Reconstructing the History of Basin and Range Extension Using Sedimentology and Stratigraphy

Edited by

Kathi K. Beratan
Department of Geology and Planetary Science
University of Pittsburgh
Pittsburgh, Pennsylvania 15260

1996

Published by The Geological Society of America, Inc.
3300 Penrose Place, P.O. Box 9140, Boulder, Colorado 80301

Printed in U.S.A.

GSA Books Science Editor Abhijit Basu

Library of Congress Cataloging-in-Publication Data
Reconstructing the history of basin and range extension using sedimentology and stratigraphy / edited by Kathi K. Beratan.
p. cm. -- (Special paper ; 303)
Includes bibliographical references (p. -) and index.
ISBN 0-8137-2303-5
1. Geology, Structural--West (U.S.)--Congresses. 2. Sedimentary basins--West (U.S.)--Congresses. 3. Geology, Stratigraphic--Neogene--Congresses. I. Beratan, Kathi K., 1957- . II. Series: Special papers (Geological Society of America) ; 303.
QE747.W47R43 1996
551.8'0978—dc20 96-805
CIP

Cover: This photograph shows the half-graben geometry typical of extensional basins. View looking to the southwest, along strike, of the Copper Basin Formation (Sequence III of Nielson and Beratan, 1990). The lake is the Gene Wash reservoir, located about 2 km west of Parker Dam in California, on the southeastern side of the Whipple Mountains. The hill in the middle distance on the right edge of the photograph marks the location of the basin-bounding high-angle normal fault along which the strata were tilted. The block tilted along an axial fulcrum, with a sedimentary basin forming near the normal fault on the down-tilted part of the block, and with the up-tilted part of the block exposed to erosion. Another high-angle normal fault is located just to the left (northeast) of the photograph. The sedimentary strata of the Copper Basin Formation were deposited in a larger half-graben basin bounded by a different set of high-angle normal faults. The Buckskin Mountains, in the background, are on an untilted block separated from the tilted Whipple Mountain blocks by a transfer fault (the Billy Mack Mountain fault of Beratan, 1991). (See Beratan and Nielson, Chapter 9, this volume.)

10 9 8 7 6 5 4 3 2 1

Contents

Plates
(in pocket)

Preface

The Mojave and Basin and Range structural provinces have experienced a complex Neogene extensional history. Extreme southwest-northeast extension in the mid-Tertiary was accommodated by large, low-angle normal fault systems known as detachment faults, and resulted in the formation of metamorphic core complexes. Lesser amounts of extension during the late Tertiary and Quaternary were accommodated by north-south–trending high-angle normal faults, forming the elongated, fault-bounded mountain ranges separated by broad valleys that characterize the Basin and Range province. Excellent exposures make this region an outstanding setting for studies of the response of continental lithosphere to these two distinct types of extensional events.

Sedimentary and volcanic strata deposited during extension preserve a record of structural events by recording changes in paleotopography through time. In the semi-arid southwestern United States, major paleotopographic changes result from structural events such as fault motion or upwarping. The synextensional strata can be dated and correlated using techniques such as radiometric dating, fossil assemblages, and magnetostratigraphy. Sedimentologic and stratigraphic studies thus can provide critical temporal and geometric constraints on crustal extension.

About 12 years ago, when I first began working in the Whipple Mountains, several people told me that correlating the Tertiary strata would be difficult, if not impossible. As one of them put it, "The Tertiary is a mess!" It *is* a mess but, thank goodness, not a hopeless one. Over the past dozen years, the combined efforts of many people in a variety of disciplines have cleaned up much of the Tertiary "mess" within the extended terranes. These multidisciplinary studies sought to gain understanding of the region's history and the mechanics of crustal extension.

To highlight recent advances in these areas, James Schmitt and I convened a symposium at the 1993 meeting of the Cordilleran and Rocky Mountian Sections of the Geological Society of America. Twenty-two talks were presented, with study areas ranging from Utah to southern California. Both detachment fault-related and classic Basin and Range–style basins were discussed. A variety of approaches was used, including field-based descriptive stratigraphic analysis, high-precision radiometric dating, detailed lithofacies analysis of depositional environments, seismic stratigraphy, computer modeling of basin geometries, biostratigraphy, and detailed field-based structural mapping. The symposium was a success; one measure of the interest in the topic was the number of people who attended the last session on the final day of the meeting. The level of interest encouraged us to propose this volume.

Some generalizations emerged from the symposium talks. Lateral facies relationships provide information on the structural geometry of synextensional basins, whereas the vertical facies changes represent datable horizons marking structural events or, rarely, climate change. Although each basin had a unique history, some features were common to all. The sedimentary basins associated with both detachment and Basin and Range faulting typically are half-graben, formed by tilting of crustal blocks along high-angle normal faults. The basin-bounding normal faults are <20 km long, and tend to be truncated by transverse structures. Individual basins thus tend to be small during active extension. Commonly, individual basins filled and, in some cases, merged with adjacent basins once active extension had slowed down. Seismic data from Nevada suggest that in some areas normal faulting was preceded by development of broad sags.

The chapters in this volume illustrate the range of variation in extensional style in the southwestern United States, as well as the variety of approaches used to study the synextensional deposits. *Dubiel and others* integrate sedimentology, molluscan paleontology and paleoecology, geologic mapping, and $^{40}Ar/^{39}Ar$ dating to study the initiation of extension in the eastern Great Basin as recorded in the early Eocene White Sage Formation, Nevada-Utah border. *Strecker and others* identify extensional structural patterns from high-resolution seismic reflection profiles of the basin fill in Goshute Valley, Nevada, a classic Basin and Range valley. *Beard* presents a revised stratigraphy for the Tertiary Horse Spring Formation in the Lake Mead area, Nevada, providing insight into complex mixed-mode deformation along kinematically linked strike-slip and extensional faults of the Lake Mead fault system.

Shifting to the central Mojave Desert, *Ingersol and others* present a stratigraphic reinterpretation of the Mud Hills and discuss implications for the structural development of this detachment-related basin. *Friedman and others* use magnetostratigraphy to refine the timing of detachment fault-related events affecting the Shadow Valley basin. *Fillmore and Walker* relate deposition in the early Miocene Pickhandle Formation to the early stage of formation of a detachment system in the central Mojave Desert, and they provide sedimentologic constraints on the magnitude of the extension.

Southeastern California and western Arizona contain a series of classic "metamorphic core complexes" formed during detachment faulting. *Dorsey and Roberts* describe sedimentation associated with the later stage of detachment faulting within the Aubrey Hills. *Lucchitta and Suneson* present a deformational history for the Castaneda Hills–Signal area of western Arizona and propose a thermally driven model for detachment faulting.

The final two chapters in the volume use stratigraphic and sedimentologic data to evaluate models for basin formation and crustal extension. *Beratan and Nielson* summarize stratigraphic relations throughout the Colorado River extensional corridor and use this information to test three different models for detachment faulting. *Schlische and Anders* describe fault and basin growth models that can refine interpretation of extensional basin architecture. Illustrative examples are drawn from Mesozoic rift basins on the East Coast of the United States and Cenozoic basins in the northern Basin and Range.

We thank Jim Schmitt for initiating this volume and providing encouragement and advice during the early stages of editing. Special thanks go to the symposium and volume participants for their contributions and to all of the reviewers for their time and careful work. Sound advice provided by Tom Anderson and Jane Nielson was greatly appreciated.

This volume is dedicated to the late Robert H. Anderson, my dissertation advisor at the University of Southern California. I am grateful to him for providing me with the opportunity to study such a wonderful topic.

Geological Society of America
Special Paper 303
1996

Reconstructing an Eocene extensional basin: The White Sage Formation, eastern Great Basin

R. F. Dubiel and C. J. Potter
U.S. Geological Survey, MS 919, Box 25046, Denver Federal Center, Denver, Colorado 80225
S. C. Good
Department of Geology, State University of New York College at Cortland, Cortland, New York 13045
L. W. Snee
U.S. Geological Society, MS 963, Box 25046, Denver Federal Center, Denver, Colorado 80225

ABSTRACT

A study integrating sedimentology, molluscan paleontology and paleoecology, geologic mapping, and $^{40}Ar/^{39}Ar$ dating of volcanic rocks indicates that the White Sage Formation north of the Deep Creek Range on the Nevada-Utah border was deposited in the early Eocene in marginal-lacustrine, lacustrine, freshwater-marsh, and minor terrestrial settings. Sedimentary facies include wave-reworked, locally derived Paleozoic carbonate-clast basal conglomerate in depositional and fault contact with bedrock; carbonate mounds; organic-rich mudstone; and laminated to medium-bedded carbonate rocks. The wave-reworked conglomerate implies the existence of a broad lake with considerable fetch to generate large waves, but one in an area with only local drainage basins and sharp relief to supply the locally derived clasts. There is a striking lack of any fluvial, deltaic, or alluvial-fan deposits that would indicate development of substantial drainage areas. The large carbonate mounds indicate a high-wave-energy shoaling environment with stable substrate and topography. The profusion of lacustrine carbonate strata indicates dominantly chemical- or biochemical-induced deposition in a carbonate-saturated lake.

Molluscan faunas include freshwater aquatic gastropods (*Valvata subumbilicata*, *Biomphilaria pseudoammonius*, and *Physa* sp.), freshwater limpets (*Acroloxus minutus*), fingernail clams (*Sphaerium* sp. indet.), terrestrial gastropods transported into the lake (*Oreoconus didacticus*, *Discus ralstonensis*, *Vertigo* sp. aff. *V. arenula*), and abundant ostracodes and charophytes. Each mollusk species is facies controlled and consistent with the depositional environment. The aquatic molluscan fauna indicates shallow, quiet lacustrine conditions with emergent vegetation. The limpets inhabited areas of rooted aquatic vegetation, and the terrestrial gastropods were transported from marshes adjacent to the lacustrine system. The molluscan assemblage constrains the age of the White Sage Formation as early Eocene, indicating a lacustrine system that is at least partially age equivalent to the Sheep Pass Formation, the Elko Formation, and outcrops near Illipah, Nevada, that have similar lacustrine facies and molluscan faunas.

The data are consistent with a model wherein the White Sage was deposited in a large lake superimposed on preexisting topography having low relief and little or no syndepositional surface-cutting faulting. White Sage strata were cut by high- and

Dubiel, R. F., Potter, C. J., Good, S. C., and Snee, L. W., 1996, Reconstructing an Eocene extensional basin: The White Sage Formation, eastern Great Basin, *in* Beratan, K. K., ed., Reconstructing the History of Basin and Range Extension Using Sedimentology and Stratigraphy: Boulder, Colorado, Geological Society of America Special Paper 303.

low-angle normal faults. New $^{40}Ar/^{39}Ar$ dates on a biotite-bearing hypabyssal rhyolitic intrusion (39.58 ± 0.10 Ma, 1 σ) and a hornblende-bearing dacite flow (39.6 ± 0.2 Ma, 1 σ) that lap onto tilted White Sage carbonate rocks require that the extension had ended by the latest Eocene.

INTRODUCTION

Much research on the Tertiary extensional history of the Great Basin focuses on structural mapping, extensional fault kinematics, and seismic data. Investigation of the stratigraphic record of sedimentation is an underutilized method for deciphering the tectonic history of these basins, in part because of limitations imposed by poor or scarce exposures and lack of chronostratigraphic control. Where suitable or fortuitous outcrops are available, sedimentary sections of extensional basins can provide information on changes in paleotopography over time, and may thus reflect otherwise subtle tectonic events (e.g., Vandervoort and Schmitt, 1990). When sedimentologic and paleontologic data are integrated, critical temporal constraints locally can be placed on the timing and tempo of extensional sedimentary basin development.

Exposures of Eocene sedimentary rocks in the eastern Great Basin are rare (Stewart and Carlson, 1976; Stewart, 1980; Stokes, 1986; Hintze, 1980, 1988). Notable exceptions include the Eocene Elko Formation in the Piñon and Adobe Ranges near Elko, Nevada (Smith and Ketner, 1976, 1978; Solomon et al., 1979; Solomon and Moore, 1982a, 1982b; Server and Solomon, 1983; Ketner and Alpha, 1992; Johnson, 1992, 1995 and references therein), part of the Late Cretaceous(?) to Oligocene Sheep Pass Formation in the Egan and Grant Ranges of east-central Nevada (Winfrey, 1958, 1960; Kellog, 1964; Fouch, 1979; Fouch et al., 1979; Good, 1987; Fouch et al., 1991), and isolated exposures of inferred partially correlative units such as the rocks in Elderberry Canyon south of Ely, Nevada (Emry, 1989; Emry and Korth, 1989), and the Kinsey Canyon Formation at Kinsey Canyon in the Schell Creek Range (Young, 1960; Fouch, 1979; Fouch et al., 1991). These scattered occurrences of Eocene conglomerate and lacustrine sequences have been intensively studied. Apart from implications for the structural evolution of the eastern Great Basin (e.g., Vandervoort and Schmitt, 1990; Thorman et al., 1991; Axen et al., 1993), the economic significance of the lacustrine rocks as oil shales and potential petroleum source rocks, particularly in producing fields in Railroad Valley and White River Valley in east-central Nevada (e.g., Winchester, 1923; Murray and Bortz, 1967; Solomon et al., 1978; Fouch, 1979), makes the Eocene section a crucial part of the early Cenozoic history disproportionate to the areal extent of its outcrops.

The White Sage Formation near Gold Hill in Tooele County, Utah (Fig. 1), provides an additional example of early Eocene conglomerate and carbonate sequences whose stratigraphic and structural relations constrain the early Cenozoic history of part of the eastern Great Basin. In contrast to numerous studies on the Eocene Elko Formation (see Johnson, 1995) and the latest Cretaceous(?) to Oligocene Sheep Pass Formation (see Fouch et al., 1991), the White Sage Formation, which was originally named and mapped by Nolan (1935), has received little attention since that first insightful study. A few reports have made tectonic inferences based on Nolan's mapping (Miller, 1990) or made passing reference to the White Sage in regional summaries of Tertiary rocks (Heylmun, 1965) and in isotopic dating of Phanerozoic magmatism (Moore and McKee, 1983). One recent study remapped the geology of the Gold Hill quadrangle just west of the area where the White Sage is exposed (Robinson, 1993). We summarized implications of the White Sage for Eocene tectonism in the eastern Great Basin in another recent paper (Potter et al., 1995).

Nolan (1935) mapped and described the White Sage Formation as part of his larger study of the stratigraphy and structural geology of the Gold Hill mining area in western Utah. He interpreted the White Sage as a series of lacustrine conglomerate and limestone strata that contained an Eocene molluscan fauna in two localities. These original interpretations are not disputed here, but they are in large part refined and expanded upon by (1) the reexamination of the sedimentology of the lacustrine carbonate rocks and conglomerate, (2) the collection of additional molluscan specimens, which were compared to molluscan faunas from other now well-known Eocene lacustrine systems on the Colorado Plateau, (3) new detailed structural mapping, and (4) new isotopic dating of overlying volcanic rocks. These new results allow more precise interpretations of the lacustrine setting of the White Sage, of the timing and structural origin of the lacustrine basin, and of the postdepositional deformation of the basin. These results also have implications for the tectonic and sedimentological development of other Eocene lacustrine basins in the eastern Great Basin and the adjacent Colorado Plateau. In addition, the interpretations have broader implications for the timing and nature of Eocene deformation in the eastern Great Basin.

GEOLOGIC SETTING

The White Sage Formation is exposed in scattered outcrops within an area of about 55 km^2 on both sides of Deep Creek, 60 km south of Wendover, Utah (Fig. 1). The study area is located 3–8 km east of the Utah-Nevada state line and about 15–20 km north of Ibapah, Utah. A small area containing similar conglomerate and lacustrine rocks that are probably correlative with the White Sage is present about 15 km to the south in the Deep Creek Range (Nutt and Thorman, 1993; R. F. Dubiel, unpublished data, 1992).

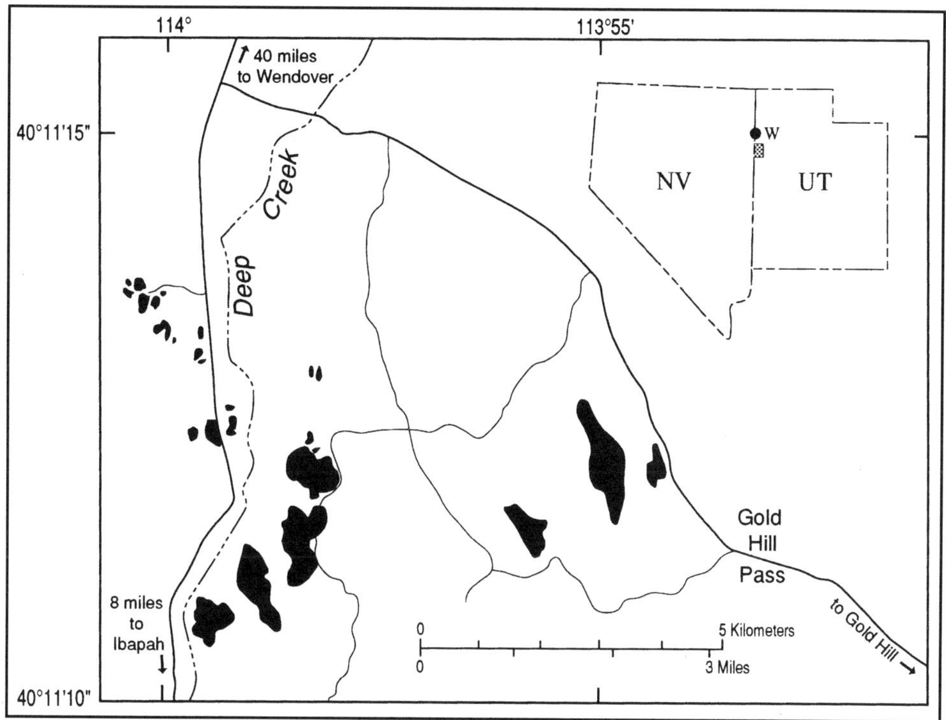

Figure 1. Map showing location of outcrops of the White Sage Formation (black areas) in the study area. Inset shows location of study area along border of western Utah and eastern Nevada. Wendover marked by W.

Within the study area, the White Sage Formation crops out in several areas, each of which contains numerous small exposures (Fig. 1). Geologic mapping of these White Sage outcrops is presented in Potter et al. (1995; see also, Nolan, 1935). One of the areas is located west of Deep Creek and about 3–5 km south of the intersection of the Ibapah road and the road that leads east to Gold Hill. A second area is located east of Deep Creek, about 1–1.5 km east of the road that leads south to Ibapah and that parallels Deep Creek. The third area lies 1–2 km west of the Gold Hill road and about 3 km northwest of Gold Hill pass.

Stratigraphic and structural setting

Locally, the White Sage Formation lies with angular unconformity upon upper Paleozoic rocks. Elsewhere, the White Sage is in fault contact with upper Paleozoic rocks and the Lower Triassic Thaynes Formation. The upper Paleozoic rocks consist of the Upper Mississippian and Pennsylvanian Ely Limestone and the Lower Permian Ferguson Mountain Formation, Arcturus Formation, and Park City Group (Hodgkinson, 1961; Wardlaw and Collinson, 1978; Hintze, 1988). These Paleozoic rocks were mapped by Nolan (1935) as the Pennsylvanian Oquirrh Formation overlain by the Permian Gerster Formation. The rocks underwent thrust faulting, folding, and normal faulting prior to 152 Ma (Robinson, 1993; Miller, 1990). In the outcrop area west of Deep Creek, the White Sage in places overlies either reddish-brown carbonate beds of the Lower Triassic Thaynes Formation that contain abundant specimens of the ammonoid *Meekoceras* or red siltstone and mudstone of the Thaynes (Potter et al., 1995). In the outcrop area east of Deep Creek, the White Sage lies in depositional contact and locally is faulted against Permian cherty carbonate rocks. In the area near Gold Hill pass, the White Sage overlies both Permian cherty carbonate rocks and red, micaceous, ripple-laminated sandstone and siltstone of the Thaynes Formation (R. F. Dubiel and E. A. Johnson, 1994, U.S. Geological Survey, unpublished measured sections and data). It is commonly difficult, but important to both structural and sedimentologic interpretations, to distinguish these red, Triassic marine sandstone and siltstone strata from similar, but slightly reddish orange, Tertiary continental sandstone beds that are part of the White Sage Formation.

The White Sage Formation is cut by numerous steep normal faults, and in two places low-angle normal faults were mapped beneath the White Sage (Potter et al., 1995). West of Deep Creek, variably dipping White Sage conglomerate and carbonate strata overlie stratigraphically attenuated and folded rocks of the Thaynes Formation that occupy a gently southeast-dipping zone. This zone is interpreted as a low-angle normal fault with the White Sage in its hanging wall. In one location east of Deep Creek, a low-angle normal fault separates the White Sage from underlying Permian strata. Potter et al.

(1995) interpret the faulting of the White Sage as part of an extensional system in which steep normal faults that cut the White Sage are listric into low-angle normal faults that underlie the White Sage.

Locally, the White Sage Formation is overlain by younger Tertiary volcanic rocks or younger gravel (Nolan, 1935; Moore and McKee, 1983; Potter et al., 1995). The volcanic rocks are essentially flat lying (dips of 0° to 5°), in contrast to dips as much as 55° in the underlying White Sage; thus, they overlie a distinct angular unconformity.

Isotopic dating

Hornblende from a dacite flow that directly overlies the faulted and tilted White Sage Formation in the area east of Deep Creek yielded an $^{40}Ar/^{39}Ar$ date of 39.6 ± 0.2 (1 σ) Ma. In addition, biotite from a hypabyssal rhyolitic intrusion that cuts the White Sage yielded an $^{40}Ar/^{39}Ar$ date of 39.58 ± 0.10 Ma (1 σ) (Dubiel et al., 1993; Potter et al., 1995). This corroborates a K-Ar date of 39.2 ± 0.6 Ma reported by Moore and McKee (1983) for a latite breccia that overlies tilted White Sage strata in the study area. These volcanic rocks are part of an extensive late Eocene volcanic field that covers much of northeastern Nevada (Thorman et al., 1993; Brooks et al., 1995). The dates on these volcanic rocks, which directly overlie tilted White Sage strata and are themselves relatively flat lying, thus indicate that essentially all of the deformation of the White Sage occurred during the Eocene.

SEDIMENTOLOGY

The White Sage Formation is locally as thick as 500 ft (150 m), although postdepositional faulting and erosion preclude an estimation of the original thickness. Whereas the faulting also precludes precise correlation of individual sections, each section contains a succession from basal conglomerate upward into bedded carbonate rocks that are succeeded by structureless dolomite. Four lacustrine facies associations are recognized in the White Sage Formation: (1) marginal-lacustrine conglomerate, (2) marginal-lacustrine to open-lacustrine limestone, (3) marginal-lacustrine carbonate mounds, and (4) stuctureless lacustrine dolomite.

Facies association 1: Marginal-lacustrine conglomerate

Conglomerate is common at the base of the White Sage Formation (Fig. 2), and it grades upward into conglomeratic limestone. The basal conglomerate is as much as 100 ft (30 m) thick. Most pebbles in the conglomerate are less than 4 in (10 cm) in diameter, but locally clasts are as large as 12 in (30 cm) across. The conglomerate consists predominantly of reworked subrounded to subangular Paleozoic carbonate clasts, with minor siliciclastic pebbles. The conglomerate occurs as medium to thick, tabular and less common lenticular beds. In places, lenticular beds have scoured contacts into underlying beds, but bed contacts are generally planar.

Figure 2. Thick conglomerate within the basal part of the White Sage Formation on the west side of Deep Creek. A section of about 15 m of conglomerate is exposed.

The conglomerate is predominantly clast supported, but matrix-supported conglomerate is locally present in the lenticular beds. The conglomerate appears to be poorly sorted on a large scale, consisting of a wide range of pebble sizes, but on closer examination, there are many beds with pebbles that show layer-selective sorting. Pebbles in many beds, and especially in the uppermost parts of the thickest conglomerate sections, are sorted selectively by layer according to both pebble size and shape (Fig. 3). In places, the uppermost beds of conglomerate are interbedded with, or grade upward into, carbonate-cemented conglomerate and conglomeratic limestone. In these beds, pebbles and granules are present as small pods and lenses in which the clasts are sorted according to both size and shape.

Conglomerate of the White Sage Formation locally contains thin beds of reddish orange, fine- to medium-grained, siliclastic sandstone and minor interbeds of gray claystone. These finer grained units are present as thin beds and small lenses within conglomerate beds, and very rarely are they interbedded with limestone higher in the section.

Interpretation. Clastic deposits in the White Sage Formation were deposited or reworked in marginal-lacustrine settings. The interpretation is based in part on lithology, clast sorting, and the presence of lacustrine carbonate and mollusks in overlying units (see discussion in a following section). The clast-supported conglomerate is interpreted to represent minor subaqueous debris flows and predominantly wave-reworked gravel deposited in a marginal-lacustrine setting. The lenticular beds of conglomerate with local scoured bases and poor internal sorting indicate channelized flow in debris flows. The conglomerate lacks the abundant scours, imbricated clasts, and dis-

Figure 3. Wave-sorted conglomerate in the White Sage Formation showing layer-selective sorting of Paleozoic carbonate clasts by both size and clast shape. Scale marked in cm.

Figure 4. Transitional facies between wave-sorted conglomerate and micrite-cemented conglomerate showing patchy distribution of clasts. Scale marked in cm.

tinctive crossbedding suggestive of organized bedforms and fluvial deposition. The layer-selective sorting by size and clast shape reflects varying wave energies that transported different classes of pebble sizes and shapes. The marginal-lacustrine interpretation is supported by the intercalation of the conglomerate with carbonate beds interpreted to be lacustrine on the basis of their included molluscan fauna (see discussion in a following section). The wave-reworked conglomerate implies the existence of a broad lake with considerable fetch to generate large waves, but one in an area of only small drainage basins and sharp relief to supply the locally derived clasts. The general lack of sandstone or mudstone deposits suggests a lack of sufficient relief, lack of extensive drainage basins, or lack of appropriate lithologies that could have supplied more abundant fine-grained fluvial or deltaic material. In the available outcrops, there is a striking scarcity of major fluvial, deltaic, or alluvial-fan deposits and only rare deposits interpreted as debris flows. However, it should be noted that the study area is small, and exposures within the area are even more localized, thus limiting precise inferences about the actual size of the original drainage basin.

White Sage conglomerate locally grades upward (Fig. 4) into conglomeratic limestone or micrite-cemented conglomerate, both of which represent lithologic and depositional gradations between the conglomerate facies and the bedded limestone facies described in a following section.

Facies association 2: Marginal-lacustrine to open-lacustrine limestone

Conglomerate of the White Sage Formation is overlain by sequences of white to tan, thin- to thick-bedded carbonate rocks that range from 1 ft (0.3 m) to 10 ft (3 m) in thickness. The bedded carbonate rocks are overlain by, or grade laterally into, irregularly thick to mound-shaped, fine-grained carbonate rocks. The bedded carbonate rocks are laterally extensive; individual beds can be traced along the entire length of several outcrops (0.5 mi or 1 km). Analysis by X-ray diffraction indicates that the bedded carbonates are primarily microcrystalline calcite (micritic), with possible trace amounts of quartz and dolomite. They are classified as micrites or carbonate mudstones (Dunham, 1962; Folk, 1980). Laminations are rare, and no desiccation features were observed in any of the limestone. Ostracodes, fragments of charophyte stems, and mollusks are locally abundant in the bedded limestone (see Molluscan Paleontology section).

The bedded limestone is interstratified with gray calcareous claystone and locally with abundant dark gray to black, organic-carbon–rich sandy mudstone as much as 2 ft (0.6 m) thick (Fig. 5). The black mudstones contain very small, 1/16 in. (1.5 mm) diameter gastropods. In the section east of Deep Creek (Fig. 1), black mudstone and thin-bedded limestone are regularly interbedded over a vertical interval of about 40 ft (10 m).

Interpretation. The bedded limestone is interpreted as marginal-lacustrine to open-lacustrine limestone that was deposited in fresh-water lacustrine conditions. Interpretation of a fresh-water but slightly alkaline lacustrine setting is supported by the presence of ostracodes, charophytes (indicate pH ±9), and mollusks. Modern charophytes commonly live in warm, shallow alkaline lakes, in depths less than 30 ft (9 m) or rarely to a maximum of 45–60 ft (13.6–18 m) (Cohen and Thouin, 1987). The profusion of lacustrine carbonate rocks indicates dominantly chemical- or biochemical-induced deposition in a carbonate-saturated lake. The general absence of lamination in the bedded carbonate rocks implies abundant bioturbation of the sediments, which in turn suggests low salinity, oxygenated bottom waters or substrate and nonpermanent lake-water stratification.

Figure 5. Black mudstone deposited in lacustrine marshes exposed in small trenches interstratified with bedded lacustrine limestone. Trenching shovel (50 cm long) for scale.

Rock-Eval pyrolysis indicates that the organic material from the black mudstone is Type III kerogen produced primarily from reeds and woody tissue. Thus, organic geochemistry supports the mollusk-based interpretations (see discussion in a following section) that organic-rich mudstone represents deposition in a marginal-lacustrine marsh. The interbedded marsh and limestone deposits imply a periodic fluctuation of lake levels and repeated transgressions and regressions of the shorelines.

Facies association 3: Lacustrine carbonate mounds

Carbonate mounds form a distinctive facies in the lower part of the White Sage Formation. Most individual mounds are less than 10 ft (3 m) high and 100 ft (30 m) across, but one mound complex is 100 ft (30 m) high and more than 1,000 ft (300 m) across (Fig. 6). The mounds are composed of tan micritic limestone similar to the bedded limestone, and they locally contain abundant ostracodes and charophyte stem fragments. Mounds commonly are developed directly over the basal conglomerate facies, and locally they drape and are formed over Paleozoic bedrock. In other areas, mounds are interbedded with and grade laterally into bedded carbonate rocks, organic-rich mudstone, minor gray mudstone, and rare red sandstone, similar to those facies interstratified with the bedded-limestone facies.

Allochemical constituents in the bedded carbonate rocks and in the mound-like carbonate rocks include abundant charophyte stem fragments, ostracodes, and mollusks. Most of the ostracodes are calcareous and represent primary preservation of the fossils, but at a few localities the fossils are silicified. The majority of the ostracode tests are articulated, indicating little evidence of transport.

Figure 6. Giant carbonate mound complex near the base of the White Sage Formation east of Deep Creek.

Interpretation. The size and the morphology of the carbonate mounds indicate carbonate precipitation in a high-wave-energy shoaling environment with stable substrate and topography. The carbonate mounds that lie stratigraphically above wave-reworked conglomerate indicate marginal-lacustrine settings in which carbonate precipitated on stabilized gravels, and the source of the mounds may have been spring flow through the porous gravels. Mounds that developed on Paleozoic units represent carbonate precipitation on bedrock near the lake margin in shallow water.

The carbonate mounds are analogous in size, morphology, and lithology to carbonate tufa mounds on the northwest margin of Pyramid Lake in western Nevada (Shearman et al., 1989) and similar tufa mounds at Mono Lake in eastern California (R. F. Dubiel, 1993, unpublished data). The giant mounds at Pyramid Lake are aligned with a major fault that may provide conduits for spring activity. Other mounds at Pyramid Lake lie above porous alluvial-fan, deltaic, and wave-sorted gravels, which served as ground-water conduits. These modern carbonate mounds provide good analogs for the mounds that overlie wave-sorted conglomerate in the White Sage. Based on the similar size and relation to underlying gravels in these modern settings, the White Sage carbonate mounds are interpreted to have been deposited in a shallow to moderately deep lacustrine basin.

Facies association 4: Lacustrine structureless dolomite

The uppermost structureless dolomite facies of the White Sage Formation consists of white, structureless to thin and very thin bedded, chippy-weathering, fine-grained carbonate rock.

Analysis by X-ray diffraction indicates that this facies is primarily very fine grained dolomite containing approximately 5%–15% quartz and minor calcite. Chert lenses and nodules as much as 1 ft (0.3 m) thick and 3–6 ft (1–2 m) across are locally present within the dolomite. The chert weathers primarily white to yellow and is black on a fresh surface. Minor bedding can be traced locally into the nodules, so that the chert appears to be a diagenetic alteration of the original dolomite. No fossils were observed in the dolomite.

Interpretation. The lack of fossils in this facies compared to abundant fossils in other facies indicates possible higher salinity, or at least higher concentrations of magnesium. The dolomite facies is interpreted to have been deposited in a lacustrine environment, possibly representing a closed-basin lacustrine setting, but the lack of obvious evaporative minerals suggests that a highly evaporative environment did not exist (Smoot and Lowenstein, 1991).

Sedimentologic discussion

The White Sage Formation was deposited in marginal-lacustrine, lacustrine, and freshwater-marsh environments (Dubiel, et al., 1992, 1993). Sedimentary facies consist of wave-reworked, locally derived Paleozoic carbonate-clast basal conglomerate, carbonate mounds, organic-rich mudstone, laminated to medium-bedded carbonate rocks, and structureless dolomite. In addition, terrestrial mollusks found in black mudstones apparently were reworked and transported into the lacustrine realm (see Molluscan Paleontology section) and provide the only direct indications of unexposed terrestrial settings that must have existed adjacent to the lake. These adjacent terrestrial settings must have included abundant leaf litter to support the terrestrial mollusk population.

Each of the sections in the White Sage exhibits the same succession of depositional facies from basal conglomerate upward into bedded limestone or locally into limestone mounds and organic-rich mudstone, overlain by structureless dolomite at the top of the section. None of the sections shows depositional contact with another overlying sedimentary facies, but they are overlain locally by Tertiary volcanic flows.

MOLLUSCAN PALEONTOLOGY

Both the carbonate mounds and the bedded limestone contain abundant ostracodes and charophyte stem fragments and locally abundant molluscan faunas (Table 1). The molluscan faunas can be grouped into associations that parallel the sedimentologic facies.

Molluscan associations

Eight taxa of nonmarine mollusks have been recovered from the carbonate mounds, the bedded limestone, and the organic-rich mudstone facies of the White Sage Formation (Table 1; Fig. 7). Mollusks locally dominate the fossil biota. Identification of the molluscan faunal assemblages facilitate (1) enhanced understanding of the White Sage depositional environments derived from the paleoecology of the molluscan biota, (2) age determination of these strata using molluscan biostratigraphy, and (3) interpretation of the molluscan paleobiogeographic relations between the White Sage lacustrine system and coeval lacustrine systems of the Western Interior, such as those represented by the Sheep Pass and Elko Formations of the Basin and Range and the Green River Formation of the central Rocky Mountains and the Colorado Plateau.

TABLE 1. MOLLUSK TAXA FROM THE WHITE SAGE FORMATION

Class Bivalvia
- Order Heterodonta
  - Family Pisidiidae
    - *Sphaerium* sp. indet.

Class Gastropoda
- Subclass Prosobranchia
  - Order Mesogastropoda
    - Family Valvatidae
      - *Valvata subumbilicata* (Meek and Hayden)
- Subclass Pulmonata
  - Order Basommatophora
    - Family Physidae
      - *Physa* sp. indet.
    - Family Planorbidae
      - *Biomphalaria pseudoammonius* (Schlotheim)
    - Family Ancylidae
      - *Acroloxus minutus* (Meek and Hayden)
  - Order Stylommatophora
    - Family Bulimulidae
      - *Oreoconus didacticus* Hanley
    - Family Endodontidae
      - *Discus ralstonensis* (Cockerel)
    - Family Pupillidae
      - *Vertigo* sp. aff. *V. arenula* (White)

Molluscan paleoecology

Two types of mollusk assemblages occur in the White Sage Formation: (1) an aquatic mollusk-dominated fauna within thin-bedded limestones, and (2) a terrestrial gastropod fauna within the organic-rich mudstones. Paleoecological interpretations (Table 2) are formulated using taphonomic analyses and taxonomic uniformitarianism (Dodd and Stanton, 1981). Taphonomic analyses were used to infer the processes responsible for producing the observed mollusk assemblage. Taxonomic uniformitarianism transfers the ecological constraints of the most closely related extant taxon to the paleoautoecology (ecologic tolerances of a single species) of the fossil assemblage. Paleo-ecological interpretations are derived from the paleosynecology (overlap of ecological tolerances of all the species) of the entire assemblage, as delimited by the combined paleoautoecologies of component species of the assemblage.

Figure 7. Molluscan fauna of the White Sage Formation. Each scale bar is 1 mm. (1). *Sphaerium* sp. indeterminate. (2, 3). *Acroloxus minutus*. (4). *Valvata subumbilicata*. (5). *Lymnaea* sp. indeterminate. (6). *Biomphalaria pseudoammonius*. (7). *Discus ralstonensis*. (8). *Vertigo* sp. cf. *V. arenula*. (9). *Oreoconus didacticus*.

Bedded limestone molluscan assemblage

The bedded-limestone molluscan assemblage is composed of aquatic molluscs, except in one locality where a single specimen of the terrestrial gastropod *Oreoconus didacticus* was found. The aquatic mollusk fauna consists of "finger-nail" clams, prosobranch gastropods, and aquatic pulmonate gastropods. Benthic mollusks that live on or in the substrate include *Sphaerium* sp. and *Valvata subumilicata*, which are freshwater Pisidiid bivalves, and prosobrach gastropods. Both of these mollusk taxa are gill-respiring organisms that prefer shallow, oxygenated, quiet-water lacustrine habitats.

The aquatic pulmonate gastropods include *Physa* sp., *Biomphalaria pseudoammonius*, and *Acroloxus minutus*. Gastropods within the genus *Physa* generally respire subaerially by bringing the pneumostome into contact with the air-water interface and bringing air into the pulmonary sac; however, some deeper water species can extract oxygen cutaneously from water

TABLE 2. PALEOECOLOGIC INTERPRETATIONS OF MOLLUSKS FROM THE WHITE SAGE FORMATION

I. Benthic Mollusks - live in or on substrate
  A. Fingernail Bivalve
    1. *Sphaerium* sp.
      a. Gill breathing, aquatic, small lakes.
      b. On or in sandy to muddy substrates.
  B. Prosobranch Gastropod
    1. *Valvata subumbilicata*
      a. Gill breathing, benthic aquatic.
      b. On gravel to muddy substrates.

II. Epibenthic Mollusks - live on emergent vegetation
  A. Basommatophorans Gastropods
    1. *Acroloxus minutus*
      a. Gill breathing.
      b. Attached to rooted vegetation in small ponds or lakes.
    2. *Biomphalaria pseudoammonius*
      a. Air breathing or gill breathing.
      b. Depths less than 6 ft (0.6 m); not below depth of rooted vegetation (15 ft; 5 m).
    3. *Lymnaea* sp.
      a. Air breathing, abundant in quiet-water habitats.
      b. Usually live on rooted aquatic vegetation.

III. Terrestrial Gastropods - live in adjacent pond margins
  A. Stylommatophoran Gastropods
    1. *Oreoconus didacticus*
      a. Air breathing.
      b. Transported from vegetated habitat into ponds.
    2. *Discus ralstonensis*
      a. Air breathing.
      b. Humid forest under dead wood and rotting leaves.
    3. *Vertigo arenula*
      a. Air breathing.
      b. Moist habitats - under fallen leaves and on sticks and stems adjacent to creeks, ponds, and marshes.

in the pulmonary sac. Extant species of *Physa* most commonly inhabit shallow, quiet-water, well-aerated aquatic habitats with firm substrates, incident sunlight, and rooted aquatic plants (Dawson, 1911). *Biomphalaria pseudoammonius* is a planorbid pulmonate, a group that is characterized by the pseudobranch (a gill for subaqueous respiration). Planorbids retain the pulmonary sac that permits them also to respire subaerially. Planorbids inhabit a wide variety of aquatic habitats, ranging from small to large, perennial to ephemeral, lotic and lentic aquatic systems. Planorbids are most abundant in water habitats less than 6 ft (2 m) deep, and they do not live deeper than the maximum depth of rooted vegetation, which is approximately 15 ft (5 m) (Baker, 1945). *Acroloxus minutus*, a freshwater limpet, is an aquatic pulmonate that respires subaqueously via an external pseudobranch gill. Freshwater limpets inhabit small lakes and small fluvial systems, typically living upon rooted aquatic vegetation (Basch, 1963).

The paleosynecology of the mollusk fauna indicates that these limestone beds were deposited in shallow aquatic habitats (definitely less than 15 ft, and probably less than 6 ft in depth), that were well oxygenated and low energy, and contained abundant rooted aquatic vegetation. Most localities represent ecologically homogeneous mollusk faunas. The one locality that contains the terrestrial gastropod is ecologically heterogeneous.

This mollusk fauna is preserved in random orientations within limestone blocks. Mollusk specimens are very finely recrystallized, with no evidence of fragmentation or abrasion. The matrix is micrite, indicating a low-energy lacustrine depositional environment. The aquatic mollusk fauna is a parautochthonous assemblage (Kidwell et al., 1986) or a disturbed neighborhood assemblage (Scott, 1970). This assemblage type indicates an ecologically homogeneous fauna that occurs in sediments deposited within the area that the taxa inhabited but in which the specimens are not in life position. The assemblage containing the terrestrial gastropod represents a mixed parautochthonous-allochthonous assemblage (Kidwell et al., 1986) or mixed assemblage (Scott, 1970) with individuals having been transported to the site of burial.

This mollusk assemblage, with the exception of the terrestrial gastropod specimen, indicates an ecological association of freshwater mollusks that cohabited as the molluscan component of an ecological community. This mollusk association inhabited shallow, quiet-water, marginal-lacustrine environments with rooted, emergent aquatic vegetation. The micritic matrix of these beds indicates weak currents, rendering winnowing ineffective, but the disturbed orientations of specimens and presence of a terrestrial gastropod indicate occasional higher energy conditions, perhaps from storm and/or flooding events.

Black-mudstone molluscan assemblage

The black-mudstone molluscan assemblage is composed of two species of terrestrial gastropods, *Discus ralstonensis* and *Vertigo* aff. *V. arenula*. The habitat of modern species of the genus *Discus* is under dead wood and among moist, rotting leaves and grass in humid forests. The paleoautoecology of the genus *Vertigo* is moist habitats adjacent to creeks, ponds, and marshes; it is typically found either on or under leaf litter and dead wood with locally high humidity. Some species of *Vertigo* require soils rich in lime (Pilsbry, 1948; LaRocque, 1970).

The black-mudstone molluscan assemblage is present in an approximately 2 ft (0.6 m) thick, black, organic-carbon–rich sandy mudstone that is intercalated with thin-bedded carbonate rocks in sequences 1–10 ft (0.3–3 m) in thickness. The mudstone represents marginal-lacustrine marsh environments. Both genera of mollusks in this bed are small, less than 3 mm in diameter, and they are very well preserved.

The black-mudstone molluscan assemblage is an allochthonous assemblage (Kidwell, et al., 1986), or transported assemblage (Scott, 1970) that inhabited terrestrial environments adjacent to the marsh. The mollusks were probably washed into the marsh during storms and high runoff. Modern representatives of these terrestrial families often are found with aquatic mollusks in the deposits of marl lakes, where they are interpreted to have been transported during storms (Nylander, 1900; Walker, 1900, 1903). The genera of terrestrial gastropods found in the White Sage Formation inhabited similar environments, and they were most likely members of a common terrestrial community. Therefore, this assemblage does form an ecological association; however, it has undergone transport from that environment and subsequent deposition in marginal-lacustrine marshes.

Molluscan biostratigraphy

The stratigraphic ranges of numerous species of nonmarine mollusks have been documented in unpublished range data collated by John Hanley (deceased; 1987, U.S. Geological Survey unpublished data). This extensive database was produced from the literature on Tertiary stratigraphy and molluscan paleontology studies from the Western Interior of the United States. The ranges of species that occur in the White Sage Formation are illustrated in Figure 8 (see also Table 3).

The date of the last appearance datum (LAD) of both *Discus ralstonensis* and *Acroloxus minutus* is the top of the Wasatchian (early Eocene). The first appearance datum (FAD) of both *Biomphalaria pseudoammonius* and *Oreoconus didacticus* is the base of the Wasatchian (early Eocene). The combination of this FAD and LAD restricts the age of the White Sage Formation to the early Eocene (Wasatchian). *Vertigo arenula* is restricted to the Bridgerian (middle Eocene); however, the specimens in this study are noted to have only affinities to this species. The specimen may represent a slightly older form than *V. arenula*, and therefore does not contradict the age determination based on the other four species of nonmarine gastropods.

Molluscan paleobiogeography

The paleobiogeographic distribution of comparable mollusk faunas throughout lower Eocene rocks of the Western Interior indicates the widespread distribution of similar lacustrine habitats and successful dispersal of mollusk species into these habitats throughout the region. Aquatic prosobranch and pulmonate gastropods reproduce by laying eggs on shallow-water vegetation. The eggs frequently become attached to the legs of water birds, and thus they are dispersed efficiently to other water bodies by the birds (Rees, 1965). Pisidiid bivalves (*Sphaerium* sp.) have been documented to be locally dispersed by insects (Rees, 1965), another effective transport mechanism between water bodies. After introduction into a new waterbody, prosobranch valvatid gastropods, pulmonate gastropods, and pisidiid bivalves rapidly populate the new habitat by hermaphroditic reproduction.

Molluscan biostratigraphy indicates that the White Sage Formation represents a lacustrine system equivalent in age to other Eocene lacustrine sequences in the eastern Great Basin, including the Elko Formation and part of the Sheep Pass Formation (Fouch, 1979; Fouch et al., 1991, 1992; Good, 1987), and to outcrops near Illipah, Nevada (Dubiel et al., 1993), that have similar lacustrine facies. The bedded-limestone molluscan assemblage of the White Sage is taxonomically very similar to the *Lymnaea-Biomphalaria* association of the siliceous and fossiliferous carbonate member (member E of Winfrey, 1960) of the Sheep Pass Formation (for discussion of Sheep Pass mollusks, see Good, 1987; Fouch et al., 1991). The bedded-limestone molluscan assemblage of the White Sage is also similar to the *Physa-Biomphalaria-Omalodiscus* association of Hanley (1976), which is found in the main body (Lysitean and Lost Cabinian ages of the Wasatchian, early Eocene) and Niland Tongue (Lost Cabinian of the Wasatchian, early Eocene) of the Wasatch Formation. The *Physa-Biomphalaria-Omalodiscus* association occurs in gray limestone and marlstone lithofacies that are interpreted to have been deposited in flood-plain ponds in the predominantly terrestrial facies surrounding the lacustrine facies of the Green River Formation (Hanley, 1976).

BASIN-FORMING PROCESSES

Regional stratigraphic patterns (Armstrong, 1968; Stewart, 1980; Hintze, 1988) strongly suggest that much of the eastern Great Basin was a region of nondeposition from Late Jurassic through Cretaceous time and that the eastern Great Basin was a region of low relief at the end of the Mesozoic. Vandervoort and Schmitt (1990) inferred that localized early Cretaceous basins were developed in front of surface-breaking thrust faults, and they speculated that localized late Cretaceous to Eocene basins were extensional in nature, accommodating gravitational collapse of the uplifted hinterland behind the frontal Sevier fold-thrust belt. These regional stratigraphic and tectonic issues invite speculation about the origin of the White Sage basin. The sedimentology and structure of the White Sage provide no strong evidence for syndepositional faulting. There is no evidence for fault scarps shedding megabreccia debris (Burchfiel, 1966), nor is there a systematic variation in the degree of tilt of the basin fill, which is a hallmark of synextensional deposition above an active normal fault. The mapped faulting and tilting of the White Sage occurred sometime during a 12 m.y. (or greater) interval between the deposition of the early Eocene White Sage Formation and the unconformably overlying 39 Ma volcanic rocks. Although it appears that the normal faults completely postdated White Sage deposition, it remains possible that they began during White Sage basin formation (which would imply that the White Sage strata accumulated in an extensional basin). Another possibility is that the White Sage Formation accumulated in a sag that developed above a zone of lithospheric or crustal thinning, in which case the basin would have been generated by deep extensional processes not linked to local surface faulting (Gibbs, 1987).

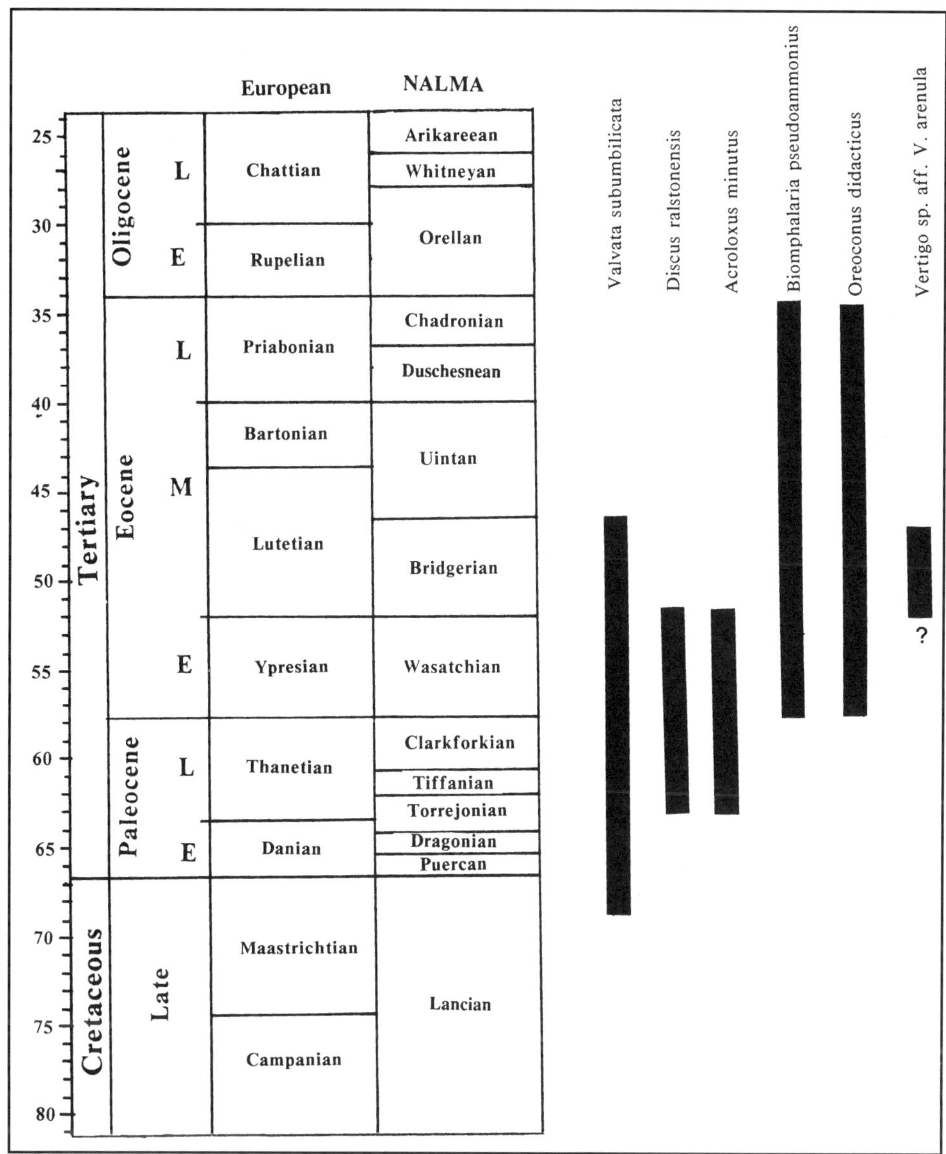

Figure 8. Diagram showing stratigraphic range of mollusk species in the White Sage Formation and comparison of European and North American land mammal ages (NALMA) (Woodburne, 1987). Ages shown on left in Ma.

TABLE 3. MOLLUSCAN BIOSTRATIGRAPHY

Valvata subumbilicata (Meek and Hayden)	Late Cretaceous to Middle Eocene
Discus ralstonensis (Cockerell)	Late Paleocene to Early Eocene
Acroloxus minutus (Meek and Hayden)	Late Paleocene to Early Eocene
Biomphalaria pseudoammonius (Schlotheim)	Early Eocene to Late Eocene
Oreoconus didacticus Hanley	Early Eocene to Late Eocene
Vertigo sp. aff. *V. arenula* (White)	Middle Eocene (Bridgerian)

CONCLUSIONS

Sedimentologic and paleontologic evidence indicates that the White Sage Formation in the eastern Great Basin was deposited in a shallow to moderately deep lacustrine system. There is no evidence of subaerial exposure or pedogenesis of the lacustrine carbonates. Depositional facies present in the White Sage can be grouped into four main associations: (1) marginal-lacustrine, wave-reworked conglomerate, (2) marginal-lacustrine to open-lacustrine bedded carbonate rocks and organic-rich marsh mudstone, (3) marginal-lacustrine carbonate mounds, and (4) shallow-water, lacustrine dolomite. There is a marked lack of alluvial-fan or deltaic deposits in the White Sage, but the study area is small and the exposures of the White Sage do not extend over a very large area, so that the lack of outcrops displaying those facies may be the result of preservational bias.

Mollusk assemblages found within the White Sage Formation provide constraints on the depositional facies, the age of the deposits, and correlations to other correlative Eocene lacustrine deposits both within the eastern Great Basin and on the adjacent Colorado Plateau to the east.

Lack of alluvial-fan and landslide debris in the White Sage may argue for low topographic relief surrounding the basin, but the presence of these types of deposits in partially correlative rocks such as the Late Cretaceous(?) to middle Eocene Sheep Pass Formation, exposed farther south in the Egan Range, indicates that rapid development of topographic relief and thus active tectonism did accompany some coeval Tertiary basin evolution. Deposits of landslide debris and alluvial fans that grade abruptly into lacustrine facies are common along Tertiary normal faults throughout the Great Basin (Leeder and Gawthorpe, 1987). The Late Cretaceous to early Tertiary record in the eastern Great Basin has been interpreted to reflect ponding of drainage systems within extensional basins situated behind an active fold and thrust belt (Vandervoort and Schmitt, 1990). Sedimentation within the internally drained system filled closed-drainage depressions to the level of the terrain. Subsequent middle Tertiary volcanic rocks were deposited over a broad region with little apparent topographic relief.

Sedimentologic, biostratigraphic, and isotopic evidence within the White Sage Formation indicates that lacustrine deposition was extant within a broad, low-relief early Eocene system that was subsequently deformed above an extensional fault system prior to volcanism at about 39 Ma. No definitive sedimentologic or structural evidence was observed to indicate that the White Sage accumulated in an actively extending basin, although it is possible that the basin developed in a sag above deeper seated lithospheric thinning. Stratigraphic and isotopic age constraints require that the normal faults mapped in the area occurred prior to the late Eocene. This dating allows the possibility that these faults were initiated as part of an extensional fault system that formed the basin. The low relief inferred from the sedimentology implies an absence of major syndepositional fault scarps that would indicate rapid extension, but it does not preclude faulting at slip rates too low to generate significant topography.

ACKNOWLEDGMENTS

Many people have contributed toward our introduction to and understanding of the geology of the eastern Great Basin. The late Tom Nolan's early work on the White Sage Formation and much of the geology of the Great Basin, in general, have provided the groundwork for many continuing studies. K. B. Ketner, F. G. Poole, C. H. Thorman, and T. D. Fouch (U.S. Geological Survey) were instrumental in our introduction to the geology of the Great Basin and the White Sage Formation. W. E. Brooks (U.S. Geological Survey) provided valuable advice and regional context for our study of the volcanic rocks. We thank T. D. Fouch, R. C. Johnson (U.S. Geological Survey), and K. Beratan (University of Pittsburgh) for reviews of the paper. We also thank K. Beratan and J. Schmitt (Montana State University) for inviting us to present our research both in the original symposium and in this publication.

REFERENCES CITED

Armstrong, R. L., 1968, Sevier orogenic belt in Nevada and Utah: Geological Society of America Bulletin, v. 79, p. 429–458.

Axen, G. J., Taylor, W. J., and Bartley, J. M., 1993, Space-time patterns and tectonic controls of Tertiary extension and magmatism in the Great Basin of the western United Sates: Geological Society of America Bulletin, v. 105, p. 56–76.

Baker, F. C., 1945, The molluscan family Planorbidae: Collation, revision and additions by Harley Jones Van Cleave: Urbana, University of Illinois Press, 530 p.

Basch, P. F., 1963, A review of recent freshwater limpet snails of North America (Mollusca: Pulmonata): Bulletin of the Museum of Comparative Zoology, v. 129, p. 399–461.

Brooks, W. E., Thorman, C. H., and Snee, L. W., 1995, $^{40}Ar/^{39}Ar$ ages and tectonic setting of upper middle Eocene volcanic rocks in northeast Nevada and adjacent Utah: Journal of Geophysical Research, Special Section on Magmatism and Extension, v. 100, no. B7, p. 10,403– 10,416.

Burchfiel, B. C., 1966, Tin Mountain landslide, southeastern California, and the origin of megabreccia: Geological Society of America Bulletin, v. 77, p. 95–100.

Cohen, A. S., and Thouin, C., 1987, Nearshore carbonate deposits in Lake Tanganyika: Geology, v. 15, p. 414–418.

Dawson, J., 1911, The biology of Physa: Behavior Monographs, v. 1, no. 4, 125 p.

Dodd, J. R., and Stanton, R. J., Jr., 1981, Paleoecology, concepts and applications: New York, John Wiley and Sons, 559 p.

Dubiel, R. F., Potter, C. J., and Good, S. C., 1992, Depositional and structural setting of the Lower Eocene White Sage Formation, Eastern Great Basin: Boulder, Colorado, Geological Society of America Abstracts with Programs, v. 24, no. 6, p. 8.

Dubiel, R. F., Good, S. C., Potter, C. J., and Snee, L. W., 1993, Sedimentologic and biostratigraphic constraints on Lower Eocene lacustrine systems, Eastern Great Basin, Nevada: Boulder, Colorado, Geological Society of America Abstracts with Programs, v. 25, no. 5, p. 32.

Dunham, R. J., 1962, Classification of carbonate rocks according to depositional texture, *in* Ham, W. E., ed., Classification of carbonate rocks: American Association of Petroleum Geologists Memoir, v. 1, p. 108–121.

Emry, R. J., 1989, A tiny new Eocene ceratomorph and comments on "tapiroid" systematics: Journal of Mammology, v. 70, p. 794–804.

Emry, R. J., and Korth, W. W., 1989, Rodents of the Bridgerian (middle Eocene) Elderberry Canyon local fauna of eastern Nevada: Smithsonian Contributions to Paleobiology, no. 67, 14 p.

Folk, R. L., 1980, Petrology of sedimentary rocks: Austin, Texas, Hemphill Publishing, 182 p.

Fouch, T. D., 1979, Character and paleogeographic distribution of Upper Cretaceous(?) and Paleogene nonmarine sedimentary rocks in east-central Nevada, *in* Armentrout, J. M., Cole, M. R., and TerBest, H., eds., Cenozoic Paleogeography of the Western United States: Society of Economic Paleontologists and Mineralogists, Pacific Section, Pacific Coast Paleogeography Symposium 3, p. 97–111.

Fouch, T. D., Hanley, J. H., and Forester, R. M., 1979, Preliminary correlations of Cretaceous and Paleogene lacustrine and related nonmarine sedimentary and volcanic rocks in parts of the eastern Great Basin of Nevada and Utah, *in* Newman, G. W. and Goode, H. D., eds., Basin and Range Symposium and Great Basin Field Conference: Denver, Colorado, Rocky Mountain Association of Geologists, p. 305–312.

Fouch, T. D., Lund, K., Schmitt, J. G., Good, S. C., and Hanley, J. H., 1991, Late Cretaceous (?) and Paleogene sedimentary rocks and extensional(?) basins in the region of the Egan and Grant Ranges, and White River and Railroad Valleys, Nevada: Their relation to Sevier and Laramide contractional basins in the southern Rocky Mountains and Colorado Plateau, *in* Flanigan, D. M. H., Hansen, M., and Flanigan, T. E., eds., Geology of White River Valley, the Grant Range, Eastern Railroad Valley, and Western Egan Range, Nevada: Reno, Nevada, Nevada Petroleum Society, 1991 Fieldtrip Guidebook, p. 15–23.

Fouch, T. D., Potter, C. J., and Dubiel, R. F., 1992, Evolution of Maastrichtian and Paleogene lake basins and associated terranes, Eastern Great Basin, Nevada: Boulder, Colorado, Geological Society of America Abstracts with Programs, v. 24, no. 6, p. 12.

Gibbs, A., 1987, Development of extension and mixed-mode sedimentary basins, *in* Coward, M. P., Dewey, J. F., and Hancock, P. L., eds., Continental extensional tectonics: London, Geological Society of London Special Publication No. 28, p. 19–33.

Good, S. C., 1987, Mollusc-based interpretations of lacustrine paleoenvironments of the Sheep Pass Formation (Latest Cretaceous to Eocene) of east-central Nevada: Palaios, v. 2, p. 467–478.

Hanley, J. H., 1976, Paleosynecology of nonmarine Mollusca from the Green River and Wasatch Formations (Eocene), southwestern Wyoming and northwestern Colorado, *in* Scott, R. W., and West, R. R., eds., Structure and classification of paleocommunities: New York, Dowden, Hutchinson and Ross, p. 235–261.

Heylmun, E. B., 1965, Reconnaissance of the Tertiary sedimentary rocks in western Utah: Utah Geological and Mineralogical Survey Bulletin 75, 38 p.

Hintze, L. H. (compiler), 1980, Geologic map of Utah: Salt Lake City, Utah Geological and Mineral Survey, scale 1:500,000.

Hintze, L. H., 1988, Geologic history of Utah: Provo, Utah, Brigham Young University Geology Studies, Special Publication 7, 202 p.

Hodgkinson, K. A., 1961, Permian stratigraphy of northeastern Nevada and northwestern Utah: Brigham Young University Geology Studies, v. 8, p. 167–196.

Johnson, R. C., 1992, A general model for the evolution of Eocene lake basins of the Elko area, northeastern Nevada: Boulder, Colorado, Geological Society of America Abstracts with Programs, v. 24, no. 6, p. 20.

Johnson, R. C., 1995, Evolution of the lower Tertiary Elko Lake basin, northeast Nevada: U.S. Geological Survey Bulletin, in press.

Kellogg, H. E., 1964, Cenozoic stratigraphy and structure of the southern Egan Range, Nevada: Geological Society of America Bulletin, v. 75, p. 949–968.

Ketner, K. B., and Alpha, A. G., 1992, Mesozoic and Tertiary rocks near Elko Nevada—Evidence for Jurassic to Eocene folding and low-angle faulting: U.S. Geological Survey Bulletin 1988-C, 13 p.

Kidwell, S. M., Fursich, F. T., and Aigner, T., 1986, Conceptual framework for analysis and classification of fossil concentrations: Palaios, v. 1, p. 228–238.

LaRocque, A., 1970, Pleistocene Mollusca of Ohio: Ohio Geological Survey Bulletin 62, 800 p.

Leeder, M. R., and Gawthorpe, R. L., 1987, Sedimentary basins for extensional tilt-block/half-graben basins, *in* Coward, M. P., Dewey, J. F., and Hancock, P. L., eds., Continental extensional tectonics: London, Geological of London Society Special Publication No. 28, p. 139–152.

Miller, D. M., 1990, Mesozoic and Cenozoic tectonic evolution of the northeastern Great Basin, *in* Shaddrick, D. R., Kizis, J. A., Jr., and Hunsaker, E. L., III, eds., Geology and ore deposits of the northeastern Great Basin: Geological Society of Nevada, p. 43–73.

Moore, W. J., and McKee, E. H., 1983, Phanerozoic magmatism and mineralization in the Tooele 1° × 2° quadrangle, Utah, *in* Miller, D. M., Todd, V. R., and Howard, K. A., eds, Tectonic and stratigraphic studies in the eastern Great Basin: Geological Society of America Memoir 157, p. 183–190.

Murray, D. K., and Bortz, L. C., 1967, Eagle Springs oil field, Railroad Valley, Nye County, Nevada: American Association of Petroleum Geologists Bulletin, v. 51, p. 2133–2145.

Nolan, T. B., 1935, The Gold Hill mining district, Utah: U.S. Geological Survey Professional Paper 177, 172 p.

Nutt, C. J., and Thorman, C. H., 1993, Eocene sedimentary and volcanic rocks and their use in dating Mesozoic and Tertiary structures in the southern Deep Creek Range, Nevada and Utah: Geological Society of America Abstracts with Programs, Cordilleran and Rocky Mountain Sections, v. 25, no. 5, p. 128.

Nylander, O. O., 1900, Shells from the marl-deposits of Aroostook County, Maine, as compared with living forms in the same locality: The Nautilus, v. 14, p. 101–104.

Pilsbry, H. A., 1948, Land mollusca of North America (North of Mexico): Academy of Natural Science of Philadelphia, Monograph 3, 1,113 p.

Potter, C. J., Dubiel, R. F., Snee, L. W., and Good, S. C., 1995, Eocene extension of early Eocene lacustrine strata in a complexly deformed Sevier-Laramide hinterland, northwest Utah and northeast Nevada: Geology, v. 23, p. 181–184.

Rees, W. J., 1965, The aerial dispersal of Molluscs: Proceedings of the Malacological Society of London, v. 36, p. 269–282.

Robinson, J. P., 1993, Provisional geologic map of the Gold Hill quadrangle, Tooele County, Utah, with a plate showing mines, prospects, and workings by H. M. Messenger, H. H. Doelling. B. T. Tripp, and M. E. Jensen: Utah Geological Survey Map 140, scale 1:24,000.

Scott, R. W., 1970, Paleoecology and paleontology of the Lower Cretaceous Kiowa Formation, Kansas: Lawrence, University of Kansas Paleontological Contributions, Article 52 (Cretaceous 1), 94 p.

Server, G. T., Jr., and Solomon, B. J., 1983, Geology and oil shale deposits of the Elko Formation, Piñon Range, Elko County, Nevada: U.S. Geological Survey Miscellaneous Field Investigations Map MF-1546, scale 1:24,000.

Shearman, D. J., McGugan, Stein, C., and Smith, A. J., 1989, Ikaite, $CaCO_3 \cdot 6H_2O$, precursor of the thinolites in the Quaternary tufas and tufa mounds of the Lahontan and Mono Lake Basins, western United States: Geological Society of America Bulletin, v. 101, p. 913–917.

Smith, J. F., Jr., and Ketner, K. B., 1976, Stratigraphy of post-Paleozoic rocks and summary of resources in the Carlin-Piñon Range area, Nevada: U.S. Geological Survey Professional Paper 867-B, p. B1–B48.

Smith, J. F., Jr., and Ketner, K. B., 1978, Geologic map of the Carlin-Piñon Range area, Elko and Eureka Counties, Nevada: U.S. Geological Survey Miscellaneous Investigations Map I-1028, scale 1:62,500.

Smoot, J. P., and Lowenstein, T. K., 1991, Depositional environments of nonmarine evaporites, *in* Melvin, J. L., ed., Evaporites, petroleum, and mineral resources: New York, Elsevier, p. 189–345.

Solomon, B. J., and Moore, S. W., 1982a, Geologic map and oil shale deposits of the Elko East quadrangle, Elko County, Nevada: U.S. Geological Survey Miscellaneous Field Investigations Map MF-1421, scale 1:24,000.

Solomon, B. J., and Moore, S. W., 1982b, Geologic map and oil shale deposits of the Elko West quadrangle, Elko County, Nevada: U.S. Geological Survey Miscellaneous Field Investigations Map MF-1410, scale 1:24,000.

Solomon, B. J., McKee, E. H., Brook, C. E., and Smith, J. W., 1978, Tertiary geology and oil shale resources of the south Elko Basin, Nevada: American Association of Petroleum Geologists Bulletin, v. 62, p. 2362–2363.

Solomon, B. J., McKee, E. H., and Anderson, D. W., 1979, Stratigraphy and depositional environments of Paleogene rocks near Elko, Nevada, *in* Armentrout, J. M., Cole, M. R., and TerBest, H., eds., Cenozoic paleogeography of the western United States: Society of Economic Paleontologists and Mineralogists, Pacific Section, Pacific Coast Paleogeography Symposium 3, p. 75–88.

Stewart, J. H., 1980, Geology of Nevada: Nevada Bureau of Mines and Geology, Special Publication 4, 136 p.

Stewart, J. H., and Carlson, J. E., 1976, Cenozoic rocks of Nevada - Four maps and brief description of distribution, lithology, age, and centers of volcanism: Nevada Bureau of Mines and Geology, Map 52.

Stokes, W. L., 1986, Geology of Utah: Utah Museum of Natural History and Utah Geological and Mineral Survey, Occasional Paper no. 6, 280 p.

Thorman, C. H., Ketner, K. B., Snee, L. W., and Zimmermann, R. A., 1991, Late Mesozoic–Cenozoic tectonics in northeastern Nevada, *in* Raines, G. L., Lisle, R. E., Schafer, R. W., and Wilkinson, W. H., eds., Geology and ore deposits of the Great Basin: Geological Society of Nevada, Symposium Proceedings, p. 25–45.

Thorman, C. H., Brooks, W. E., Snee, L. W., Potter, C. J., Dubiel, R. F., and Ketner, K. B., 1993, Late middle Eocene Nanny Creek calc-alkaline volcanic field, NE Nevada and NW Utah: Age, extent, and implications for Eocene tectonics: Geological Society of America Abstracts with Programs, v. 25, no. 5, p. 155.

Vandervoort, D. S., and Schmitt, J. G., 1990, Cretaceous to early Tertiary paleogeography in the hinterland of the Sevier thrust belt, east-central Nevada: Geology, v. 18, p. 567–570.

Walker, B., 1900, The fossils of the marls of Huron County, Michigan: Geological Survey of Michigan, v. 7, p. 2, p. 246–252.

Walker, B., 1903, Contributions to the natural history of marl, appendix on the shells of marls: Geological Survey of Michigan, v. 8, no. 3, p. 97–102.

Wardlaw, B. R., and Collinson, J. W., 1978, Stratigraphic relations of the Park City Group (Permian) in eastern Nevada and western Utah: American Association of Petroleum Geologists Bulletin, v. 62, p. 1171–1184.

Winchester, D. E., 1923, Oil shale of the Rocky Mountain region: U.S. Geological Survey Bulletin 729, 204 p.

Winfrey, W. M., 1958, Stratigraphy, correlation, and oil potential of the Sheep Pass Formation, east-central Nevada: American Association of Petroleum Geologists Bulletin, Rocky Mountain Section of Geological Records, p. 77–82.

Winfrey, W. M., 1960, Stratigraphy, correlation, and oil potential of the Sheep Pass Formation, east-central Nevada, *in* Boettcher, J. W., and Sloan, W. W., eds., Guidebook to the geology of east-central Nevada, Intermountain Association of Petroleum Geologists, 11th Annual Field Conference, p. 126–133.

Woodburne, M. O., 1987, Cenozoic mammals of North America—Geochronology and biostratigraphy: Berkeley, University of California Press, 336 p.

Young, J. C., 1960, Structure and stratigraphy in north-central Schell Creek Range, *in* Guidebook to the geology of east-central Nevada: Salt Lake City, Utah, Intermountain Association of Petroleum Geologists and Eastern Nevada Geological Society, 11th Annual Field Conference, p. 158–172.

Manuscript Accepted by the Society April 21, 1995

Printed in U.S.A.

Geological Society of America
Special Paper 303
1996

Cenozoic basin extension beneath Goshute Valley, Nevada

Uwe Strecker, Scott B. Smithson, James R. Steidtmann
Department of Geology and Geophysics, University of Wyoming, Laramie, Wyoming 82071

ABSTRACT

High-resolution seismic reflection profiles across Goshute Valley, Nevada, record a continuum of protracted Cenozoic basin extension. A basinwide sequence boundary separates two different groups of syntectonic Tertiary seismic sequences. Seismic sequences below this seismic boundary are interpreted to record a sag stage of the basin (sequences 1–5). In contrast, the younger group of sequences is characterized by expansion of stratigraphic section across the intrabasin graben (sequences 6–8). Thickness trends of sequences suggest that extension migrated toward the eastern basin margin but culminated in the intrabasin graben. Analysis of seismic facies relative to the timing of structural basin evolution emphasizes the role of tectonism on spatial and temporal seismic facies occurrence. The sag stage is characterized by sediment transport parallel to the long axis of the basin (axial subsidence increased to the basin center) with minor sediment contributions from the eastern footwall of the basin. Transverse, east–west subsidence in the form of listric growth-faulting along the eastern intrabasin margin resulted in asymmetric sequence formation and changes in sediment dispersal patterns, probably because of decrease in axial (north-south) and increase in transverse (east-west) sediment transport slopes. This second phase of sedimentation resulted in more subdued reflectivity of Cenozoic basin fill.

INTRODUCTION

Only a few Cenozoic basins in eastern Nevada provide sufficient outcrop exposure to determine complete basin evolution from field studies alone (e.g., Solomon et al., 1979). Consequently, much basin investigation relies on geophysical techniques, usually seismic reflection profiling (e.g., Effimoff and Pinezich, 1981; Anderson et al., 1983). Previous workers used seismic profiles and line drawings made from the profiles to infer the overall geometry of basin-fill strata and style of extension (Effimoff and Pinezich, 1981; Anderson et al., 1983). Gans et al. (1985) used seismic character and unconformities to divide the basin-fill of Spring Valley, Nevada, into stratigraphic packages. In recent years workers have increasingly used sequence stratigraphic concepts in seismic interpretations of modern lacustrine basins of the East African Rift system and structural basins elsewhere (Cohen, 1990; Hanneman and Wideman, 1991; Scholz et al., 1990; Brocher et al., 1993). High-quality seismic reflection data from Goshute Valley, Nevada, provide an excellent opportunity for the application of sequence stratigraphic principles to resolution of tectonic and sedimentary features of the basin's Cenozoic evolution (Fig. 1). In addition, we use the first vertical derivative of Bouguer gravity anomalies to outline the flanks of the basin in the subsurface (e.g., Evjen, 1936; Saltus, 1988) (Fig. 2).

Contrasting seismic sequence geometries beneath Goshute Valley are interpreted to record a continuum of protracted basin subsidence and migration of extensional deformation. Analysis of seismic facies relative to the timing of structural basin evolution emphasizes the role of tectonism on the spatial and temporal occurrence of seismic facies.

GEOLOGIC SETTING

Goshute Valley, northeastern Nevada, is located between the Pequop Mountains on the west and the Toano Range on the east (Fig. 1). Both mountain ranges are mainly comprised of Paleozoic carbonate strata. Goshute Valley trends north and is

Strecker, U., Smithson, S. B., and Steidtmann, J. R., 1996, Cenozoic basin extension beneath Goshute Valley, Nevada, *in* Beratan, K. K., ed., Reconstructing the History of Basin and Range Extension Using Sedimentology and Stratigraphy: Boulder, Colorado, Geological Society of America Special Paper 303.

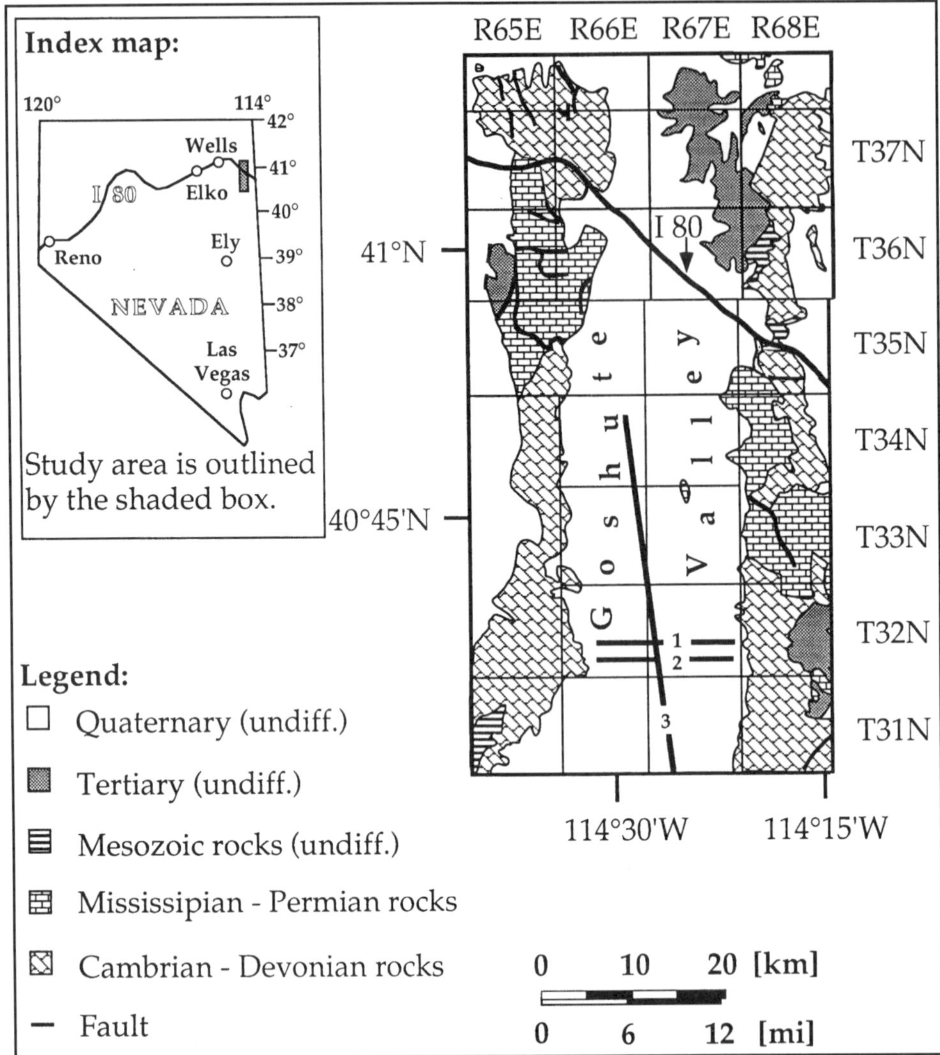

Figure 1. Simplified tectonostratigraphic map of Goshute Valley. Modified from Coats (1987) and Camilleri (1992). Because of licensing restrictions the exact locations of seismic lines 1, 2, and 3 (Plates 1, 2, and 3) cannot be revealed. Numbers 1, 2, and 3 denote seismic lines and their approximate locations.

approximately 90 km long and 19 km wide. The valley is floored by a structural basin that formed during Cenozoic Basin and Range extension (Effimoff and Pinezich, 1981; Anderson et al., 1983). Based on stratigraphic relationships, Effimoff and Pinezich (1981) suggested an upper Tertiary age for basin-fill deposits. Integration of geophysical and drilling data indicates that a maximum of 2.7 km of Cenozoic strata directly overlie Permian carbonate and shale beneath Goshute Valley (Effimoff and Pinezich, 1981). At 1,600 m depth the Shell Goshute Unit No. 1 well (Section 19, T32N R67W; API Serial No. 27-007-05207) encountered the top of the Paleozoic section represented by lower Wolfcampian strata (age analysis performed by H. R. Downs; S. Reeves, personal communication, 1993).

An east-west–trending, migrated seismic profile across Goshute Valley was previously interpreted as showing an eastward-thickening wedge of late Tertiary sediment (Effimoff and Pinezich, 1981; Anderson et al., 1983). The asymmetry of the basin fill was interpreted as having resulted from the rotation of strata by motion on a basin-bounding, westward-dipping listric master normal fault that flanks the eastern basin margin (Effimoff and Pinezich, 1981; Anderson et al., 1983). Anderson et al. (1983) suggested that a steeper dip to the listric master fault must be present at greater depth to account for the overall gentle easterly tilt of the basin-fill wedge. Using the horizontal scale given in Effimoff and Pinezich (1981) and the vertical depth scale shown in Anderson et al. (1983), an approximate dip of fault segments can be computed by connecting the fault tips to obtain a chord. In both interpretations the upper segment of the fault dips approximately 60°. However, in Effimoff and Pinezich (1981) fault dip decreases to less than 25° after the basin-bounding fault is interpreted to flatten out into a shallow detachment, whereas in Anderson et al. (1983) a steep fault dip of nearly 50°

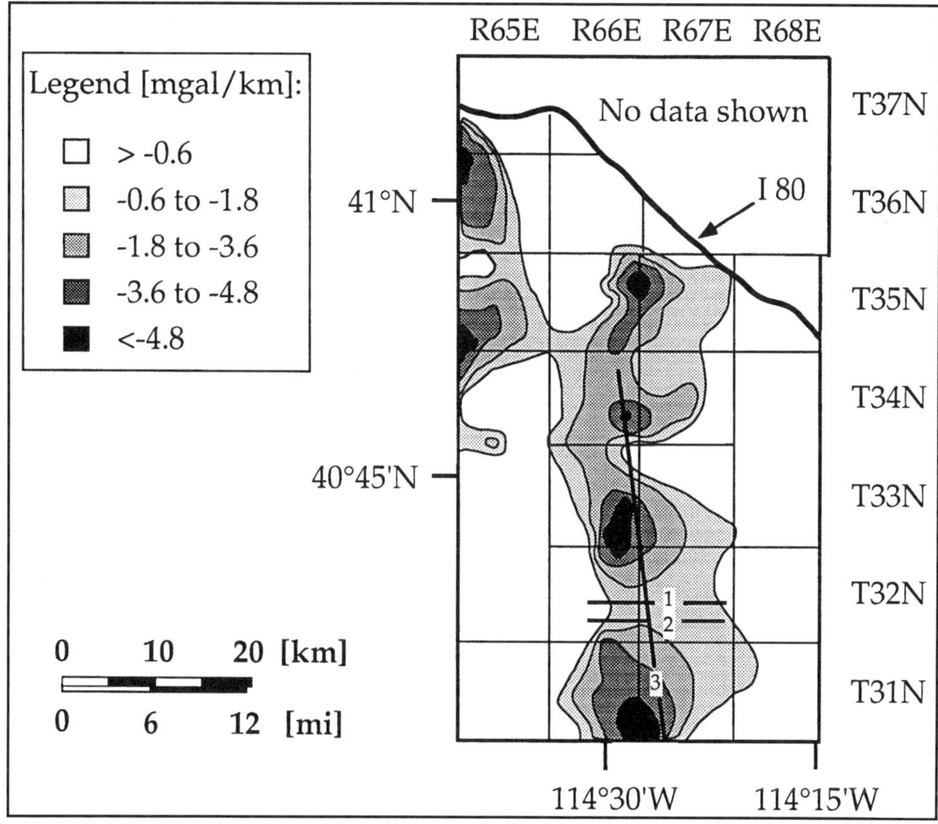

Figure 2. Map of the vertical derivative of Bouguer gravity (modified from Saltus, 1988). The first derivative of Bouguer gravity amplifies the gravitational effect of near-surface density distributions (e.g., Evjen, 1936). Negative gravity anomalies are interpreted as constraining the overall geometry of structural subbasins beneath Goshute Valley. The northern boundary of townships T33N, R66E, and R67E crosses an east-west–trending transverse bedrock high between the small northern half graben and the southern subbasin. Numbers 1, 2, and 3 denote seismic lines and their approximate location.

is maintained. Effimoff and Pinezich (1981) interpreted migration of upper Tertiary depocenters from west to east as the sedimentary response to movement along the listric fault.

SEISMIC DATA ACQUISITION AND PROCESSING

Seismic data were provided by Shell Western Exploration & Production Incorporated. The data consist of two east-west–trending profiles (lines 1 and 2; Plates 1A, 2A) that intersect a north-south line (line 3, Plate 3A). Only one migrated seismic profile of the northern east-west–trending section is available (line 1, Plate 1A). In all figures only the general location or an approximate horizontal scale is indicated due to licensing restrictions (Figs. 3–7, Plates 1–3).

The seismic data were acquired in 1974 using a 96 channel split-spread geometry and 10 pound dynamite energy sources. Each geophone array consisted of 12 GSC-11D geophones. Field instrumentation filtered out low frequencies (the exact filter range is not available for publication). All original seismic data were recorded for 4 s. Replacement velocities ranged from 1,370 m/s (4,500 ft/s for line 3 to 1,800 m/s (5,900 ft/s) for lines 1 and 2. A sloping datum (topography) was used for static corrections on all lines. After application of normal moveout corrections, seismic traces were stacked twelvefold during data processing. However, this twelvefold multiplicity of the seismic data tapers off from 1,200% to 100% over a range of about 35 stations at the ends of each seismic profile.

METHODS OF INTERPRETATION

Geological studies of Cenozoic basin-fill deposits in northeastern Nevada use unconformable or disconformable stratigraphic contacts to define formation boundaries (e.g., Solomon et al., 1979; Snoke and Lush, 1984). Similarly, we use reflection-termination patterns to delineate seismic sequence boundaries in subsurface Cenozoic deposits (e.g., Hanneman and Wideman, 1991). The pattern of truncated reflections typically includes onlap and downlap, as well as toplap stratal signatures (e.g., Mitchum and Vail, 1977).

Geometric patterns (tilt, continuity, position, and shape of

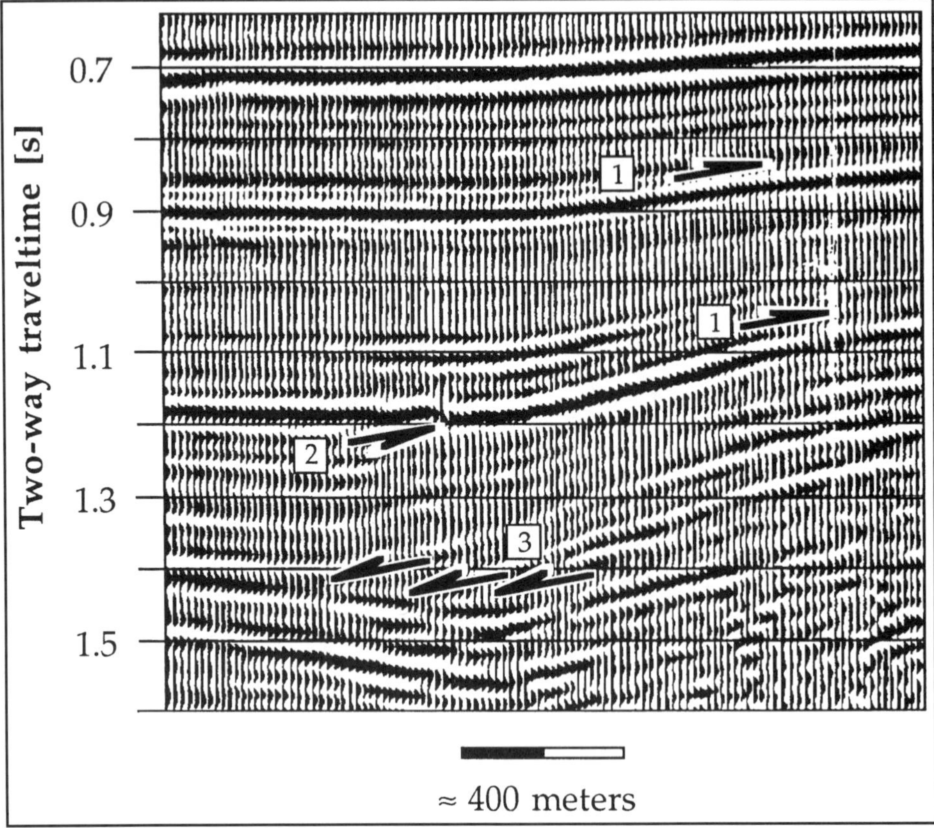

Figure 3. Interpreted detail of line 3 depicting the southern margin of the axial depocenter of the southern subbasin. Various reflection-termination patterns are shown: 1 identifies onlap, 2 denotes toplap, and 3 shows downlap. Termination patterns of stratal reflectors define sequence boundaries displayed on Plate 3B.

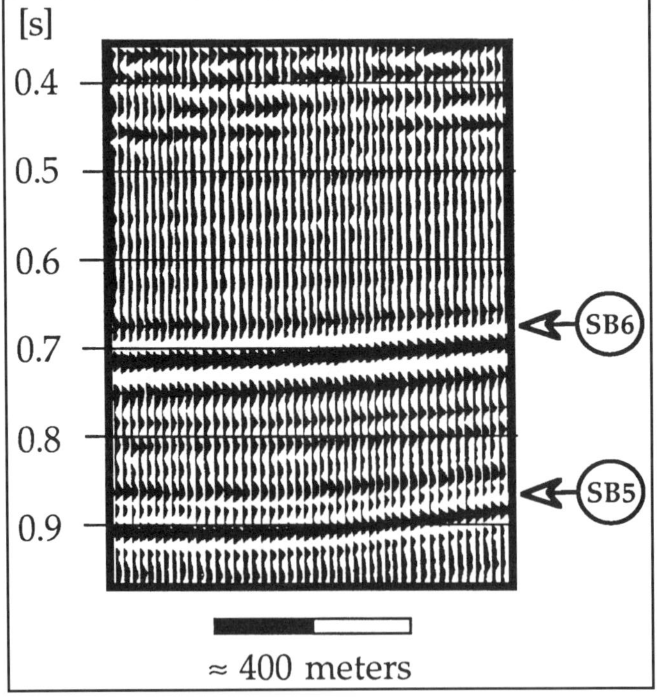

Figure 4. Detail of line 3 within the axial depocenter of the southern subbasin. Vertical scale is two-way traveltime (s). SB 6 at ≈ 0.7 s (upper arrow) separates sequence 5 (below) from sequence 6 (above). Acoustic transparency of sequence 6 is higher, because of slightly lower reflector continuity compared with sequence 5. Both reflection units in these sequences are interpreted as lacustrine facies (e.g., Anderson et al., 1983). Higher acoustic transparency within sequence 6 could be the result of a change in sediment dispersal concomitant with an increased sand/shale ratio. Alternatively, this seismic signature could indicate a higher degree of lithification of sequence 5, or a combination of both mechanisms.

reflectors) and abrupt transitions in reflection character (reflection configuration, amplitude, continuity, frequency, and interval velocity) result in a variety of distinctions that define a three-dimensional unit of seismic reflections, termed seismic facies (Brown and Fisher, 1985). Thus, a seismic facies is primarily the "recognition of lesser reflection units within a sequence" inherited from depositional processes (Brown and Fisher, 1985, p. 12; Hubbard et al., 1985). Depositional processes characterized by progradation result in time-transgressive facies patterns (e.g., Liro and Pardus, 1990). Beneath Goshute

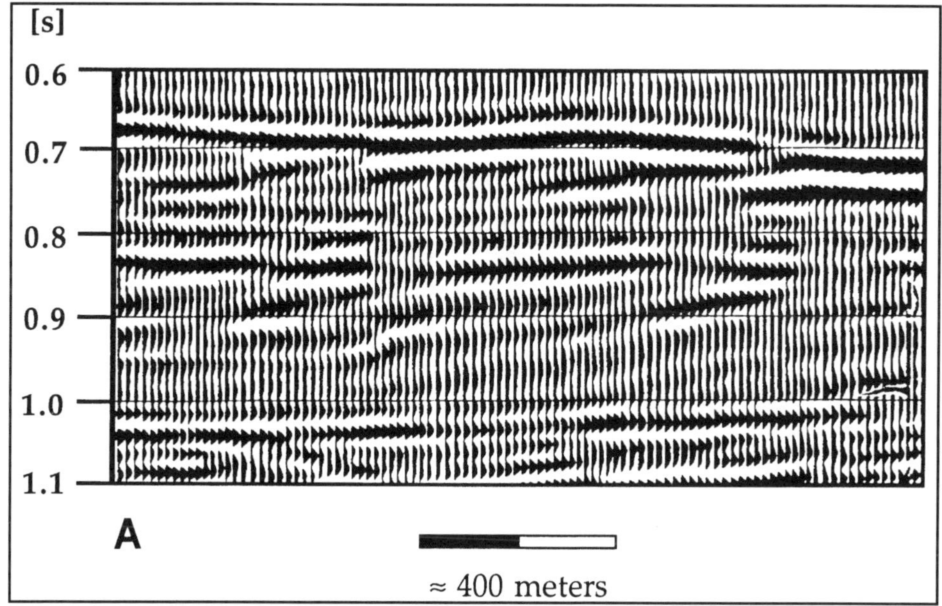

Figure 5A. Uninterpreted detail of line 3. Vertical scale is two-way traveltime (s).

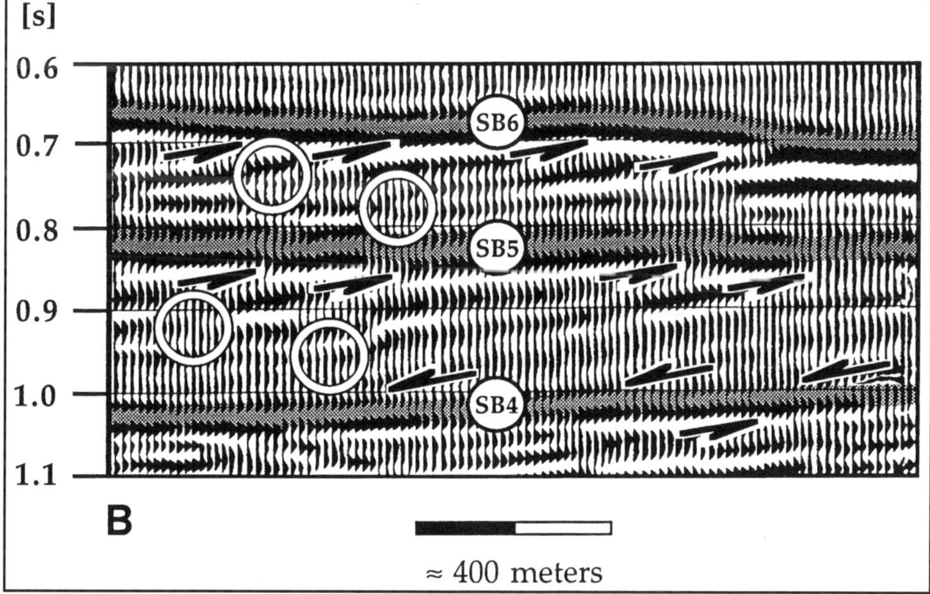

Figure 5B. Interpreted detail of line 3 at the southern margin of the axial depocenter of the southern subbasin, showing clinoforms of oblique progradational facies (OPF). Vertical scale is two-way traveltime (s). Termination patterns of stratal reflectors define interpreted sequence boundaries (gray-shaded bands). Sequence boundaries 4, 5, and 6 are identified by white solid circles labeled SB4, SB5, and SB6, respectively. Toplap against upper sequence boundaries is oblique, and downlap onto lower sequence boundaries is tangential. Where OPF clinoforms are not terminating against either sequence boundary, they are intercalated downdip with reflectors of lacustrine facies (white open circles).

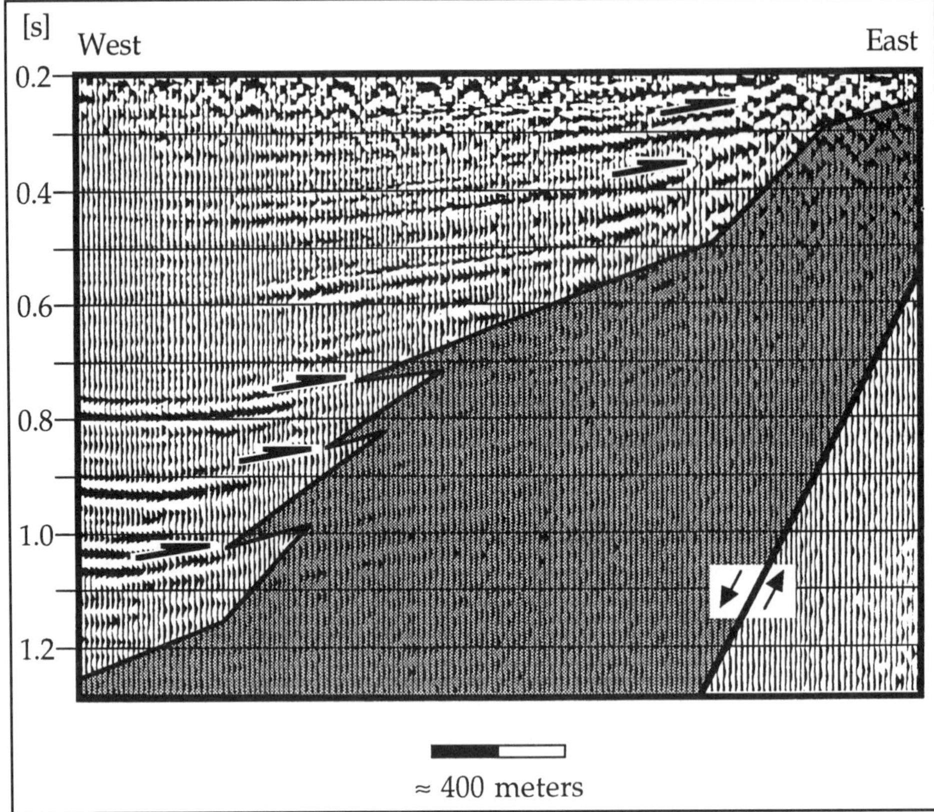

Figure 6. Interpreted detail of line 1 at the eastern basin margin of the southern subbasin. Black half-arrows (white background) denote terminations of seismic reflectors by onlap. Acoustically transparent unit (gray background) is interpreted as coarse-grained fault front facies (FFF) (e.g., Anderson et al., 1983). Normal displacement of basin-fill deposits is indicated by full arrows in box adjacent to the eastern margin fault. Vertical scale is two-way traveltime (s).

Valley, the most prominent progradations of seismic facies occur at the eastern basin margin (Plate 1, Fig. 7), and at the southern margin of the axial depocenter (Plate 3).

Similar external and internal attributes of seismic sequences define a sequence group; external attributes include criteria such as the basin location, geometry, and symmetry of a sequence, whereas type and migratory pattern of seismic facies constitute internal attributes (e.g., Hubbard et al., 1985).

STRUCTURAL STYLE

The structural configuration of the basin fill must be interpreted prior to sequence stratigraphic analysis to avoid stratal miscorrelations. On all sections, faulting in the seismic data is interpreted on the basis of structural offset of reflectors that are seismically similar and by the presence of subvertically aligned diffractions on unmigrated seismic profiles. In a zone of poor seismic reflectivity, we used the easternmost termination of interpreted seismic sequence boundaries as seed points to approximate the trace of the eastern margin fault (EMF) in the subsurface (discussed in a following section).

Line 3 (Plates 3A, 3B) suggests that two subbasins coalesced across an approximately east-west–trending, transverse, pre-Cenozoic bedrock high (Fig. 2) (e.g., Effimoff and Pinezich, 1981; Anderson et al., 1983). This report focuses on the larger subbasin to the south.

The southern subbasin decreases in depth both toward the north against the transverse bedrock high and toward the south (Plate 3A). The axial depocenter of this subbasin is depicted in gravity data as an oval-shaped, NNE-trending low more than 7 km long and 6 km wide (minimum ≤-4.8 mgal/km) (Fig. 2). North of the transverse bedrock high a half graben developed. On line 3 (Plates 3A, 3B), the deepest portion of that basin faces south.

Based on the two east-west–trending seismic profiles, the southern subbasin can be divided into three major fault blocks of differing geometry: (1) a gently east-tilted fault block in the west, (2) an intrabasin graben, and (3) an eastern fault block adjacent to the Toano Range. Each one of these structures within the basin fill is several kilometers wide (Plates 1, 2).

The western fault block is downfaulted against a horst composed of miogeoclinal Mesozoic and Paleozoic rocks (Coats,

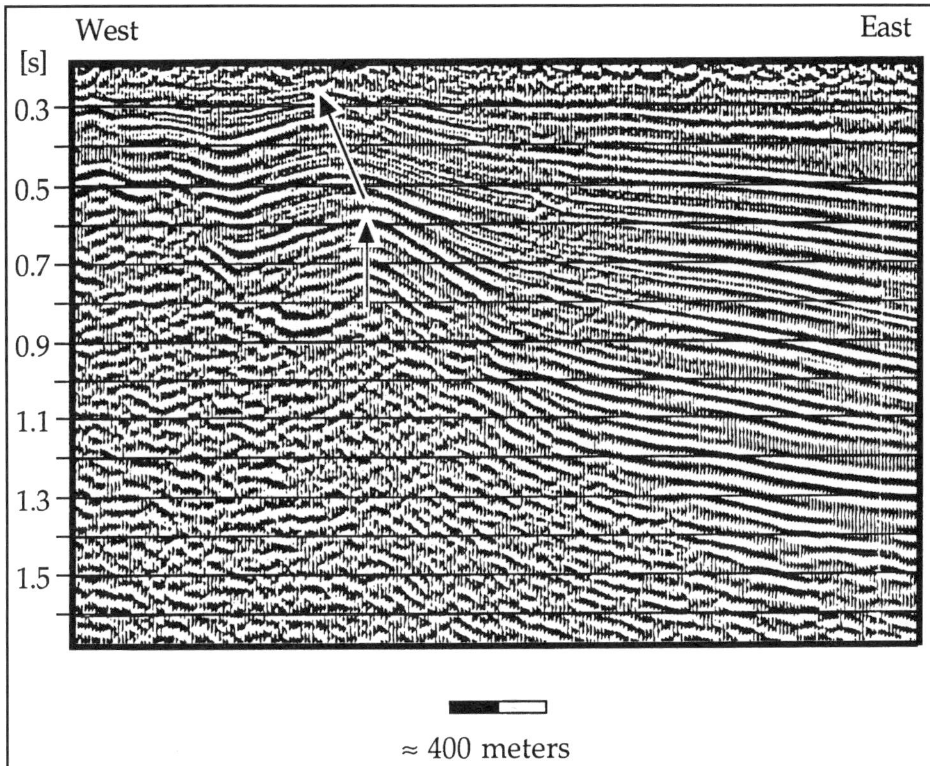

Figure 7. Detail of east-west–trending migrated section at the western margin of the southern sub-basin (line 1). Arrows connect successive fold crests of anticlinal strata. Fold crests in the lower portion of the anticline align vertically, whereas the axial plane in the upper portion of the anticline dips east. Basin-fill deposits become stratigraphically attenuated toward the anticline, suggesting that onlap toward a syndepositionally active structure occurred. Vertical scale is two-way traveltime (s).

1987; Camillieri, 1992). This horst, the subcrop of the Pequop Mountains, is internally dissected by several normal-fault splays that combine to form horst-and-graben topography buried beneath Cenozoic sediments. Normal faulting within the western fault block itself is negligible. Near the western basin margin an anticline is overlain by a thin veneer of subhorizontal strata (Plate 1A). In the lower part of the anticline the axial plane is vertical, whereas in the upper part of the anticline the axial plane dips east (see discussion in a following section).

In contrast to its bounding fault blocks, the intrabasin graben is internally dissected by numerous normal faults (Plates 1A, 1B). From west to east, the key structural elements of the intrabasin graben are formed by a growth-faulted, small crestal collapse graben (e.g., Faure and Chermette, 1989), a horst-and-graben series near the center, and two fault blocks at the eastern margin. Major faults within the intrabasin graben are synthetic to the bounding faults with the exception of a series of horsts and grabens near the center and the crestal collapse graben (Plate 1). The small crestal compensation graben is growth faulted and roots antithetically into the western intrabasin graben fault. The pair of fault blocks, each several hundred meters wide, and the horst-and-graben series record only minor structural offsets. Farther to the east, the intrabasin graben is downthrown against the eastern fault block along the eastern intrabasin fault.

The eastern fault block, several kilometers wide, bounds the basin fill against the eastern basin margin. Although patterns resembling structural offsets along the small horsts and grabens in the intrabasin graben may be present in the upper section, no significant faulting of this block is interpreted from seismic data. The eastern fault block is dominantly comprised of basin fill.

Except for the northern boundary of the transverse intrabasin high, faulting on line 3 (Plate 3B) is subdued. However, faulting of the intrabasin graben is interpreted from structural offset of several basin-fill reflectors south of the axial depocenter (Plates 3A, 3B).

SEISMIC STRATIGRAPHY

Termination patterns of stratal reflections

Updip terminations of stratal reflections are seen along the boundary between Pre-Cenozoic bedrock and Tertiary basin fill (Plates 1B, 2B, 3B) and at several locations within the Tertiary section (Plates 1B, 2B, 3B; Fig. 3). This interpreted onlap occurs near both basin margins on lines 1 and 2 (Plates 1B, 2B) and in the vicinity of the southern margin of the axial depocenter on line 3 (Plate 3, Fig.3). On line 3, updip termination of clinoform-shaped stratal reflections is not associated with the development of paleotopography and is interpreted as toplap. The most prominent toplap reflector signatures occur at the southern margin of the axial depocenter (see discussion in a following section). Downdip termination of stratal reflections is observed along the southern margin of the axial depocenter and farther south (line 3; Plate 3B). Apparent dip of toplap and downlap reflectors is generally steeper than that of onlapping reflectors.

Diffraction apices present between 1.1 s and 1.6 s on unmigrated line 3 (Plate 3A) are interpreted to originate from the erosional truncation of pre-Cenozoic bedrock stratal reflections (e.g., Sheriff and Geldart, 1986). Thus, connecting the crests of neighboring diffractions defines the interpreted base of Tertiary basin fill (Plate 3B) (e.g., Snyder, 1988). Where diffractions are not prominent or are absent on seismic profiles, the base of the Tertiary section is defined by the onset of discontinuous seismic reflections (Plate 3A, 3B) (e.g., Hastings, 1979).

Unconformities

Angular stratal seismic reflections, exemplified by toplap, onlap, and downlap, define interpreted unconformable surfaces constituting seismic sequence boundaries. Seismic sequence boundaries (SBs) are informally designated as SB 1–SB 9, with SB 1 being the interpreted boundary between Paleozoic bedrock and Tertiary basin fill, and SB 9 being the present-day valley floor. Because of licensing restrictions, the topography of the valley floor (i.e., SB 9) is not superposed onto the seismic sections.

On the east-west–trending sections, sequence boundaries appear to span the entire southern subbasin (Plates 1B, 2B). However, SB 3 merges with SB 2 at the base of the southern slope of the axial depocenter on line 3 (Plate 3B, Fig. 3).

Compared to most stratal reflections, interpreted sequence boundaries are generally represented by low-frequency and high-amplitude seismic reflections. For instance, on all three seismic sections onlap, downlap, and toplap delineate SB 6, a prominent low-frequency reflector of basinwide continuity (dominant frequency ≈ 20 Hz) (Plates 1B, 2B, 3B). However, on line 3 (Plate 3B), acoustic contrast across SB 7 and SB 8 appears decreased compared to SB 6. Moreover, reflection character of SB 7 and SB 8 varies laterally in distinction to older sequence boundaries, in particular across the axial depocenter (Plate 3B). In the depocenter, SB 7 and SB 8 appear concordant and nearly indistinguishable from stratal reflections.

Seismic sequences

On lines 1 and 2 (Plates 1B, 2B), S3–S9 denote interpreted seismic sequences 3–9. Additional sequences (S1–S2) are shown on line 3 (Plate 3B). Fewer sequences are discernible on lines 1 and 2 (Plates 1B, 2B) because close fault spacing in the deepest parts of the basin inhibits reliable cross-correlations of basin-fill deposits.

All sequences are faulted. When structurally balancing basin-fill deposits in Plate 1B, reflectors contained in sequences 1–7 match across the eastern intrabasin fault, but only sequence 8 is stratigraphically expanded adjacent to the eastern intrabasin fault. However, sequences 6 and 7 thicken across the central intrabasin graben on lines 1, 2, and 3 (Plates 2B, 3B).

Two seismic sequence groups are defined based on similarity of geometry, seismic attributes, and seismic facies. A discussion of differences between the two sequence groups regarding thickness and seismic attributes (frequency content, reflectivity, reflector continuity, etc.) follows.

Sequences 1–5. Seismic sequences 1–5 thin toward both basin margins on lines 1 and 2 (Plates 1, 2), and no thickening of stratal reflections or sequences into the eastern margin fault is observed (Plates 1, 2). On line 3 (Plate 3B), maximum thickness of this sequence group is in the axial depocenter. Farther north, toward the transverse intrabasin high, these sequences once again become stratigraphically attenuated.

Sequences 6–8. All of these sequences also thin toward the basin margins. However, in contrast to sequences 1–5, the stratigraphic section of these sequences is expanded across the intrabasin graben. On the east-west–trending lines, the thickness of sequence 8 is at its maximum adjacent to the eastern intrabasin fault. Sequence 8 is thus characterized by a pronounced asymmetry resulting from the preferred accumulation of basin fill in the eastern portion of the intrabasin graben (Plates 1B, 2B). Excess sediment thickness here amounts to 150 ms two-way traveltime (140 m). However, across the eastern fault block, the geometry of sequence 8 resembles that of previously deposited sequences. Sequences 6 and 7 appear expanded across the intrabasin graben on line 3 (Plates 3A, 3B). Sequences 6 and 7 thicken considerably in the central portion of the intrabasin graben on lines 1 and 2 (Plate 2B).

Seismic facies

Although seismic facies within the depocenter occur in the same spatial position, subtle differences in seismic attributes emerge across sequence boundary 6 (Fig. 4). Stratal reflections above SB 6 are less continuous than reflectors below. However, reflectors above and below this unconformity are similar, and reflections appear conformable and continuous (Plate 3, Fig. 4).

Both reflection units are probably related to well-layered, relatively fine grained strata deposited in shallow depressions. Continuous reflections, like those observed above and below SB 6, have been related to quiet-water sedimentation (e.g., Anderson et al., 1983). A lacustrine origin is compatible with this seismic signature and paleotopographic position. Higher continuity of stratal reflections below SB 6 is interpreted as the result of a higher degree of lithification or a decreased sand/shale ratio.

A second distinct seismic facies occurs near the southern edge of the depocenter (Plate 3, Fig. 5). SB 8 is the youngest seismic boundary below which this seismic facies is seen. Reflectors within sequences 1–7 display progradation in an axial direction. Stratal reflections are characterized by an oblique termination at the upper sequence boundary (toplap against SB 6) and a tangential contact with the lower sequence boundary (downlap onto SB 5; Fig. 5). Moreover, clinoform reflectors that do not terminate against either sequence boundary intercalate downdip with reflectors of interpreted lacustrine origin (Fig. 5). The continuity of progradational reflectors is high in the upper part of the clinoform, but becomes increasingly lower in the mid- and lower parts of the clinoform (Fig. 5). Additionally, reflector frequency and amplitude generally decrease downdip (e.g., Brown and Fisher, 1985). Compared to lacustrine facies, clinoform reflectors are characterized by lower frequencies and display slightly lower interval velocities.

This observed stratal reflector geometry is compatible with the oblique progradational clinoform of Sangree and Widmier (1977), herein termed OPF. Beneath Goshute Valley, OPF is found near the southern edge of the axial depocenter approximately where a gradient in the first vertical derivative of Bouguer gravity occurs, indicating that this facies accumulated on the southern flank of this basin (Fig. 2). Stratigraphic position and seismic attributes of the oblique progradational facies comply with the depositional environment of a delta (e.g., Brown and Fisher, 1985).

Another seismic facies primarily found within sequences 1–5 is characterized by semicontinuous to discontinuous low-frequency reflections defining an area of acoustic near-transparency adjacent to the eastern basin margin (Anderson et al., 1983). The maximum basinward extent of this poorly reflective unit occurs along the lower sequence boundary (Fig. 6). Overlying seismic sequences bordering the EMF also lack good reflections, resulting in a pattern of saw-tooth-shaped, poorly reflective, vertically stacked wedges (Fig. 6). Low-frequency, semicontinuous to discontinuous reflectors are not well represented in sequences above SB 6.

Similar seismic signatures have been related to the heterogeneity of sedimentologic processes on alluvial fans (Effimoff and Pinezich, 1981; Anderson et al., 1983; Okaya and Thompson, 1985). The fact that alluvial-fan deposits are often poorly layered explains the paucity of continuous reflectors in this coarse-grained fault-front seismic facies (FFF) (e.g., Anderson et al., 1983). Fault-front facies is found in the sequence succession above SB 6, but constitutes a smaller proportion of the footwall-sourced sediments.

DISCUSSION

Crustal-scale deformation is commonly expressed in the fill geometry of basins (e.g., Okaya and Thompson, 1985). We have applied seismic stratigraphy using reflection termination patterns, reflection character of seismic sequence boundaries, geometry of seismic sequences, and occurrence and migration of seismic facies to detect changes in the tectonic history of this continental basin, recorded as changes in fill geometry.

Earlier structural interpretations of Goshute Valley based on conventional line drawings and seismic profiles suggested (1) an asymmetric basin fill in response to growth faulting, (2) a listric master normal fault bounding the basin fill to the east, and (3) a temporal shift of depocenters toward the master fault (Effimoff and Pinezich, 1981; Anderson et al., 1983). Anderson et al. (1983) observed that the overall gentle tilt of the basin fill was not compatible with the high degree of curvature of a listric normal fault shown in Effimoff and Pinezich (1981). To account for the low easterly dip of basin-fill strata, Anderson et al. (1983, p. 1064) suggested "the need for deeply penetrating strands of the listric fault system"; i.e., these authors advocated the presence of a steeply dipping, listric master fault. Alternatively, we suggest that the overall gentle easterly tilt of Tertiary basin fill is the result of a continuum of protracted extension expressed in two contrasting groups of seismic sequences.

Syntectonic sequences and style of faulting

Several workers recognized that the dip of basin fill can be used "to infer selected aspects of subsurface fault geometry" (Anderson et al., 1983, p. 1056; Okaya and Thompson, 1985). In contrast to well-imaged stratal reflections within the basin, the structural geometry of the basin margins in Goshute Valley is poorly constrained. However, the angularity of reflections at sequence boundaries can be used to reconstruct depositional history and to infer paleoenvironmental conditions (e.g., Hardage, 1987). One such reflection pattern, toplap, is commonly interpreted to have been "caused by nondeposition along the top of a prograding wedge of sediments" (e.g., Hardage, 1987, p. 31). Thus, with sedimentation rates high enough to allow sequence aggradation to base level throughout most of the basin (see nondepositional upper sequence boundary of OPF), we use thickness changes of syntectonic sequences to infer the basinwide distribution of tectonic subsidence and to constrain master fault geometry. For example, equal sequence thickness attests to uniform subsidence, whereas symmetric or asymmetric sequence geometries reflect differential subsidence.

Thinning of sequences 1–5 toward the basin margins, particularly the eastern basin margin, is not compatible with a growth-faulting pattern of sedimentation indicative of a listric normal fault (e.g., Effimoff and Pinezich, 1981; Anderson et al., 1983). However, thickness trends of sequences in this group (sequences 1–5) suggest increased differential subsidence toward the east through time (Effimoff and Pinezich, 1981) (Plates 1B, 2B). Except for sequence 3, which is not basinwide, sequences are at maximum thickness in the axial depocenter (Plate 3B).

We interpret the observed external geometry of sequences 1–5 as the result of sagging with increased subsidence near the basin axis and maximum subsidence at the axial depocenter. The seismic data suggest that this sequence group probably formed in response to displacement along a steeply dipping segment of a north-south–trending normal fault bounding the basin to the east. Whether the curvature of the EMF increases with greater depth (i.e., becomes listric) cannot be determined because of insufficient velocity information. However, the interpretation of a near-planar normal fault segment overlying a (sub)horizontal detachment is supported by two independent lines of analogous inference. First, sandbox analog modeling of Ellis and McClay (1988, p. 118, fig. 5b) bears a striking resemblance to the structural geometry of the Goshute Valley. In this model, a central graben is downfaulted against a footwall block and subsidiary hanging wall fault blocks. Sheared, pre-extensional, hanging-wall horizons of the sandbox analog model suggest that attenuation of synrift strata near the eastern basin margin of Goshute Valley may be partly the result of fault plane drag. Second, a seismic reflection profile across a basin in the Bay of Biscay identifies a graben that closely compares in size, scale, and structural configuration to the basin-fill geometry of Goshute Valley (Faure and Chermette, 1989, p. 464, fig. 6). Faure and Chermette (1989) used the term "pseudo-rollover" to denote a syntectonic basin-fill geometry resembling that of a listric growth-faulted basin which instead accumulated adjacent to a planar fault.

Three fundamental changes in seismic properties occur across sequence boundary 6: (1) a stratigraphic thickening of sequences 6, 7, and 8 across the intrabasin graben (Plates 1B, 2B, 3B), (2) variations of seismic attributes (frequency content, reflectivity, reflector continuity, etc.), and (3) resulting seismic facies changes (Plates 1A, 1B, 2A, 2B, 3A, 3B; Fig. 5A, 5B). Stratigraphic expansion of sequences 6 and 7 and east-west asymmetry of sequence 8 attests to a shift of differential subsidence from an axial subsidence pattern to a transverse one. The crestal collapse graben at the western end of the intrabasin graben is interpreted as a structure accommodating hanging-wall strain associated with the rollover geometry of sequence 8.

Lack of paleotopographic relief along sequence boundaries supports their nondepositional origin, and implies closed drainage. If acoustic contrast across sequence boundaries is related to lithification, then the strong impedance contrast across sequence boundary 6 may indicate a period of "tectonic doldrums" prior to deformation of the intrabasin graben. In contrast, correlative conformity of sequence boundaries 7 and 8 in the depocenter and their decreased acoustic contrast elsewhere in the basin possibly indicate a lesser degree of induration and may reflect an acceleration of tectonism and sedimentation.

Extensional deformation

Review of the above data reveals that the site of maximum subsidence migrated east during initial extension, as recorded by the eastward shift of depocenters through time (Effimoff and Pinezich, 1981), but that subsidence parallel to the long-axis of the basin was dominant. Subsequent deformation is indicated by the initiation and collapse of the rollover geometry of basin fill. This pattern is observed in rifted basins elsewhere (Strecker et al., 1990). This pattern is probably related to the presence of a (sub)horizontal detachment horizon, or, possibly, a shallowing of the master fault at depth.

Onlap of strata and thinning of sequences toward the anticline near the western basin margin suggest the presence of a syndepositionally rising structural block (Fig. 7). An east-dipping axial plane in the upper portion of the structure suggests updip migration of fold crests above a fault with an undulation that creates a space problem (e.g., Billingsley, 1982). However, neither migrated seismic nor gravity data clearly support the presence of a fault, although diffractions are present on the unmigrated profile. A similar structure, located in the vicinity of a basin-bounding fault, is depicted on two seismic profiles across Railroad Valley, Nevada (Bortz and Murray, 1979, fig. 9; Foster, 1979, fig. 2).

In summary, the application of sequence stratigraphy to the Goshute basin identifies an asymmetric graben containing two distinctly different successions of synextensional sequences that record sagging of the entire basin and growth faulting of the intrabasin graben. Differences in fault angles influence sequence geometries and may be the result of differing mechanical response to contrasting lithologies. During early extension, faults cut pre-Cenozoic bedrock, whereas later listric normal faults dominantly cut poorly consolidated Tertiary sediments resulting in stronger fault curvature. Similar mechanical behavior has been modeled and is reported for listric faults in sediments of the Gulf Coast (Bradshaw and Zoback, 1988).

Tectonic stages and seismic facies

OPFs mainly prograde north, toward the main depocenter of the southern subbasin (Plate 3). Although onlap/downlap patterns represent only apparent dips, they appear to be maximum in the direction of the basin axis. Coarse-grained fault-front sediments are contributed to the basin near the eastern basin margin even during axial sedimentation (Anderson et al., 1983). The onset of growth faulting of the intrabasin graben appears to diminish and eventually terminate coarse-grained fault-front sedimentation.

Explanations for the waning importance of both OPF and FFF above SB 6 must remain purely speculative because of the limited number of seismic sections available. A possible explanation for the observed decrease in progradational clinoform dip could be the onset of growth faulting that reduces depositional slopes in an axial direction. Slightly shallower dips of OPF clinoforms higher in the section support a decrease in topographic slopes in an axial direction (Plate 3, line 3). However, delta-lobe shift, or a combination of both mechanisms may be equally valid. Likewise, various mechanisms could explain the reduced importance of FFF: depletion of a coarse-clastic source, shift of the fan conduit caused by an increasing strike-slip component in basin displacement, climate change, depositional fan lobe switching, or some combination of these mechanisms. The most likely cause, however, is the onset of growth faulting of the intrabasin graben that reduced the amount of tectonic subsidence along the eastern basin margin and thus decreased topographic relief required for fan development. Listric growth faulting resulted in change in sediment dispersal so that the eastern fault block, instead of the Toano Range, was the sediment source. Although this fault block remained below depositional base level, cannibalization of previously deposited, finer grained sediment from this newly formed footwall would have almost certainly inhibited alluvial fan development.

CONCLUSIONS AND IMPLICATIONS

(1) Application of sequence stratigraphy to a structural subbasin beneath Goshute Valley allowed identification of an asymmetric graben that contains two distinctly different groups of synextensional sequences. The first sequence group records sag along an essentially planar fault, and the second sequence group formed dominantly in response to listric growth faulting of the intrabasin graben.

(2) Axial (parallel to the north-south–trending long axis of the basin) and transverse (east-west) components of differential subsidence influence sequence geometry. Sequences that accumulated during the sag stage thin toward both basin margins on the east-west–trending sections. Sequences that accumulated during the deformation of the intrabasin graben are characterized by expansion of stratigraphic section and thicken asymmetrically. This second phase of sedimentation resulted in more subdued reflectivity of basin fill. Thickness trends of sequences suggest that extension migrated toward the eastern basin margin, but culminated in the intrabasin graben.

(3) Partitioning of differential subsidence into axial and transverse directions may affect the type and distribution of seismic facies contained in a sequence. Therefore, knowledge of prevailing deformational styles facilitates the prediction of tectonic basin maturation and of seismic facies (e.g., Lambiase, 1990).

(4) Seismic reflection data indicate that the north-south–trending subbasin coalesced with another subbasin to the north across a transverse, approximately east-west–trending intrabasin high.

(5) Seismic reflection and gravity data indicate that these subbasins are two of four structural basins that floor Goshute Valley. Three of these subbasins trend approximately north-south; the subbasin north of the transverse intrabasin high appears to trend east-west.

(6) In a regional context, seismic sequence stratigraphy can be used to demonstrate extensional stages and tectonic gradients for elements of the Basin and Range Province. In conjunction with previously documented geologic evidence, this information can serve to identify Basin and Range Province crustal-scale deformation expressed in the fill geometry of superposed basins (Okaya and Thompson, 1985).

ACKNOWLEDGMENTS

This work was supported in part by Geological Society of America Student Research Grants to Strecker. The authors would like to thank Shell Western Exploration & Production Incorporated for the opportunity to publish these data. Strecker would like to thank Edwin R. Goter, Rick J. Marinko, Scott Reeves, and Dan M. Worrall at Shell Western, and Alan Ramelli at the Nevada Bureau of Mines and Geology, Reno. The authors thank Kathi Beratan, Phyllis Camilleri, William Clement, Manfred Strecker, and three anonymous reviewers for providing invaluable comments on earlier versions of this manuscript.

REFERENCES CITED

Anderson, R. E., Zoback, M. L., and Thompson, G. A., 1983, Implications of selected subsurface data on the structural form and evolution of some basins in the northern Basin and Range Province, Nevada and Utah: Geological Society of America Bulletin, v. 94, p. 1055–1072.

Billingsley, L.T., 1982, Geometry and mechanisms of folding related to growth-faulting in Nordheim Field Area (Wilcox), De Witt County, Texas: Transactions, Gulf Coast Association of Geological Societies, v. 32, p. 263–274.

Bortz, L. C., and Murray, D. K., 1979, Eagle Springs Oil Field Nye County, Nevada, *in* Newman, G. W., and Goode, H. D., eds., Rocky Mountain Association of Geologists–Utah Geological Association 1979 Basin and Range Symposium: Denver, Colorado, Rocky Mountain Association of Geologists, p. 441–453.

Bradshaw, G. A., and Zoback, M. D., 1988, Listric normal faulting, stress refraction, and the state of stress in the Gulf Coast basin: Geology, v. 16, p. 271–274.

Brocher, T. M., Carr, M. D., Fox, K. F., Jr., and Hart, P. E., 1993, Seismic reflection profiling across Tertiary extensional structures in the eastern Amargosa desert, southern Nevada, Basin and Range Province: Geological Society of America Bulletin, v. 105, p. 30–46.

Brown, L. F., Jr., and Fisher, W. L., 1985, Seismic stratigraphic interpretation and petroleum exploration (fourth printing): Tulsa, Oklahoma, American Association of Petroleum Geologists Continuing Education Course Note Series # 16, 56 p.

Camilleri, P. A., 1992, Mesozoic structural and metamorphic features in the Wood Hills and Pequop Mountains, Northeastern Nevada, *in* Wilson, J. R., ed., Field guide to geologic excursions in Utah and adjacent areas of Nevada, Idaho, and Wyoming: Utah Geological Survey Miscellaneous Publication 92-3, p. 93–105.

Cohen, A. S., 1990, Tectono-stratigraphic model for sedimentation in Lake Tanganyika, Africa, *in* Katz, B. J., ed., Lacustrine basin exploration—Case studies and modern analogs: Tulsa, Oklahoma, American Association of Petroleum Geologists Memoir 50, 340 p.

Coats, R. R., 1987, Geologic map of Elko County, Nevada: Nevada Bureau of Mines and Geology Bulletin 101, Plate 1, scale 1:250,000.

Effimoff, I., and Pinezich, A. R., 1981, Tertiary structural development of selected valleys based on seismic data: Basin and Range Province, northeastern Nevada: Royal Society of London, Philosophical Transactions, Series A, v. 300, p. 435–442.

Ellis, P. G., and McClay, K. R., 1988, Listric extensional fault systems—Results of analogue model experiments: Basin Research, v.1, p. 55–70.

Evjen, H. J., 1936, The place of the vertical gradient in gravitational interpretations: Geophysics, v.1, p. 127–136.

Faure, J.-L., and Chermette, J.-C., 1989, Deformation of tilted blocks, consequences on block geometry and extension measurements: Bulletin de la Société géologique de France, no. 3, p. 461–476.

Foster, N. H., 1979, Geomorphic exploration used in the discovery of Trap Spring Oil Field, Nye County, Nevada, *in* Newman, G. W., and Goode, H. D., eds., Rocky Mountain Association of Geologists–Utah Geological Association 1979 Basin and Range Symposium: Denver, Colorado, Rocky Mountain Association of Geologists, p. 477–486.

Gans, P. B., Miller, E. L., and McCarthy, J., and Ouldcott, M. L., 1985, Tertiary extensional faulting and evolving ductile-brittle transition zones in the northern Snake Range and vicinity: New insights from seismic data: Geology, v. 13, p. 189.

Hanneman, D. L., and Wideman, C. J., 1991, Sequence stratigraphy of Cenozoic continental rocks, southwestern Montana: Geological Society of America Bulletin, v. 103, p. 1335–1345.

Hardage, B., ed., 1987, Seismic stratigraphy, *in* Helbig, K., and Treitel, S., eds., Handbook of geophysical exploration, section I, seismic exploration, v. 9: London, Geophysical Press, 432 p.

Hastings, D. D., 1979, Results of exploratory drilling Northern Fallon Basin, Western Nevada, *in* Newman, G. W., and Goode, H. D., eds., Rocky Mountain Association of Geologists–Utah Geological Association 1979 Basin and Range Symposium: Denver, Colorado, Rocky Mountain Association of Geologists, p. 515–522.

Hubbard, R. J., Pape, J., and Roberts, D. G., 1985, Depositional sequence mapping as a technique to establish tectonic and stratigraphic framework and evaluate hydrocarbon potential on a passive continental margin, *in* Berg, O. R, and Woolverton, D. G., eds., Seismic stratigraphy, Volume 2: Tulsa, Oklahoma, American Association of Petroleum Geologists Memoir 39, 276 p.

Lambiase, J. J., 1990, A model for tectonic control of lacustrine stratigraphic sequences in continental rift basins, *in* Katz, B. J., ed., Lacustrine basin exploration—Case studies and modern analogs: Tulsa, Oklahoma, American Association of Petroleum Geologists Memoir 50, 340 p.

Liro, L. M., and Pardus, Y. C., 1990, Seismic facies analysis of fluvial-deltaic lacustrine systems—Upper Fort Union Formation (Paleocene), Wind River Basin, Wyoming, *in* Katz, B. J., ed., Lacustrine basin exploration—Case studies and modern analogs: Tulsa, Oklahoma, American Association of Petroleum 340 p.

Mitchum, R. M., and Vail, P. R., 1977, Seismic stratigraphy and global changes of sea level, part 7: Seismic stratigraphic interpretation procedure, *in* Payton, C. E., ed., Seismic stratigraphy—applications to hydrocarbon exploration: Tulsa, Oklahoma, American Association of Petroleum Geologists Memoir 26, 516 p.

Okaya, D. A., and Thompson, G. A., 1985, Geometry of Cenozoic extensional faulting: Dixie Valley, Nevada: Tectonics, v. 4, p. 107–125.

Saltus, R. W., 1988, Regional, residual, and derivative gravity maps of Nevada: Reno, Nevada Bureau of Mines and Geology Map 94B, 4 sheets, scale 1:1,000,000.

Sangree, J. B., and Widmier, J. M., 1977, Seismic stratigraphy and global changes of sea level, part 9: Seismic interpretation of clastic depositional facies, *in* Payton, C. E., ed., Seismic stratigraphy—applications to hydrocarbon exploration: Tulsa, Oklahoma, American Association of Petroleum Geologists Memoir 26, 516 p.

Scholz, C. A., Rosendahl, B. R., and Scott, D. L., 1990, Development of coarse-grained facies in lacustrine rift basins: Examples from East Africa: Geology, v. 18, p. 140–144.

Sheriff, R. E., and Geldart, L. P., 1986, Exploration seismology: History, theory, and data acquisition, Volume 1: Cambridge, United Kingdom, Cambridge University Press, 253 p.

Snoke, A. W., and Lush, A. P., 1984, Polyphase Mesozoic-Cenozoic deformational history of the northern Ruby Mountains–East Humboldt Range, Nevada, *in* Lintz, J., Jr., ed., Western geological excursions: Geological Society of America, 1984 Annual Meeting Field Trip Guidebook: Reno, Nevada, Mackay School of Mines, v. 4, p. 232–260.

Snyder, D., 1988, Foreland crustal geometries in the Andes of Argentina and the Zagros of Iran from seismic reflection and gravity data [Ph.D. dissertation]: Ithaca, New York, Cornell University, 196 p.

Solomon, B. J., McKee, E. H., and Andersen, D. W., 1979, Paleogene rocks near Elko, Nevada, *in* Armentrout, J. M., Cole, M. R., and TerBest, H., Jr., eds., Cenozoic paleogeography of the Western United States: Society of Economic Paleontologists and Mineralogists, Pacific Coast Paleogeographic Symposium 3, Anaheim, California: Santa Fe Springs, California, Society of Economic Paleontologists and Mineralogists, p. 75–89.

Strecker, M. R., Blisniuk, P. M., and Eisbacher, G. H., 1990, Rotation of extension direction in the central Kenya Rift: Geology, v. 18, p. 299–302.

Manuscript Accepted by the Society April 21, 1995

Printed in U.S.A.

Geological Society of America
Special Paper 303
1996

Paleogeography of the Horse Spring Formation in relation to the Lake Mead fault system, Virgin Mountains, Nevada and Arizona

L. Sue Beard
U.S. Geological Survey, 2255 North Gemini Drive, Flagstaff, Arizona 86001

ABSTRACT

The Tertiary Horse Spring Formation in southeast Nevada records the development of strike-slip and extensional tectonism that formed the Lake Mead left-lateral fault system. Stratigraphic, geochronologic, and mapping studies in the Virgin and south Virgin Mountains clarify the details of that evolution. Exposures of the Horse Spring Formation in the Virgin Mountains area consist of only its lower two members, in ascending order, the Rainbow Gardens and Thumb Members. The Horse Spring deposits are faulted and variably tilted by normal and oblique-slip faults. The Rainbow Gardens Member (ca. 26–18 Ma) was deposited in a broad shallow basin with positive areas to the northwest and southeast; facies changes are gradual and show no influence of active faulting. In contrast, complex facies relations in the Thumb Member are the result of deposition during complex mixed-mode deformation along kinematically linked strike-slip and extensional faults of the Lake Mead fault system, about 16–14 Ma. Tilting of Tertiary and older rocks about horizontal axes apparently occurred late in the extensional episode, at about 14 Ma or later.

The Rainbow Gardens Member consists of three units that each represent separate pre-extensional depositional systems. The basal conglomerate is a braid plain or pediment deposited by northeast paleoflow. The middle unit is a mixture of tuffaceous fluvial and lacustrine deposits that are the result of a reversal of drainage to the southwest, into the Rainbow Gardens lake. The change in paleoflow direction is attributed to extensive volcanism to the north that blocked the earlier northeastward flow. The upper unit consists of pedogenically altered, palustrine carbonate rocks that formed during cessation of lacustrine deposition and waning of volcanic activity to the north.

A previously unrecognized unconformity that records the onset of faulting separates the Rainbow Gardens Member from the overlying Thumb Member and can be traced throughout the Virgin Mountains area. The unconformity separates the underlying pedogenically altered carbonate rocks from overlying well-bedded lacustrine limestones. Locally the unconformity is marked by conglomerate and megabreccia deposits derived from the underlying Rainbow Gardens carbonate rocks across nearby fault scarps. Carbonate rocks above the unconformity grade laterally and vertically into lacustrine gypsum and fine-grained sandstone of the Thumb Member. These rocks in turn intertongue laterally and vertically with marginal lacustrine and alluvial fan facies. Abrupt influx of megabreccia and coarse conglomerate into lacustrine deposits occurred northward from both the Gold Butte and Lime Ridge faults. At approximately the same time, megabreccia and coarse conglomerate were shed

Beard, L. S., 1996, Paleogeography of the Horse Spring Formation in relation to the Lake Mead fault system, Virgin Mountains, Nevada and Arizona, *in* Beratan, K. K., ed., Reconstructing the History of Basin and Range Extension Using Sedimentology and Stratigraphy: Boulder, Colorado, Geological Society of America Special Paper 303.

southward from the uplifting Virgin Mountains. Rocks of the uppermost part of the Thumb above the megabreccia horizons everywhere in the south Virgin Mountains are conglomerates of prograding alluvial fan and braid plain facies. The conglomerates record decreasing tilt and grade upward and laterally eastward to mostly younger fine-grained clastic rocks widely exposed in the western Grand Wash trough. By 14 Ma, the Virgin Mountains area became a source region for younger strata deposited to the east in the Grand Wash trough, to the west in the Overton Arm, and to the north in the Virgin River depression.

INTRODUCTION

Recent debate on the extensional history of southeastern Nevada has centered on the relative roles of strike-slip versus detachment faulting and on the magnitude of extension (Anderson and Barnhard, 1993a, 1993b; Anderson and Bohannon, 1993; Axen et al., 1993; Campagna and Aydin, 1991; Duebendorfer and Wallin, 1991; Duebendorfer and Simpson, 1995). Because syn-extensional strata can record the evolution of both strike-slip and extensional structures, they provide an opportunity to separate, both temporally and spatially, the influence of various structures on depositional systems. The evolution of these depositional systems can in turn yield information on deformation processes (Bohannon, 1984; Beratan, 1993; Fedo and Miller, 1992; Fillmore, 1993; Sherrod and Nielson, 1993). The data and conclusions presented here support a model of kinematically linked strike-slip and extensional faulting, with the dominance of one style of faulting over the other being a function of location and timing within the regional extensional deformation.

Late Oligocene to middle Miocene strata of the Horse Spring Formation in the south part of the Virgin Mountains include pre- and syndeformational deposits. Nearly continuous north-south and discontinuous east-west exposures of these strata between strands of the Lake Mead fault system allow correlation of Tertiary strata across the various faults. The Lake Mead fault system in the south Virgins is comprised of northeast-striking left-separation faults that bound domains cut by low- to high-angle normal faults and minor right-separation faults. Horizontal axis tilting of both fault blocks and their bounding normal faults, up to 60°, is common within the domains. Correlation of the Tertiary strata provides constraints on the timing of both fault movement and horizontal-axis tilting.

Stratigraphic studies of the lower two members of the Horse Spring Formation in the south Virgin Mountains were undertaken during regional geologic mapping (Beard and Campagna, 1991, Beard, 1992, 1993a, 1995). This chapter revises the Tertiary stratigraphy of the south Virgins and presents new $^{40}Ar/^{39}Ar$ ages of intercalated tuffs. These data allow construction of paleogeographic scenarios before, during, and after peak regional Miocene extension from about 16 Ma to 12 Ma.

The Virgin Mountains are just west of the Colorado Plateau margin, in northwest Arizona and southeast Nevada (Fig. 1A). The Virgins form an elongate north-south uplift from Gold Butte north to Bunkerville Ridge, at which point the range bends to the east before continuing northward to the Virgin River Gorge. In this chapter, the Gold Butte to Bunkerville Ridge section will be referred to informally as the south Virgin Mountains, and the remainder as the north Virgin Mountains. To the east, the south Virgin Mountains are physically separated from the Colorado Plateau by the topographically lower Grand Wash trough. However, the Grand Wash trough ends northward so that the north Virgin Mountains merge topographically with the Colorado Plateau. To the north, the Virgin River depression (Bohannon et al., 1993) marks the northern extent of the mountains. To the west, Overton Arm of Lake Mead fills a north-south linear trough separating the south Virgins from the Muddy Mountains. Overton Arm is used in this chapter to informally divide the region into the western and eastern Lake Mead areas.

There are three main conclusions drawn from the data presented in this chapter: (1) The oldest Tertiary strata, the Rainbow Gardens Member (26–18 Ma) of the Horse Spring Formation, were deposited in a pre-extensional sag basin that was the result of shallow ponding caused by extensive volcanism to the north. (2) Surface manifestation of extension was first reflected in complex facies changes in the Thumb Member of the Horse Spring at about 16 Ma. The mixed-mode deformation included normal faulting and oblique slip on major segments of the Lake Mead fault system, causing segmentation of the early Rainbow Gardens sag basin into subbasins but not large-scale disruption of the paleosurface by horizontal axis tilting of crustal blocks. (3) By 14 Ma, large-scale tilting and relative uplift occurred during structural differentiation of basins and ranges. The entire south Virgin Mountains area was most likely uplifted as one single block, albeit internally deformed, and deposition of syn-extensional strata shifted into large linear graben or half-graben type basins to the west, east, and north.

GEOLOGIC SETTING

The Virgin Mountains are in the southeasternmost Great Basin where it meets the southern Basin and Range province. Tertiary deformation includes large-magnitude normal faulting, strike-slip faulting, and north-south shortening (Anderson and Barnhard, 1993b). The northwest-striking, right-lateral Las Vegas Valley shear zone and northeast-striking, left-lateral Lake Mead fault system intersect in the western Lake Mead area (Fig. 1B) accompanied by south-directed structural crowding (Anderson and Barnhard, 1993a, 1993b). These large fault systems interrupt the linear north-south–oriented basin-range

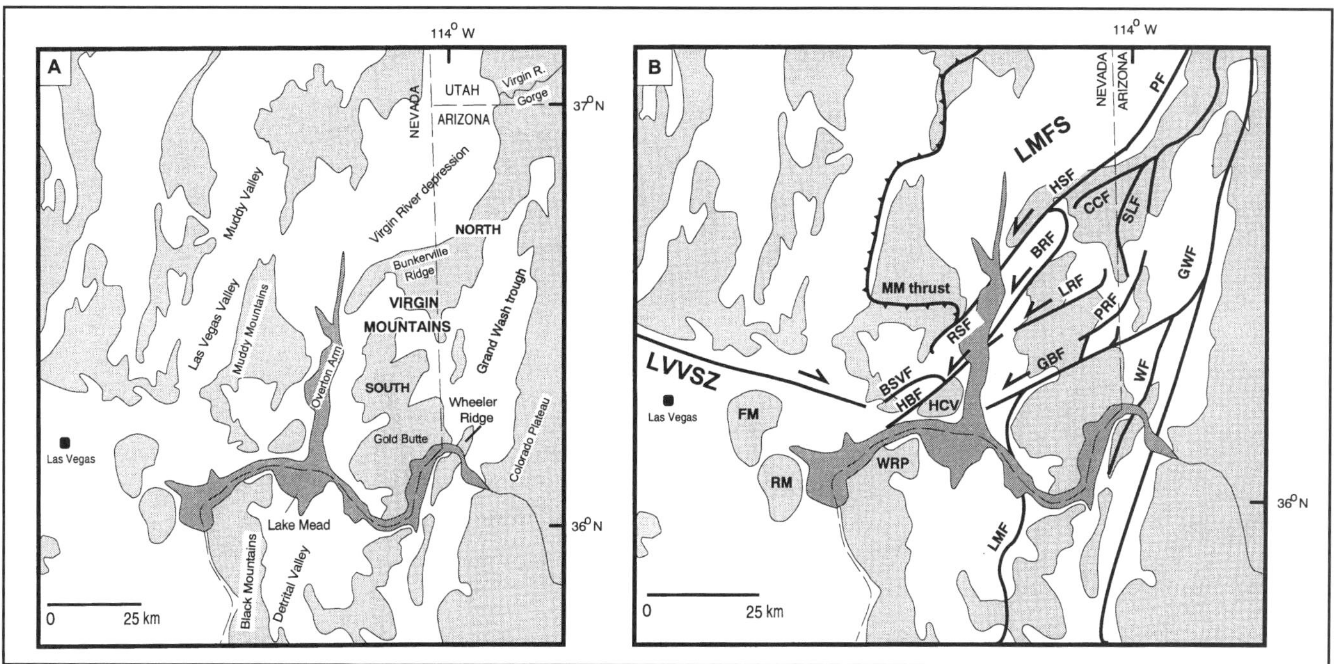

Figure 1. (A) Location map of southeastern Nevada, southeastern Utah, and northwestern Arizona showing major physiographic features in relation to Virgin Mountains. (B) Slight enlargement of map A, showing geologic features mentioned in text. Includes location of Las Vegas Valley Shear Zone (LVVSZ) and Lake Mead fault system (LMFS), as well as the Sevier-age Muddy Mountain (MM) thrust. Strands of the LMFS and related normal faults shown on figure and mentioned in text are as follows: BRF—Bitter Ridge fault, BSVF—Bitter Spring Valley fault, CCF—Cabin Canyon fault, GBF—Gold Butte fault, GWF—Grand Wash fault, HBF—Hamblin Bay fault, HSF—Hen Spring fault, LMF—Lakeside Mine fault, LRF—Lime Ridge fault, PF—Piedmont fault, PRF—Pakoon Ridge fault, RSF—Roger Spring fault, SLF—State Line fault, and WF—Wheeler fault. Also shown are the location of Frenchman Mountain (FM), River Mountains (RM), Hamblin-Cleopatra stratavolcano (HCV), and Wilson Ridge pluton (WRP).

structural fabric and separate terrane to the north containing thick sequences of Paleozoic and Mesozoic strata from a southerly terrane of Tertiary volcanic rocks underlain by Proterozoic crystalline rocks.

The Lake Mead fault system, which extends northeast through the Virgin Mountains to the Colorado Plateau margin, consists of several strands each with its own movement history. It was active chiefly from about 16 Ma to post-10 Ma, as well as some Quaternary slip. Longwell (1971, 1974) recognized lithologic similarities between proximal megabreccia clasts in Tertiary rocks at Frenchman Mountain (Fig. 1B) to a source in Proterozoic basement rocks in the south Virgin Mountains. He suggested 60 km of tectonic offset along the Las Vegas Valley shear zone (LVVSZ). Anderson (1973) recognized that the offset was along a left-lateral system, subsequently named the Lake Mead fault system by Bohannon (1979). Of the 60 km of estimated slip, only part can be uniquely identified with specific fault strands. Weber and Smith (1987) documented 20 km of slip between volcanic rocks of the River Mountains and the subvolcanic Wilson Ridge pluton to the east (Fig. 1B), occurring by a combination of transfer and extensional faulting sometime after 13.4 Ma. The Hamblin Bay fault, which is probably the youngest segment of the Lake Mead fault system, displaces half of the Hamblin-Cleopatra stratovolcano (11.7–10 Ma) about 20 km in a left-lateral sense (Anderson, 1973) (Fig. 1B).

The Lake Mead fault system forms a fairly narrow zone near its junction with the LVVSZ, dominated by the Hamblin Bay and Bitter Spring Valley faults. It widens eastward into a broad zone of fault strands in the Virgin Mountains (Fig. 1B). The northernmost strand of the broad zone is the linked Bitter Ridge–Hen Spring fault system which correlates southwestward with the Hamblin Bay–Roger Spring fault system (Campagna and Aydin, 1994). The Bitter Ridge–Hen Spring fault system slipped episodically in the Miocene, Pliocene, and Pleistocene, and most likely connects eastward with the Piedmont fault (Moore, 1972). The Piedmont fault forms the eastern and southeastern boundary of the Virgin River depression and also has Quaternary motion.

The kinematic significance of the Lake Mead fault system within the overall extensional deformation is controversial. Wernicke et al. (1988) pointed out that although the 60 km transport of the Frenchman Mountain block relative to the Gold Butte block is widely accepted, the relative roles of crustal-scale strike-slip versus normal faulting are unknown. Anderson (1990) and Anderson and Barnhard (1993a) suggested that strike-slip faulting in southeastern Nevada accommodated and is genetically linked to north-south shortening during Tertiary extension. An alternative interpretation has been that the Lake Mead and Las Vegas Valley fault systems developed as transfer faults between differentially extended terranes (Wernicke et al., 1982, 1984; Duebendorfer and Black, 1992). Ron et al. (1986) suggested that strike-slip faulting has been the dominant mode of deformation in the western Lake Mead area since 11 Ma, based on paleomagnetic and structural data.

The data presented here indicate a dominance of mixed mode deposition along the Lake Mead fault system (LMFS) from at least 16 to14 Ma, as the result of kinematically linked strike-slip and normal faulting (Beard, 1994). Although the data presented here do not resolve the magnitude of slip along the LMFS, future studies like this one that reconstruct the paleogeography of the Horse Spring Formation west of Overton Arm should provide further contraints on displacement amounts and vectors.

South Virgin Mountains

The south Virgin Mountains contain Proterozoic, Paleozoic, Mesozoic, and Tertiary rocks (Fig. 2) exposed in structural blocks that are variably tilted, mostly to the east but also southwesterly. The structural blocks are bounded by northeast-striking, left-separation faults of the Lake Mead left-lateral fault system as well as north-striking normal separation faults with low- to high-dip angles. Subordinate right-lateral faults strike northwest and are transpressional. Southeast-directed thrusts

Era	Period	Epoch	Units
CENOZOIC	TERTIARY	Miocene	Muddy Creek Formation Rocks of the Grand Wash trough Red sandstone unit Horse Spring Formation Lovell Wash Member Bitter Ridge Limestone Member Thumb Member Rainbow Gardens Member
		Oligocene	
MESOZOIC	Cretaceous		Baseline Sandstone Willow Tank Formation
	Jurassic		Navajo Sandstone
	?		
	Triassic		Kayenta and Moenave Formations, undivided Chinle Formation Moenkopi Formation
PALEOZOIC	Permian		Kaibab Limestone Toroweap Formation Hermit Shale Esplanade Sandstone Pakoon Limestone of McNair, 1951
	Pennsylvanian		Callville Limestone
	Mississippian		Redwall Limestone
	Devonian		Temple Butte Limestone
	Cambrian		Nopah Formation Undivided dolomites Muav Limestone Bright Angel Shale Tapeats Sandstone
PROTEROZOIC			Crystalline rocks

Figure 2. Generalized stratigraphic column of Proterozoic to Tertiary strata exposed in region.

and folds, exposed in the northwest part of the area, are cut by the strike-slip and normal faults. Five major east-northeast– to northeast-striking faults partition the south Virgin Mountains into separate domains, which are here termed, from south to north, the Gold Butte, Tramp Ridge, Lime Ridge, Virgin Peak, and Bunkerville Ridge domains (Fig. 3). The three southerly domains are separated by the east-northeast-striking Gold Butte and Lime Ridge faults. The Virgin Peak domain is bounded to the south by a reverse fault and separated from the Bunkerville Ridge block to the north by the Hen Spring fault. The fault bounding the north side of the Bunkerville Ridge block is buried beneath late Tertiary and Quaternary alluvial fan deposits in the Virgin River depression.

The various left-lateral faults in the Virgins are kinematically linked to the normal and right-lateral faults. Displacement along the left-lateral fault strands is accommodated by north-striking normal faults whose displacements decrease away from their junction with the left-lateral faults (Campagna, 1990). These normal faults essentially transfer slip between left-lateral faults in a left-stepping sense. South from a pinning point in the central part of the Virgins around Whitney Pocket, down-to-the-west normal faults splay northward from the left-lateral faults so as to increase slip along the left-lateral faults southwestward. To the north and east of Whitney Pocket, down-to-the-east normal faults splay southward from the left-lateral faults and increase slip on the left-lateral faults to the northeast. The net result is that the broad zone of the LMFS in the Virgin Mountains functions as a large, complex, left-stepping system that youngs northward, with minimum slip centered in the Whitney Pocket area.

The two southerly faults, the Gold Butte and Lime Ridge faults, strike east-northeast and dip about 50° north; both faults exhibit down-to-the-north and left-lateral oblique slip and splay at their east ends into small north-striking, west-dipping normal faults. Two other left-separation faults, the Pakoon Ridge and Bitter Ridge faults, are nearly vertical and strike more northerly than the Gold Butte and Lime Ridge faults. The Pakoon Ridge fault starts near the eastern side of the Tramp Ridge block and heads northeast to bound Pakoon Ridge on the west side. The Bitter Ridge fault and the southern extent of the Hen Spring fault bound a small, northeast-elongate pull-apart basin (Overton Arm pull-apart basin) formed on the northwest side of the Virgin Mountains (Campagna and Levandowski, 1991). The Bitter Ridge fault forms the southeast margin of the basin; it is the northeast continuation of the Hamblin Bay fault exposed west of the Overton Arm of Lake Mead. Slip on the Bitter Ridge fault is transferred to the Hen Spring fault on the northwest side of the basin by a series of normal faults that form the pull-apart basin. The Hen Spring fault bends into a more easterly strike through bedrock exposed at Bunkerville Ridge before connecting with the Piedmont fault in the Virgin River depression.

The five structural domains within the south Virgins exhibit contrasting structural styles (Beard, 1991). These contrasting styles are at least partly a function of the amount of east-west

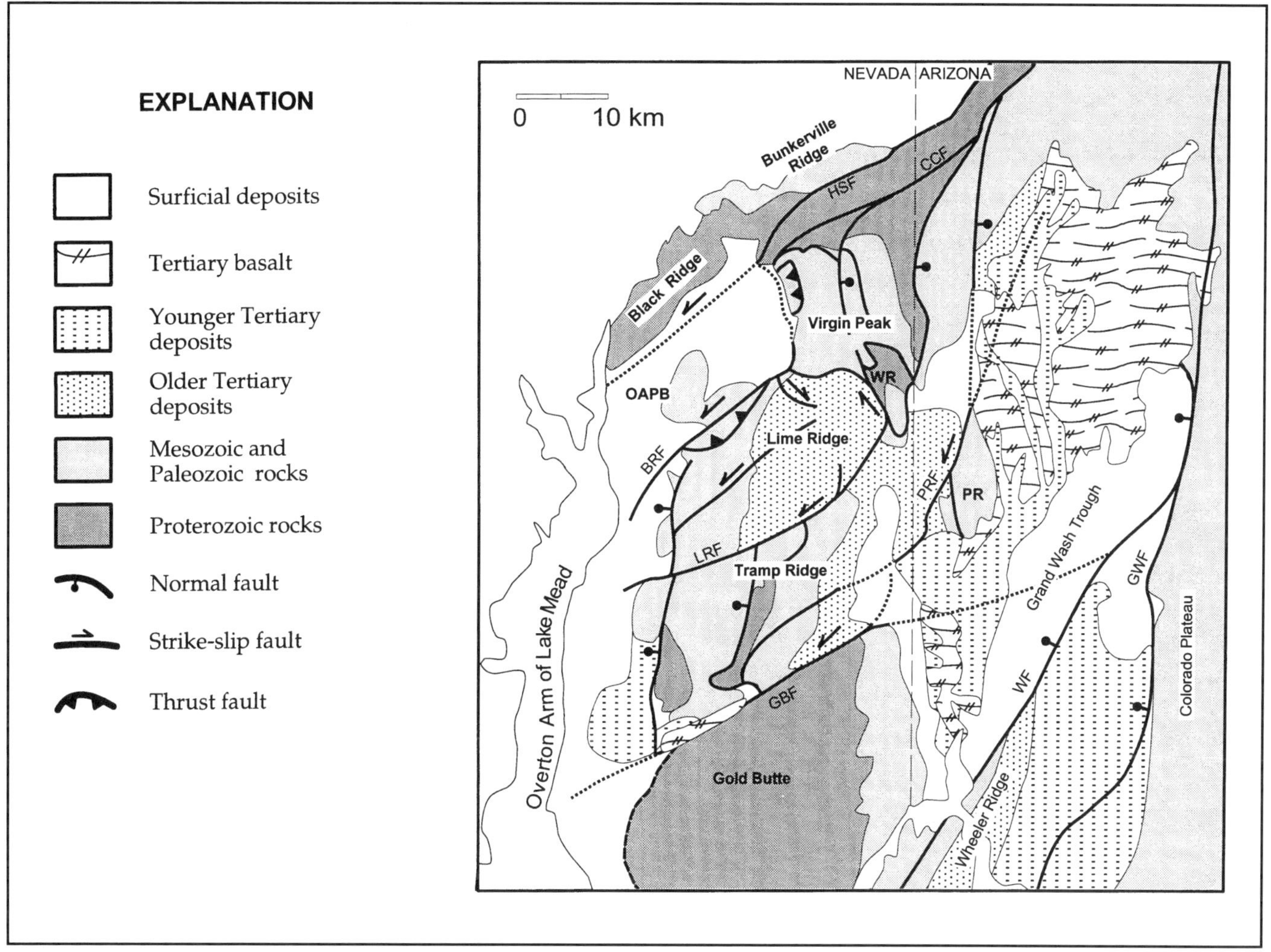

Figure 3. Generalized geologic map of the Virgin Mountains and Grand Wash trough showing segmentation of Virgin Mountain area into structural domains by various strands of the LMFS. Fault strand names as in Figure 1. Major structural domains discussed in text are Virgin Peak, extending from the Hen Spring fault (HSF) to Wilson Ridge (WR); Lime Ridge, extending from Bitter Ridge fault (BRF) and WR to Lime Ridge fault (LRF); Tramp Ridge, extending from LRF to Gold Butte fault (GBF) and from Overton Arm east to Pakoon Ridge fault (PRF); and the Gold Butte domain south of the GBF. Other features mentioned in text are the Overton Arm pull-apart basin (OAPB), and Pakoon Ridge (PR)

extension each domain has undergone. The Tramp Ridge domain exhibits the greatest amount of extension, whereas domains to the north and south exhibit lesser amounts. To the south, the eastern part of the Gold Butte domain contains Paleozoic rocks that dip 50°–70° eastward and are cut by several low-angle and high-angle normal faults. Extensive exposures of Proterozoic rocks west of the sub-Paleozoic unconformity are mostly unfaulted (Fitzgerald et al., 1991; Fryxell et al., 1992). High-grade metamorphic Proterozoic rocks at the west end of the block are mylonitic and chloritically altered; these features are considered evidence for a low-angle, west-dipping detachment fault (Fitzgerald et al., 1991; Fryxell et al., 1992). The Gold Butte block has been interpreted as a single east-tilted crustal block exposing as much as 14–17 km of crust that has been isostatically uplifted during tectonic denudation (Wernicke and Axen, 1988; Fitzgerald et al., 1991; Fryxell et al., 1992). An alternative interpretation that requires less extension is presented in the discussion section.

The Tramp Ridge domain is dominated by moderate- to low-angle normal faults and accompanying east-tilted fault blocks (Morgan, 1968). At least three major east-tilted blocks composed of mostly Paleozoic rocks, but locally including Proterozoic to Tertiary rocks, are exposed in the Tramp Ridge domain. The fault blocks are bounded by shallowly to moderately west-dipping (30°–45°) normal faults. Strata within the blocks dip about 40°–50° and are cut internally by numerous small displacement low- to high-angle normal faults and minor strike-slip faults (Morgan, 1968).

Strike-slip and high-angle normal faults dominate the Lime Ridge domain. One major fault block occupies most of the domain. Mesozoic and Tertiary rocks are exposed throughout most of the block, although Cambrian through Permian rocks are found at the west end. Rocks in the domain are only slightly tilted (5°–10° south) in the center of the block; southward toward the Lime Ridge fault the rocks tilt about 25°–30° east, and northward toward the Virgin Peak domain strata dip about 25°–30° southeast and south (Beard, 1991, 1992, 1993).

In contrast, the Virgin Peak domain is tilted to the southwest and is bounded on the south by an east-west–striking reverse fault (Seager, 1966, 1970; Beard, 1993b, 1995). Proterozoic rocks are exposed to the north near Bunkerville Ridge. Paleozoic rocks are cut by numerous north- to northwest–striking normal faults that have minor, down-to-the-east displacement. Only a few large displacement faults cut into Proterozoic basement. These faults end northward against the Cabin Canyon fault, an east-northeast–striking strike-slip fault of probable limited displacement south of the Hen Spring fault (Anderson and Bohannon, 1993; Beard, 1993, 1995; Campagna, 1990).

Whitney Ridge, a narrow prong on the southeast side of the Virgin Peak domain, is a south-tilted block of Proterozoic through Pennsylvanian rocks. A major low-angle normal fault system separates Whitney Ridge from the bulk of the Virgin Peak domain, placing Cambrian through Pennsylvanian rocks down to the north against basement rocks of Whitney Ridge.

Paleozoic through Mesozoic rocks at the west sides of both the Virgin Peak and Lime Ridge blocks are deformed into folds and thrusts of probable Sevier age. The fold and thrust system is cut internally by younger north-striking normal faults and is terminated on the west side by the Bitter Ridge strike-slip fault.

The Bunkerville Ridge domain, which lies north of the Hen Spring fault, has been mapped by Seager (1966), Beal (1965), and more recently by Bohannon (Williams et al., 1995; see also Anderson and Bohannon, 1993). Proterozoic rocks are extensively exposed in the southern and western parts of the block, including the Black Ridge prong that extends southwest from Bunkerville Ridge. Steeply dipping to overturned Paleozoic and Mesozoic strata form Bunkerville Ridge on the north side of the block.

TERTIARY STRATIGRAPHY OF SOUTHEAST NEVADA

The principal Tertiary stratigraphic elements in southeastern Nevada are, from older to younger, the Horse Spring Formation, the red sandstone unit and the rocks of the Grand Wash trough, both defined by Bohannon (1984), and the Muddy Creek Formation (Fig. 2). These stratigraphic units each result from unique depositional systems reflecting spatial and temporal differences in extensional deformation. Horse Spring strata, widely exposed in southeastern Nevada, overlie Paleozoic and Mesozoic rocks with little or no angular discordance, indicating that they mostly pre-date tilting. In turn, red sandstone unit strata in the western Lake Mead area overlie Horse Spring rocks with little to no angular discordance (Duebendorfer and Wallin, 1991). However, in the eastern Lake Mead area, the rocks of the Grand Wash trough, which are similar in age to the red sandstone unit, are angularly discordant on the Rainbow Gardens Member of the Horse Spring. Muddy Creek strata are mostly flat lying except locally near faults or salt domes and mostly overlap major extensional structures.

The Horse Spring Formation crops out from Las Vegas on the west to the Grand Wash trough in the east. The Horse Spring was originally named by Longwell (1921, 1922) for exposures in the Horse Spring area in the southern Virgin Mountains. Revision of the Tertiary stratigraphy by Bohannon (1984) designated a principal reference area for the Horse Spring Formation in White Basin and Bitter Spring Valley, west of the Overton Arm of Lake Mead. Bohannon divided the Horse Spring into four members, in ascending order, the Rainbow Gardens, Thumb, Bitter Ridge Limestone, and Lovell Wash Members. The reader is referred to Bohannon (1984) for an excellent summary of the regional distribution and facies of the Horse Spring Formation. The lower two members are described in detail later in this chapter; the upper two members are of only peripheral interest to this chapter and are described only briefly.

The lower members of the Horse Spring Formation, the Rainbow Gardens and Thumb Members (26–13.5 Ma), are exposed from the Las Vegas area east to the Grand Wash Trough, but the upper two members are restricted to the western Lake Mead area. The Bitter Ridge Limestone Member (13.5–13.0 Ma) is mostly confined to the area north of the Las Vegas Valley Shear zone. Deposits are dominantly lacustrine limestone, but include a distinctive alluvial fan facies derived from the south. The Lovell Wash Member (13.0–12.0 Ma) occurs only between Frenchman Mountain and White Basin and includes lacustrine carbonate strata with interbedded fallout tuffs and tuffaceous sandstones. Bohannon (1984) interpreted coarse-grained terrigenous rocks in both the Bitter Ridge Limestone and the Lovell Wash Members as recording activity on the Bitter Spring Valley fault, a major strand of the Lake Mead fault system.

The red sandstone unit of Bohannon (1984) (11.9–10.6 Ma) is a diverse package of conglomerate, fine-grained clastic strata, and fallout tuffs deposited in several separate basins. Bohannon (1984) described restricted outcrops in the White Basin area that include alluvial fan and playa deposits and mapped a few local exposures in the north Muddy Mountains (Bohannon, 1983). Duebendorfer and Wallin (1991) recognized extensive outcrops of the red sandstone unit south of the Las Vegas Valley shear zone between Frenchman Mountain and Hamblin Mountain. They documented linkage between deposition of the red sandstone unit and simultaneous motion along the Las Vegas Valley shear zone and the Saddle Island detachment.

The interior-basin deposits filling the Grand Wash trough were correlated by Longwell (1936) with the type locality of the Muddy Creek Formation to the west. The deposits have been studied and described by Lucchitta (1966, 1972, 1979) and are only briefly summarized later in the chapter. Bohannon (1984) reclassified the deposits as the rocks of the Grand Wash trough, arguing that the deposits were not physically connected to the Muddy Creek Formation, were of a different age, and were not similar lithologically.

The name "Muddy Creek Formation" has been applied widely to deposits that are mostly flat lying and postdate major extension. Bohannon (1984) restricted the term "Muddy Creek Formation" to rocks that are demonstrably continuous with those at the type locality north of Glendale. The type Muddy Creek fills lower elevation valleys and basins from the Virgin River depression and the Muddy Valley south to Detrital Valley and Las Vegas Valley. The age of the Muddy Creek Formation is poorly constrained, but it includes a basalt dated at 8 Ma (Eberly and Stanley, 1978) and is capped by a 5.9 Ma basalt on Fortification Hill (Damon et al., 1978) in the Boulder Basin area. Bohannon (1984) considers the Muddy Creek Formation to be no older than 10.6 Ma, the age of the red sandstone unit near Frenchman Mountain, and no younger than 5.9 Ma. The Hualapai Limestone interfingers with and overlies the rocks of the Grand Wash trough at Grapevine Mesa, and it also overlies rocks called the Muddy Creek Formation to the west in Detrital Valley. Blair and Armstrong (1979) reported an age of 8.44 Ma from a tuff intercalated with the Hualapai Limestone; its upper age relative to the basalt on Fortification Hill is unknown.

PRE-MIOCENE PALEOGEOGRAPHY

The Late Cretaceous to middle Cenozoic paleogeographic configuration of southeastern Nevada and western Arizona places important constraints on reconstruction of Miocene extensional deformation. A highland upheld by allochthonous rocks of the Sevier orogenic belt lay to the west and northwest of the Virgin Mountains. Cretaceous deposits were shed eastward from thrust nappes into the Sevier foreland basin. Laramide basement uplifts disrupted the foreland basin, from latest Cretaceous through Eocene time, creating internally drained basins such as the Claron basin centered in southwestern Utah (Dickinson et al., 1988). Sedimentologic and stratigraphic data, summarized herein, indicate that the Late Cretaceous to Oligocene paleodrainage direction of southeastern Nevada and western Arizona was northeastward (Goldstrand, 1992; Graf et al., 1987).

South and east of Lake Mead

Just east of the frontal thrust, Bohannon (1984) recognized southward beveling of autochthonous Permian and Mesozoic strata at the sub-Tertiary unconformity at Frenchman Mountain and in the south Virgin Mountains. He attributed the beveling to erosion of the north flank of a north-plunging gentle arch east of the Sevier thrust belt. The western flank of the arch and its relationship with the Sevier orogenic belt is problematic and poorly understood; for more discussion the reader is referred to Bohannon (1984).

Eastward, similar southwestward beveling of Proterozoic to middle Paleozoic rocks has long been recognized on the western Colorado Plateau margin in northwestern Arizona (Lucchitta, 1966, 1979; Young, 1966, 1979, 1982). This beveling corresponds to the eastern flank of the arch which was named the Kingman uplift by Lucchitta in Goetz et al. (1975) (Fig. 4). Eocene or older deposits ("rim gravels") filled northeastward-draining paleocanyons cut into the beveled surface on the Hualapai Plateau (Young, 1966, 1979), indicating that the Kingman uplift was a Laramide or older structural feature. The "rim gravel" deposits include well-rounded, far-traveled Proterozoic clasts as well as Laramide intrusive clasts that could have been derived only from the Kingman uplift to the west (Young, 1966, 1979).

The gently beveled surface was interrupted to the northeast by a major northeast-retreating scarp, created by the resistant limestone cliffs of the Permian Kaibab Limestone and Toroweap Formation (Young, 1985). Above the scarp, a higher, gently bevelled surface was developed on Permian and Mesozoic strata. Northeast-flowing drainages on the lower beveled surface were deflected into strike valleys by the paleoscarp, flowing northwest or southeast to major reentrants into the scarp, such as along the Hurricane monocline (Fig. 4). Major northeast drainages at these reentrants transported material northward onto the higher surfaces and into the Claron basin in southern Utah (see Graf et al., 1987, fig. 6).

The landscape created largely in the early Cenozoic by the northeast paleodrainage is still visible today on the western Colorado Plateau, only slightly modified by the cutting of the Grand Canyon by the Colorado River. Young (1985) showed that the Kaibab scarp on the western Plateau has retreated no more than 8 km northward since the Eocene strata were deposited. The ancient beveled surfaces and the Kaibab paleoscarp trend northwestward into the extended terrane west of the Plateau margin (Lucchitta and Young, 1986, fig. 7), and must have wrapped around the north-plunging nose of the Kingman uplift. Within what is now the extended terrane, Bohannon (1984) envisioned an even erosional surface beneath the sub-Tertiary unconformity at Frenchman Mountain and the south Virgins. However, to the south of the beveled Permian rocks, the paleogeography was most likely similar to that on the western Colorado Plateau. Therefore, rather than an even erosional surface, there must have been a south-facing Kaibab paleoscarp, and farther south, the lower beveled surface underlain by Cambrian to Mississippian rocks. Within a few kilometers south of the scarp, Proterozoic rocks were exposed at the surface as evidenced by their widespread occurrence beneath Tertiary volcanic rocks.

This geometry is well preserved in Wheeler Ridge, an east tilted block just west of the Grand Wash fault (Fig. 3) (Lucchitta, 1966; Lucchitta and Young, 1986). At the north end of the ridge, the Rainbow Gardens Member of the Horse Spring Formation rests on Permian rock, but just a few kilometers southward the strata are beveled down to Cambrian rocks (Matthews,

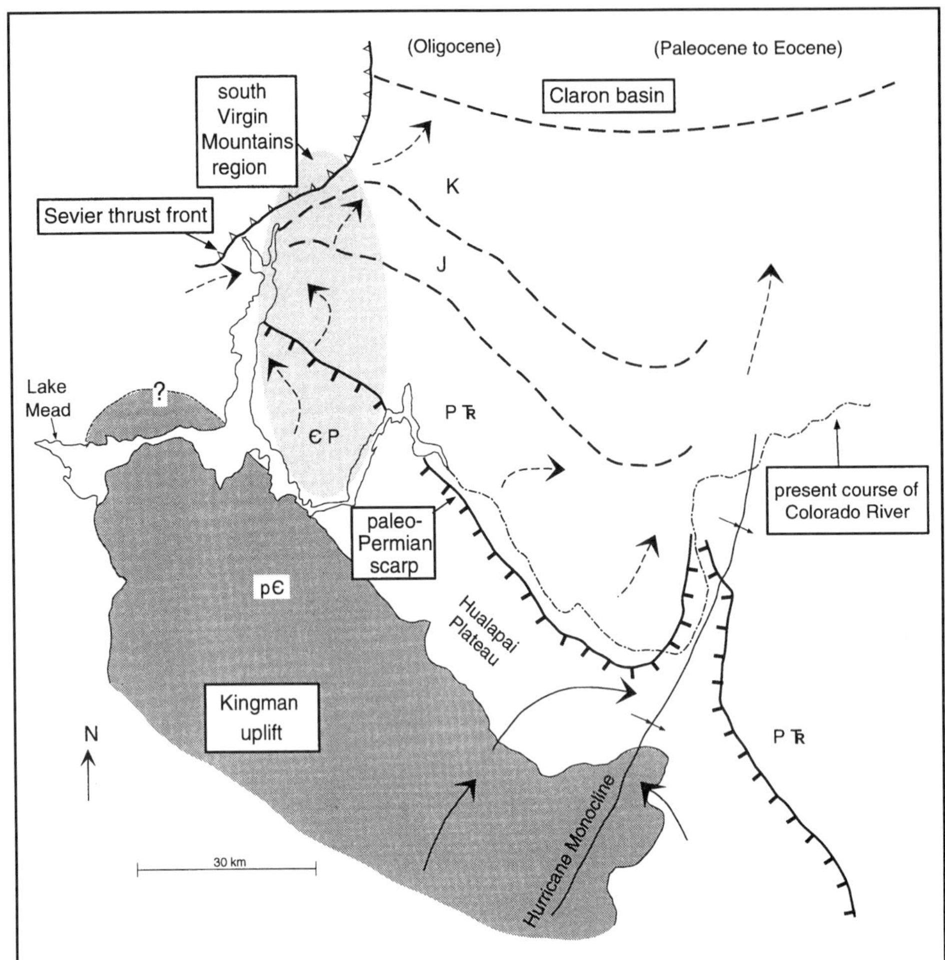

Figure 4. Simplified paleogeographic sketch map of southeast Nevada and northwest Arizona during the early Cenozoic. Inferred location of Kingman uplift from Lucchitta (1966) and Bohannon (1984). Paleo-Permian scarp inferred from Young (1985). Southwest erosional limit of strata shown by line separating Є P—Cambrian through Permian strata; paleo-Permian scarp; and dashed lines separating P TR—Permian to Triassic strata; J—Jurassic strata; and K—Cretaceous strata. Projection of Cretaceous strata eastward from Sevier thrust front is speculative, as is southern margin of Claron basin. Westward shift of the Paleocene to Eocene Claron basin (Taylor, 1993) denoted by location of word "Oligocene" on figure. Paleoflow directions shown by solid arrows based on preserved canyons and deposits (see text); directions shown by dashed arrows are inferred from regional geologic relations.

1976). Both the Permian paleoscarp and the lower beveled surface are found in Wheeler Ridge and project directly toward the Gold Butte block at the southern end of the south Virgin Mountains. Therefore, the entire Paleozoic section was most likely eroded from part of the Gold Butte block prior to extension.

The late Eocene or older deposits preserved within paleocanyons on the western Colorado Plateau are overprinted by a thick terra-rosa soil, indicative of a long period of weathering and nondeposition (R. A. Young, personal commununication, 1993). A relatively thin but widespread deposit, the Buck and Doe Conglomerate of Gray (1964), unconformably overlies the deeply weathered Eocene rocks. The Buck and Doe, in contrast to the older deposits, was locally derived, containing mostly early Paleozoic clasts in its lower part and arkosic sandstone and locally derived crystalline clasts in its upper part (Young, 1966). To the west, prior to extension, volcanic strata as old as 21 Ma were deposited on early Paleozoic rocks on the northern nose of the arch and on Proterozoic rocks within the core of the arch. In the latter setting, arkosic sediments similar to the upper part of the Buck and Doe are found locally beneath the volcanic rocks (Faulds, 1992).

The Buck and Doe Conglomerate and the volcanic strata deposited on the lower beveled surface are about the same age as the Rainbow Gardens Member of the Horse Spring Formation that is deposited on the upper beveled surface. However, the Permian paleoscarp probably formed a physical barrier between the two depositional systems. The two deposits consequently were probably not connected, although early Paleozoic clasts and rare clasts of crystalline rocks in the Rainbow Gardens indicate limited transport of sediment eroded below the Permian paleoscarp. In contrast, by the time strata of the Thumb Member of the Horse Spring were deposited, crystalline source areas were widely exposed.

North of Lake Mead

The Claron Formation, composed of sandstone, conglomerate, and lacustrine limestone, is exposed sporadically from Bryce Canyon National Park to the northern Beaver Dam Mountains. Claron deposition began in an interior basin that formed as a response to Laramide deformation in southern Utah (Goldstrand, 1990a, 1990b). In detail, the Claron basin is a composite of a main Paleocene to Eocene basin centered in southwestern Utah (Goldstrand, 1992) and a smaller Oligocene synvolcanic basin centered in southeastern Nevada (Fig. 4) (Taylor, 1993). The use of the name "Claron" for the younger synvolcanic deposits is controversial (Anderson and Kurlich, 1989; Hintze, 1986; Taylor, 1993) but it will be used here for simplicity. Possible unconformable relationships between the two basins are also in question.

Taylor (1993) documented a westward shift in the western margin of the Claron basin between Paleocene and Oligocene time and suggested that the shift may have been caused by the onset of volcanism and extension to the north of the Claron basin at about 33–31 Ma. The younger sedimentary rocks are intercalated with volcanic rocks, including the 29 Ma Wah Wah Springs Formation near the base and the 25–26 Ma Isom Formation near the top (Taylor, 1993). Small local deposits of lacustrine strata persist until about 22 Ma (Hintze, 1986).

Extensive volcanism, which formed the Caliente caldera complex and the Pine Valley Mountains destroyed most of the younger Claron basin by about 25 Ma (Anderson, 1987). Large volumes of ash-flow tuff were erupted during the main phase of volcanism from about 25–18 Ma; lesser amounts of silicic rocks were erupted from about 16–12 Ma (Best et al., 1993).

HORSE SPRING FORMATION

Introduction

Only the Rainbow Gardens and Thumb Members of the Horse Spring Formation occur in the south Virgin Mountains (Fig. 5). In the Grand Wash trough, only the Rainbow Gardens Member has been identified, although the Thumb probably exists in the subsurface. Rainbow Gardens and Thumb strata are also inferred to exist in the subsurface in Overton Arm and the Virgin River depression (Bohannon et al. 1993; Carpenter, 1988). Bohannon's (1984) definition of the Rainbow Gardens Member consists of rocks comprising the lowest unit of the Horse Spring Formation of Longwell (1921, 1922) in the Virgin Mountains, the formation at Tassai Wash of Lucchitta (1966) in the central Grand Wash Trough (Pigeon Wash in Bohannon, 1984), and the Cottonwood Wash Formation of Moore (1972) exposed in Cottonwood Wash at the north end of the Grand Wash Trough. The Thumb Member is widespread in the south Virgins; locally some strata were misidentified as the Muddy Creek Formation (Morgan, 1968; Longwell et al., 1965), but geologic mapping (Beard and Campagna, 1991; Beard, 1992, 1993) and new geochronology presented in the following section correctly place these strata into the Thumb.

Geochronology

Previous ages. Based on fission-track ages, as well as published K-Ar ages, Bohannon (1984) considered the Rainbow Gardens Member to range from about 20 Ma to 17.2 Ma in age. Bohannon also reported a fission-track age of 15.1 ± 0.8 Ma for

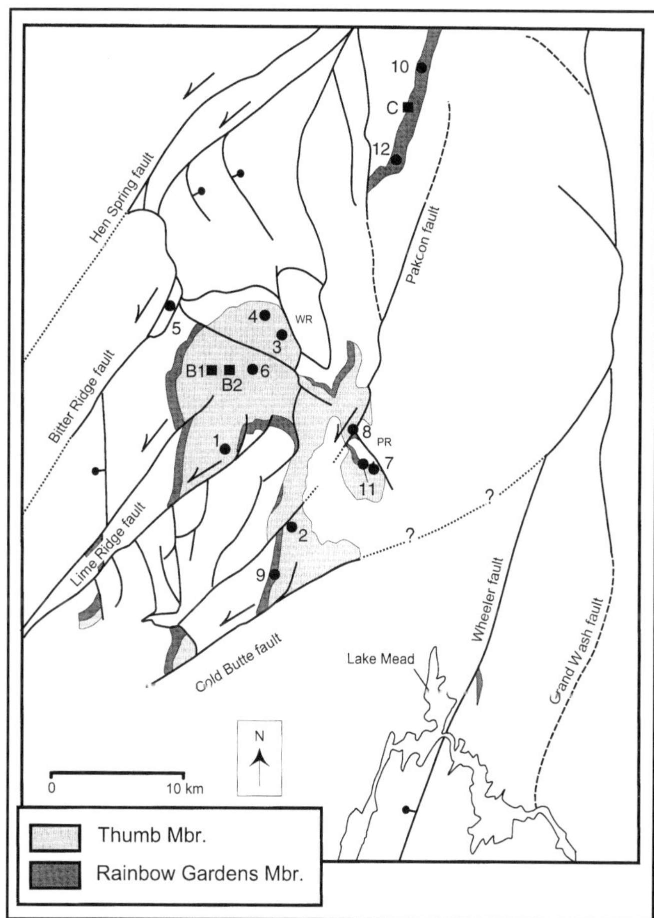

Figure 5. Map showing location of geochronologic samples discussed in text, superimposed on outcrop pattern of Thumb and Rainbow Gardens Members of Horse Spring Formation in south Virgin Mountains. Strands of LMFS, Wheeler and Grand Wash faults, Pakoon Ridge (pr), and Whitney Ridge (wr) shown for location of outcrops relative to structural domains (see Fig. 3). Location of geochronologic samples shown by numbered circles for new data presented in Table 1. Squares show locations for previously published dates: B1 is 15.6 Ma and B2 is 15.4 Ma fission-track ages from Bohannon (1984); C is 24.3 Ma K-Ar age presented in Carpenter et al. (1989).

the Rainbow Gardens at Horse Spring in the south Virgin Mountains, but he considered the age too young (a conclusion supported by data from this study). However, Carpenter et al. (1989) reported a K-Ar age on biotite of 24.3 ± 1.0 Ma from a vitric tuff in the Rainbow Gardens Member sampled at Tom and Cull Wash in the northern Grand Wash trough (Table 1). In addition, two K-Ar ages of 21.3 ± 0.4 and 19.6 ± 0.8 Ma (Anderson et al., 1972), a fission-track age of 20.7 ± 1.2 Ma (Bohannon, 1984), and a K-Ar age of 20.0 ± 0.8 Ma (Shafiqullah et al., 1980) were reported from an outcrop near Glendale, Nevada, that contains unwelded ash-flow tuff overlain by carbonate rocks (Table 1). Bohannon thought that these rocks were too old for the Horse Spring Formation, but these rocks are probably equivalent to the Rainbow Gardens Member based on the Carpenter et al. (1989) date and the $^{40}Ar/^{39}Ar$ ages from this study.

Bohannon (1984) obtained a suite of 11 fission-track ages for the Thumb Member of the Horse Spring Formation. Based on these ages and published K-Ar dates (Table 1), he considered the Thumb to be from 17.2 Ma to possibly 13.5 Ma in age. The older limit is based on a poorly constrained K-Ar age (Anderson et al., 1972) obtained from a pyroxene-olivine andesite lava at 17.2 ± 3 Ma (old constant); a second date from the same rock is 15.6 ± 3 Ma (old constant). Four fission-track ages by Bohannon (1984) from the Thumb strata just east of Frenchman Mountain suggest a range from 16.2 ± 0.8 to 13.2 ± 0.9 Ma; however, he considered the latter age anomalously young and viewed the upper limit of the Thumb as no younger than 13.5 Ma. Three dates from the Bitter Spring Valley–Echo Hills area and two dates from Overton Ridge range from 12.5 ± 0.9 to 15.6 ± 1.0 Ma. However, the younger age was discordant and is also considered too young. Bohannon obtained two fission-track dates of 15.9 ± 1.0 and 16.3 ± 1.9 Ma from Thumb sediments in Wechech Basin in the south Virgin Mountains (Table 1).

TABLE 1. GEOCHRONOLOGIC DATA FOR THUMB AND RAINBOW GARDENS MEMBERS OF HORSE SPRING FORMATION IN LAKE MEAD AREA*

Published Dates (Ma)	New $^{40}Ar/^{39}Ar$ Dates (Ma)
THUMB MEMBER	
12.5 ± 0.9 f.t. (B)	1. 14.2 ± 0.5 biotite
13.2 ± 0.9 f.t. (B)	2. 14.4 ± 0.3 biotite
14.8 ± 1.4 f.t. (B)	3. 14.4 ± 0.5 biotite
15.0 ± 2.0 f.t. (B)	4. 15.1 ± 0.1 sanidine
15.3 ± 2.0 f.t. (B)	16.2 ± 0.5 biotite
15.4 ± 0.8 f.t. (B)	5. 15.2 ± 0.2 anorthoclase
15.6 ± 1.0 f.t. (B)	6. 15.3 ± 0.1 biotite
15.6 ± 3.0 f.t. K/Ar (B)	15.4 ± 0.1 sanidine
16.2 ± 0.8 f.t. (B)	7. 15.4 ± 0.1 biotite
16.3 ± 1.9 f.t. (B)	15.4 ± 0.04 sanidine
17.2 ± 3.0 K/Ar (A)	8. 15.7 ± 0.1 biotite
	15.7 ± 0.2 sanidine
RAINBOW GARDENS MEMBER	
15.1 ± 0.8 f.t. (B)	9. 18.8 ± 0.2 anorthoclase
19.6 ± 0.8 K/Ar (A)	18.8 ± 1.0 biotite
20.0 ± 0.8 K/Ar (S)	10. 24.0 ± 0.1 sanidine
20.7 ± 1.2 f.t. (B)	11. <25 sanidine
21.3 ± 0.4 K/Ar (A)	12. <26 sanidine
24.3 ± 1.0 K/Ar (C)	

*Column on left shows published data from A = Anderson et al., 1972; B = Bohannon, 1984; C = Carpenter et al., 1989; and S = Shafiqullah et al., 1980. Column on right shows unpublished data (L. S. Beard and L. W. Snee, 1995). Locations of new dated samples and some of the previously published samples shown on Figure 5.

New $^{40}Ar/^{39}Ar$ ages. Seventeen mineral separates were dated by $^{40}Ar/^{39}Ar$ techniques at the U.S. Geological Survey Argon laboratory in Denver (L. S. Beard and L. W. Snee, unpublished data, 1993). The data, which include four samples from the Rainbow Gardens Member and eight from the Thumb Member, are presented in Table 1; locations are shown on Figure 5. The Rainbow Gardens ranges from older than 24.0 ± 0.1 Ma to less than 18.8 ± 0.2 Ma, based on two argon dates, from sanidine and anorthoclase separates, respectively. The older date was obtained from a rhyolitic fall-out tuff sampled about 70 m above the base of the Rainbow Gardens in the northern Grand Wash trough (see Bohannon, 1992). The younger age is from a similar tuff sampled about 175 m above the base of the Rainbow Gardens in the Horse Spring area, where Bohannon obtained the 15.1 Ma fission-track age. A biotite separate from the same tuff has a total fusion age of 18.8 Ma. Two other sanidine separates, from Tom and Cull Wash and Pakoon Ridge, contained excess argon; best constraints on the ages are less than 26 Ma and less than 25 Ma, respectively. A total fusion age on biotite from the Pakoon Ridge sample is 26 Ma. Therefore, the age of the Rainbow Gardens Member in the South Virgin Mountains and Grand Wash trough is bracketed between 26 Ma and 18.8 Ma.

Four sanidine, one anorthoclase, and seven biotite $^{40}Ar/^{39}Ar$ ages were obtained for the Thumb Member in the south Virgin Mountains. Two samples were collected near the base of Thumb strata at two different outcrops in the Pakoon Ridge area. The northern sample (90-33) yielded an age of 15.1 ± 0.1 (sanidine) and 16.2 ± 0.5 Ma (biotite) and the southern sample (88-28a) yielded ages of 15.7 ± 0.1 (sanidine) and 15.7 ± 0.2 Ma (biotite). Two tuffs sampled at separate locations just below an upper alluvial fan sequence in the Thumb are nearly identical in age—15.4 ± 0.04 Ma (sanidine) and 15.4 ± 0.1 Ma (biotite) for sample 89-46, and 15.4 ± 0.1 Ma (sanidine) and 15.3 ± 0.1 Ma (biotite) for sample DC-2. A tuff sampled near the probable northern margin of the Thumb basin, in the northeast part of Wechech Basin, yielded an age of 15.2±0.2 Ma (anorthoclase).

Tuff beds within the upper alluvial fan sequence were sampled at three locations. Biotite from a tuff on the west side of Whitney Ridge yielded an age of 14.4 ± 0.5 Ma (89-83). Biotite from a tuff low in the upper sequence on the north side of the Lime Ridge fault yielded an age of 14.2 ± 0.5 Ma (89-52). An age of 14.4 ± 0.3 Ma was obtained from a tuff sandwiched between a lower distinctive Proterozoic-clast conglomerate and megabreccia deposit (the Gold Butte conglomerate deposit) and the upper alluvial fan sequence north of the Gold Butte fault. In summary, the Thumb Member in the South Virgin Mountains ranges in age from slightly older than 16.2 Ma to slightly younger than 14.2 Ma. The change from finer grained facies of the Thumb to the upper alluvial fan sequence (see discussion in a following section) can be bracketed between 15.4 Ma and 14.4 Ma in age.

Rainbow Gardens Member

Detailed measured sections in the Rainbow Gardens Member show systematic stratigraphic relations without rapid facies changes (Figure 6). Two sections, Tom and Cull Wash (TCW) and Pigeon Wash (PW), were measured at the north and south ends of the Grand Wash trough (in the Cottonwood Wash Formation of Moore [1972], and the formation at Tassai Wash of Lucchitta [1966], respectively). The other sections are all located in the south Virgin Mountains. The Wechech Basin (WB) and TCW sections were measured a few hundred meters north of the sections measured by Bohannon (1984) whereas the Horse Spring (HS) section is at more or less the same location as Bohannon's section. The section at Pigeon Wash was field inspected and slightly modified from a section measured by Lucchitta (1966). Clast lithology and clast size were measured in the basal conglomerate at seven of the sections and maximum clast size was recorded for all sections. No systematic measurement was made of paleocurrent data because of poor exposures.

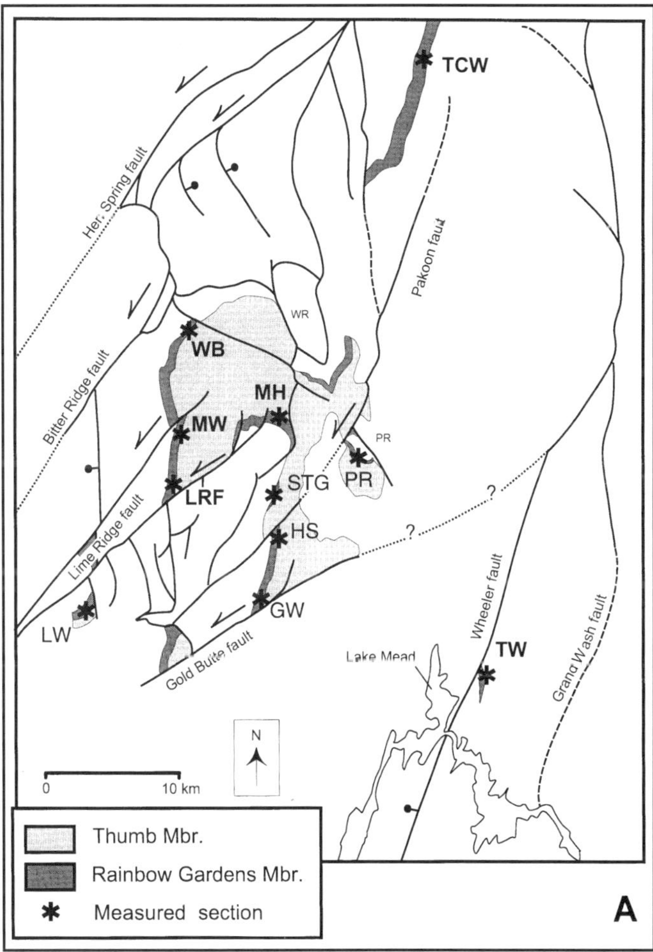

Figure 6 (on this and following two pages). Measured stratigraphic sections of Rainbow Gardens Member of Horse Spring Formation in south Virgin Mountains region. (A) Map showing location of measured sections. See text for descriptions. From north to south, TCW—Tom and Cull Wash, WB—Wechech Basin, MW—Mud Wash, LRF—Lime Ridge fault, MH—Mud Hills, STG—St. Thomas Gap, PR—Pakoon Ridge, HS—Horse Spring, GW—Garden Wash, LW—Lime Wash, TW—Tassai Wash.

Bohannon (1984) defined the type section of the Rainbow Gardens Member at Rainbow Gardens east of Frenchman Mountain. He included in the type section everything from the top of the underlying Moenkopi Formation to the top of a resistant limestone that forms a well-developed hogback in the Rainbow Gardens area. Bohannon (1984) indicated that the rocks of the upper part of the Rainbow Gardens Member made an abrupt lithofacies change northward into rocks of the Thumb Member, but based on data presented in this chapter, the section probably contains an unconformity, and so these rocks are probably the Thumb Member.

A typical section of the Rainbow Gardens Member in the Virgin Mountains region consists of basal conglomerate overlain by a complex slope-forming middle unit, which in turn is capped by an upper unit of cliff or ledge-forming carbonate rocks. A previously unrecognized unconformity lies within the carbonate rocks unit included within the Rainbow Gardens Member by Bohannon (1984). This unconformity separates pedogenically altered carbonate strata from overlying algal laminated limestones that intertongue laterally with lacustrine deposits of the Thumb Member. The underlying carbonate rocks are herein included in the Rainbow Gardens and the overlying algal limestones in the Thumb.

The northern edge of the basin was probably a short distance north of the Wechech Basin and Tom and Cull Wash sections. The thickest part of the basin in which these strata were deposited is in the Horse Spring area. The southern margin of the basin was probably not far south of the Pigeon Wash section. There, Longwell (1936) reports a deposit of sedimentary breccia (the basal conglomerate) exposed at Wheeler Ridge, where the ridge is transected by the Colorado River (the deposit is now beneath Lake Mead). The breccia is deposited, with no obvious angular discordance, on an erosional surface cut into the lower part of Permian Kaibab Limestone and is composed mostly of clasts of the Kaibab. Prior to faulting, the Pigeon Wash section most likely was positioned to the south of the Garden Wash section, which rests on the Moenkopi Formation (Fig. 6B).

Basal conglomerate. The basal conglomerate (Figs. 6B, 6C) is laterally persistent except in the northeastern Tom and Cull Wash section, where the conglomerate is preserved in broad shallow channels. Measured thicknesses range from 2–25 m and are thicker locally. The conglomerate is typically composed

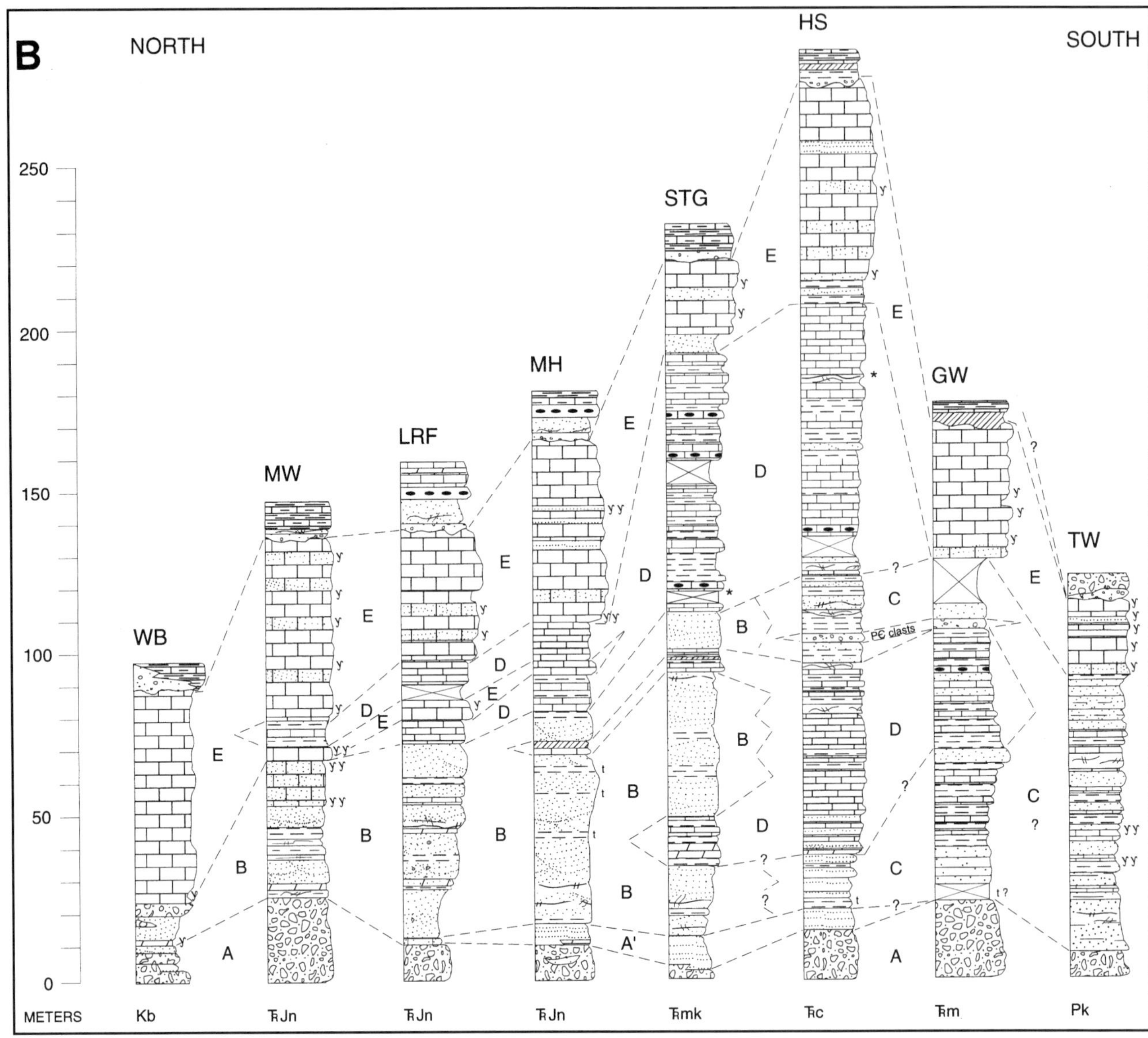

Figure 6B. North to south arrangement of stratigraphic sections, from Wechech Basin south to Tassai Wash, after restoring left-lateral displacement on Lime Ridge and Gold Butte faults. Sections represent cross-sectional view of Rainbow Gardens depositional system that is approximately perpendicular to paleoflow direction of basal conglomerate and fluvial sandstone facies. Strata underlying Rainbow Gardens Member of Horse Spring Formation shown beneath each section: Kb—Cretaceous Baseline Sandstone, TRJn-Late Triassic (?) and Jurassic Navajo Sandstone, TRmk—Triassic Moenave and Kayenta Formation, undivided TRc—Triassic Chinle Formation, TRm-Triassic Moenkopi Formation, Pk-Permian Kaibab Limestone. Legend for symbols shown on Figure 6C. Correlation lines between sections show facies relationships (see text for complete description). Lower unit: A—basal conglomerate; A′—hematitic, carbonate-cemented sandstone recycled from Navajo Sandstone. Middle unit: B—tuffaceous fluvial sandstone; C—calcareous clastic rocks of marginal lacustrine origin; D—lacustrine facies rocks. Upper unit: E—Palustrine carbonate facies rocks. Above E is unconformity below Thumb Member of the Horse Spring Formation and lowest few meters of Thumb rocks.

of carbonate and chert clasts encased in a sandy matrix with a strong carbonate cement. Some clasts have a rusty orange siliceous rind that can be as much as 0.5 cm thick. At some localities the conglomerate grades upward into a hematitic, carbonate-cemented sandstone with subrounded carbonate and chert grains as well as abundant well-rounded quartz grains which were probably recycled from the Late Triassic (?) and Jurassic Navajo Sandstone (Figs. 6B, 6C).

Systematic areal variations in maximum clast size and clast type aid in delineating source areas and provide constraints on paleoflow directions (Fig. 7). Clast types can be divided into several suites: monolithologic gray limestone (Permian Kaibab

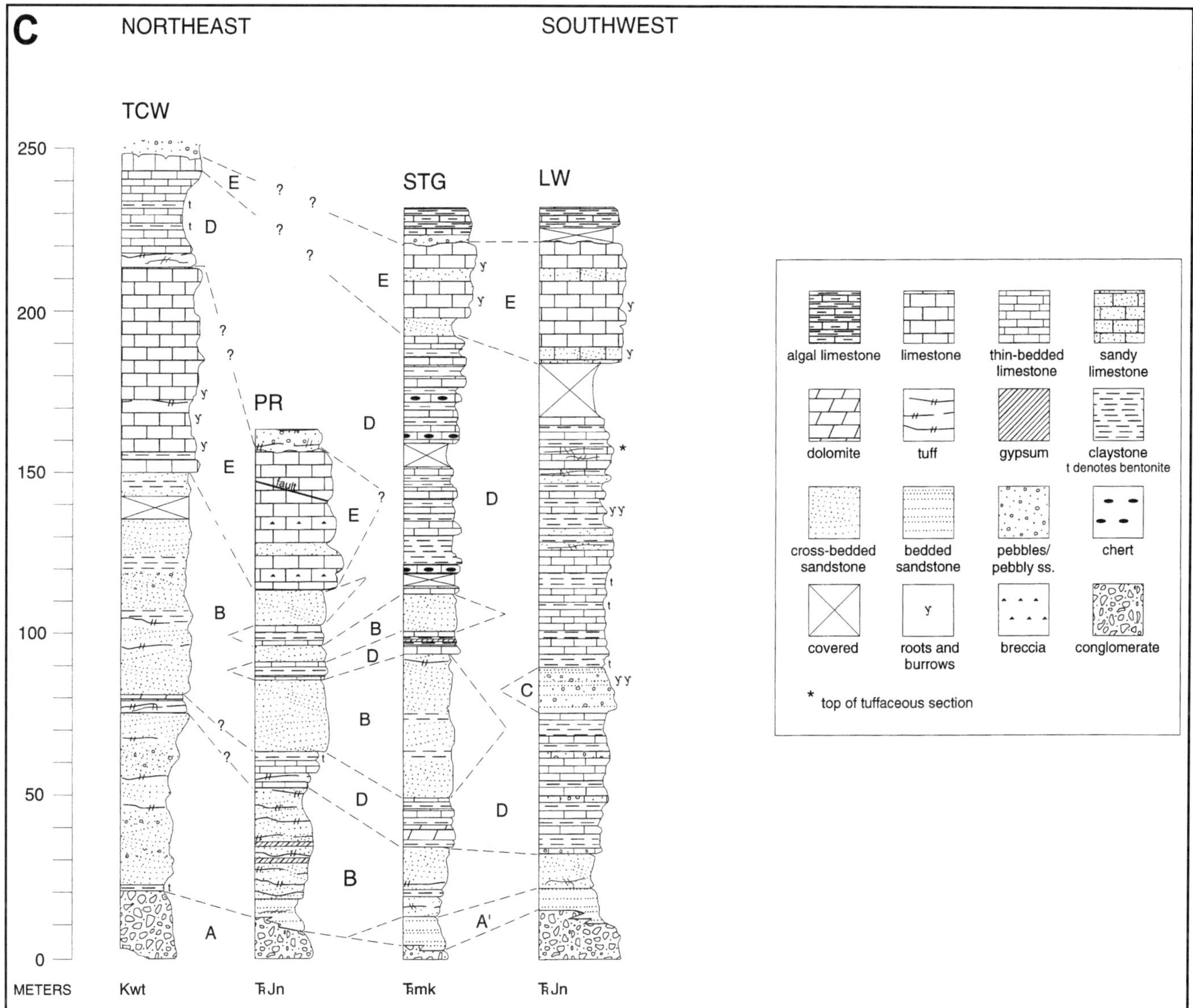

Figure 6C. Northeast-southwest stratigraphic sections arranged from Tom and Cull Wash southwest to Lime Wash, with St. Thomas Gap section providing tie to north-south section. Relative positions of sections prior to faulting uncertain, but probably approximate a northeast-southwest cross sectional view that is parallel to paleoflow directions. Unit symbol: Kwt—Cretaceous Willow Tank Formation. A–E as in Figure 6B.

Limestone and Toroweap Formation), rounded pebbles and cobbles recycled from Mesozoic rocks, mixed middle and upper Paleozoic limestone, and mixed lower Paleozoic dolomite.

In the section measured at the north end of Wechech Basin, the basal conglomerate of the Rainbow Gardens Member is composed of subangular, monolithologic gray limestone clasts derived from either the Permian Kaibab or Toroweap Formations (Fig. 8). The monolithologic conglomerate intertongues laterally with channel fill comprised of lithic sandstone containing rounded pebbles reworked from the underlying Cretaceous Willow Tank Formation. These channels of pebbly lithic sandstone within the basal conglomerate occur only in the vicinity of the Wechech Basin section. Southward from the Wechech Basin section, the basal conglomerate loses its monolithologic character and is dominated by a more diverse population of middle to late Paleozoic clasts; the gray Permian carbonate clasts, however, persist and are consistently among the largest found.

The monolithologic limestone-clast deposits at the north end of Wechech basin are sheet-like, clast-supported conglomerates. The coarsest deposit, at the base of the section, is unsorted and has an average clast size of about 5 cm; larger clasts are common and the maximum clast size is more than 14 cm. This basal deposit, as well as similar beds higher in the section,

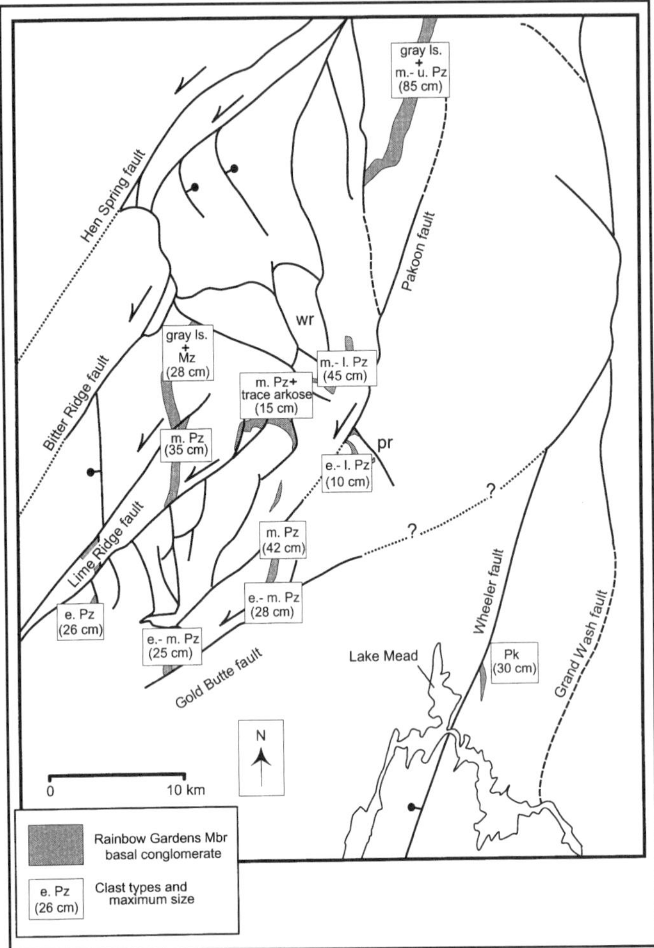

Figure 7. Map showing distribution of basal conglomerate of Rainbow Gardens Member of the Horse Spring Formation and variation in clast composition and size. Compositions include gray limestone (gry ls.) (probably Permian strata—see text); Permian Kaibab Limestone (Pk), recycled Mesozoic conglomerate clasts (Mz), and mixed Paleozoic (Pz) strata (e. = early, m. = middle, and l. = late). Maximum size given at each outcrop is long diameter of largest clast within 10 m^2 area. Faults and geographic locations same as Figure 5.

Figure 8. Photograph showing monolithologic conglomerate, composed of angular clasts of Permian gray limestone, exposed at Wechech Basin section. Fossils within clasts include bryozoans, brachiopods, and crinoids that are most likely Permian in age (G. H. Billingsley, 1993, personal communication).

appear to coarsen northward. Imbrication at several exposures indicates paleoflow from the north to northwest. The monolithologic conglomerates persist upward in the section to the contact with the upper carbonate rocks unit. No carbonate clasts other than Permian limestone were found.

The pebbly lithic sandstone is dark red to orange and hematitic, with clasts of well-rounded quartzite, and carbonate rocks. Two distinctive clasts, black radiolarian-bearing argillite and reddish orange to orange-brown litharenite, which are indicative of provenance, are also present; they occur in the Willow Tank Formation and are also well-known components of other Cretaceous foreland basin deposits in the region (cf. Goldstrand, 1992). The pebbly lithic sandstone was deposited in broad channels that appear to trend east to west. Planar and trough cross-stratification are common and indicate easterly paleoflow. Within the channels, conglomerate beds are locally bimodal, containing gray carbonate clasts and small rounded pebbles recycled from Mesozoic sandstones.

Clast compositions of the basal conglomerate are quite diverse in sections measured from south of Wechech Basin (Fig. 7). Most of the sections contain limestone, dolomite, and sandstone clasts that represent much of the Paleozoic sequence, including clasts of Cambrian Undivided Dolomites, Mississippian Redwall Limestone, Pennsylvanian Callville Limestone, Permian Pakoon Limestone of McNair (1951), and Permian Esplanade (?) Sandstone. Many sections also contain clasts of the Late Triassic (?) and Jurassic Navajo Sandstone. The only crystalline clasts found were in one small channel of arkosic sandstone at the Mud Hills section. This outcrop is enigmatic because of its isolation from similar arkosic deposits, but the source almost certainly was from the south. The Garden Wash section, just north of the Gold Butte fault, also contains clasts of the Cambrian Tapeats Sandstone, a few well-rounded quartzite cobbles, and a distinctive pebbly sandstone block probably derived from the Shinarump Member (Triassic) of the Chinle Formation. Only in the westernmost section, at Lime Wash, is the basal conglomerate dominated by subrounded dolomite clasts derived from early Paleozoic strata. Here, even the largest clasts are dolomite except for one sandstone block derived from the underlying Navajo Sandstone. No gray carbonate clasts like those seen in sections to the northeast were found. The dolomite clasts are unlike typical early Paleozoic dolomites in the Virgin Mountain area and may have been derived from allochthonous rocks in the upper plate of the Muddy Mountain thrust to the west (Bohannon, 1983) or from the lower beveled surface of the Kingman uplift.

At the northeasternmost section, Tom and Cull Wash, the basal conglomerate is the coarsest found in the Virgin Mountains. Clast types are quite varied, ranging from Cambrian through Permian lithologies, but the largest clasts (as much as 90 cm diameter) are gray Permian limestone like those seen at Wechech Basin. In contrast, the Pakoon Ridge section south of Tom and Cull Wash is much finer grained and lacks the larger gray limestone clasts. This observation, combined with the northward coarsening of the monolithologic conglomerate at Wechech Basin, indicates that Permian rocks were exposed to the north during Rainbow Gardens time.

Middle unit. The basal conglomerate is overlain by a complex intertonguing sequence of clastic and carbonate lithofacies. This middle unit typically forms a slope between the resistant basal conglomerate and overlying ledge- to cliff-forming carbonate rocks. Deposition of the middle unit was coeval with extensive silicic volcanism in the region, as shown by the abundant volcaniclastic and pyroclastic material in both clastic and lacustrine rocks. Measured sections show that tuffaceous fluvial clastic rocks dominate northern and northeastern outcrops and that these intertongue south and southwest with white tuffaceous lacustrine and marginal lacustrine deposits (Figs. 6B, 6C). The southernmost exposure at Pigeon Wash contains calcareous clastic rocks of probable marginal lacustrine origin that intertongue northward with the lacustrine deposits (Figs. 6B, 6C). In contrast to the northern clastic facies, these rocks are noticeably less tuffaceous.

Fluvial facies rocks include pebbly lithic sandstone, epiclastic rocks, and rare claystone-micrite horizons. The lithic sandstone beds commonly exhibit planar or trough cross-stratification, although some beds are structureless, unsorted, and matrix-supported. The rocks are gray to pale green and commonly very tuffaceous. Welded pumice clasts as much as 1 cm are common in the Tom and Cull Wash and Pakoon Ridge sections, and sand-size pumice clasts are a common constituent in sandstones at sections farther west and south. Other clasts include polished black and gray chert, quartzite, and carbonate rocks, ranging from sand-size to several centimeters in diameter. These clasts, which are most abundant and coarsest at the Tom and Cull Wash section, are similar to clasts found regionally in Cretaceous rocks.

At Wechech Basin and Mud Wash, the upper 6–10 m of tuffaceous lithic sandstone is heavily overprinted by probable calcrete soils exhibiting features such as thick rinds on pebbles, abundant root casts (calcite-filled tubes of Bohannon, 1984) and burrows, and a downward decrease in carbonate cement accompanied by an upward obliteration of primary features. Claystone, commonly bentonitic, occurs locally within the lithic sandstone beds forming soft, poorly exposed green, pale red, or gray slopes from about 0.5–2 m thick. The claystone horizons locally contain thin (2–10 cm) interbedded sandy micrites or calcareous sandstones that are laterally discontinuous. The combined claystone-micrite horizons are rarely more than a meter thick, and relative proportions of claystone to micrite can vary widely. Some micrite beds contain wispy laminations and small-scale cross-stratification that are outlined by rounded quartz sand grains that were probably derived from the Triassic(?) and Jurassic Navajo Sandstone. Other beds exhibit mud cracks and contain small (0.5–2 cm) pebbles of rounded carbonate rock and black and gray chert. Very minor amounts of gypsum occur within a distinctive claystone-micrite horizon at the Mud Hills and St. Thomas Gap sections (Figs. 6B, 6C). About 50 m below the gypsiferous horizon at St. Thomas Gap is a 10 m thick section of structureless to laminated dolomitic limestone with a few calcareous bioturbated mudstone horizons; the limestones are cherty and possibly algal. The origin of the claystone and micrite horizons is uncertain. In the north and northeastern sections, they may represent overbank deposits. However, the gypsiferous and dolomitic horizons described above are more likely marginal lacustrine deposits interfingering with the fluvial sandstones (this is the only section where limited interfingering between fluvial-facies rocks and overlying lacustrine deposits is observed).

The fluvial-facies rocks dominate most of the northern and northeastern sections (TCW, PR, MH, LRF, MH, WB) with the thickest section at Tom and Cull Wash. At St. Thomas Gap the fluvial facies forms the lower half of the middle unit. The southwesternmost clastic deposit of the fluvial facies is found at the Horse Spring and Lime Wash sections where about 12 m of cross-bedded to structureless tuffaceous sandstones occur just above the basal conglomerate and below lacustrine-facies rocks.

Lacustrine-facies rocks are dominantly thin-bedded calcareous mudstone to micrite that alternate with recessive claystone. Bedding is typically parallel and even, and ranges from about 5–15 cm thick. The micrite and calcareous mudstone beds are commonly structureless or bioturbated. Fossils are rare; however, a thin micritic bed with 0.5 cm long gastropods occurs at the base of the white lacustrine sequence at the LRF section, whereas bivalve and ostracode(?) shell fragments were observed in a thin section from a micritic bed slightly upsection. Grains of rounded quartz and chert form thin discontinuous wispy laminae or "float" randomly. A few beds contain 1–2 cm black or gray pebbles floating near the top.

Poorly or nontuffaceous lacustrine rocks typically weather gray or light tan; however, the bulk of the lacustrine rocks is very tuffaceous, weathering to distinctive chalky, light gray to bright white slopes. Thin, gray, crystal-lithic fallout tuff horizons, typically exposed in lenticular beds throughout the tuffaceous lacustrine facies, are cross-stratified or structureless with bioturbated tops. A few structureless fallout tuff layers are composed of delicate glass shards. At least two coarse biotite-feldspar tuffs, possibly distal unwelded pyroclastic flows, form distinctive horizons near the tops of the lacustrine internals at the Horse Spring and Lime Wash sections. Bentonitic claystone beds are typically green and range from a few centimeters to 1 m in thickness.

White, tuffaceous, lacustrine rocks constitute most of the middle unit at the Horse Spring and Lime Ridge sections. At Horse Spring, these deposits are especially tuffaceous from about 80–180 m above the base of the Rainbow Gardens Member (Fig. 6B); an 18.8 Ma $^{40}Ar/^{39}Ar$ age was obtained near the top of the tuffaceous interval. The lowermost 40 m and uppermost 15 m of the white lacustrine rocks are distinctively less tuffaceous. The less tuffaceous lower part of the lacustrine sequence at Horse Spring probably correlates northward with a thin section of poorly tuffaceous lacustrine rocks in the lower part of the section at St. Thomas Gap.

To the north, white lacustrine deposits at the south end of Whitney Ridge and at the Tom and Cull Wash section are very cherty, thicker bedded, and exhibit evidence of subaerial exposure such as burrowing and root casts. These deposits overlie fluvial-facies rocks and show limited interfingering with beds of the upper (carbonate rocks) unit (Fig. 6C).

To the south, the white lacustrine deposits overlie calcareous clastic rocks at Garden Wash. These underlying rocks are distinctively non- or poorly tuffaceous, orange-weathering sandstone, siltstone, and claystone. The lowermost 14 m of section is interbedded reddish orange sandstone and siltstone is overlain by interbedded claystone, thin sandy micrites, and rare sandstone beds. The sandstones are typically structureless and contain medium-size, well-rounded, quartz grains and subangular coarse grains of carbonate rock fragments and chert. The claystones form poorly exposed slopes from 10 cm to 1 m thick. Micritic layers, from 10–20 cm thick, are structureless except for local thin discontinuous laminae and local bioturbated horizons. The micritic layers typically contain floating sand-size grains of quartz, carbonate rock fragments, and chert.

A clastic bed, 1–2 m thick, overlies the white tuffaceous lacustrine rocks at Garden Wash. The bed is a distinctive calcareous sandstone and conglomerate unit containing crystalline clasts of granite, quartzite, and schist, as well as minor carbonate clasts (Fig. 6B). The same clastic bed, including the crystalline-clast conglomerate, is found to the north at the Horse Spring section, to the west at an unmeasured section midway to the Lime Wash section, and at the Lime Wash section. The most likely source for the crystalline rocks was far to the south, where Proterozoic rocks was widely exposed during Rainbow Gardens time.

The middle unit at the southeasternmost section, Pigeon Wash (Fig. 6B), is composed of a monotonous sequence of calcareous sandstone, siltstone, and thin-bedded impure carbonate beds (Lucchitta, 1966). Some beds contain root casts or burrows, but are mostly structureless and mottled. The section is notably nontuffaceous, except for a thin altered biotite tuff horizon near the base and a 10-m-thick horizon of tuffaceous siltstone, sandstone, and micrite near the top of the middle unit. This tuffaceous horizon most likely correlates with the white tuffaceous lacustrine rocks at Garden Wash. Middle unit rocks of the Pigeon Wash section were probably deposited in a marginal lacustrine setting that lacked the fluvial input of tuffaceous material seen to the north.

Upper unit. The upper unit which is composed of carbonate rocks forms a distinctive cap to the Rainbow Gardens Member everywhere in the Virgin Mountains, ranging from about 60 m thick at northern outcrops (WB and TCW) to as little as 20 m at Pigeon Wash (Fig. 6). The upper unit contains irregular cycles of white and pink mottled carbonate rocks and thin interbeds of orange-weathering sandy carbonate rocks to calcareous sandstone. Beds of the upper unit are commonly structureless, ranging from 30 cm to 1 m thick. Poorly preserved gastropods were observed in only one thin section.

The micritic to sparry textured carbonate beds locally contain 1–2 cm of banded calcite veins (onyx) and vug fillings; locally the beds are also quite cherty. Some beds exhibit diffuse, discontinuous horizontal laminae, formed by thin sandstone stringers or by alternating micrite and spar. The horizontal laminae are discontinuous and uneven to continuous and wavy. These laminae are poorly understood; the irregular and discontinuous laminae may be related to pedogenic processes, whereas more continuous laminae may be algal in origin.

Other beds are vuggy, irregularly fractured, and locally brecciated. Weathered surfaces show abundant diffuse white "spots," from 0.5 to 3 mm diameter, in a pink micritic matrix. The vugs and associated fractures are commonly filled with pale orange sandy micrite or with coarse spar. The spots range from angular to rounded in shape and are commonly associated with brecciation. The spots and breccia fragments are embedded in a micritic matrix. Although the spots superficially resemble ooids, petrographic examination suggests that both the spots and breccia fragments have a pedogenic origin. Some spots exhibit irregular laminae and dessication followed by continued calcitic precipitation, suggesting that they could be calcrete or vadose glaebules that were formed during pedogenesis (see fig. 3c in Platt and Wright, 1992). The breccia fragments are also probably the result of intermittent dessication.

Interbedded calcareous sandstone and sandy carbonate beds are orange to white, have nodular textures, and contain vertical to horizontal calcite-filled tubes formed by both root casts and burrows (Fig. 9). The sandstone is well sorted and is composed of well-rounded quartz grains that are recycled from Jurassic Navajo Sandstone. In the calcareous sandstone, the root casts and burrows are preserved in the lower, poorly cemented part of

Figure 9. Photograph of vertical root casts, encased in a network of smaller root casts and worm burrows. Host rock is a sandy limestone of palustrine origin.

the beds, which grade upward to nodular or locally brecciated chalky carbonate rock with floating sand grains. The sandy limestone beds usually have calcite tubes either throughout the beds or concentrated near their tops. The sandstone may represent eolian deposits formed during periods when the lake was dry.

At Pakoon Ridge and at the south end of Whitney Ridge, most of the upper (carbonate rocks) unit is eroded at the sub–Thumb Member unconformity. At Pakoon Ridge within immediately overlying Thumb beds, a locally derived conglomerate facies contains clasts of carbonate rock fragments that look like the typical vuggy, root-cast carbonate material of the upper (carbonate rocks) unit. The Tom and Cull Wash section includes several white tuffaceous lacustrine facies rocks of the middle unit that intertongue with the upper (carbonate rocks) unit.

Interpretation of deposits of the Rainbow Gardens Member. Bohannon (1984) and Rice (1987) interpreted the basal conglomerate of the Rainbow Gardens Member as braided stream deposits, most likely in a gentle downwarp with low-relief margins formed by the relict Sevier thrust belt highlands to the northwest and the Kingman uplift to the south. The thin, regionally uniform distribution of the basal conglomerate suggests a pediment gravel deposit or bajada-style alluvial plain (Bohannon, 1984, Rice, 1987). Because the basal conglomerate was probably formed by coalescence of many pediment deposits, it could represent a time span of millions of years, with the endpoint of deposition extending somewhere beyond the present-day outcrops. Based on regional paleogeography presented above, I infer that the sedimentary materials comprising the basal conglomerate were eventually deposited to the north or northeast in the Oligocene Claron basin (Anderson, 1987; Taylor, 1993).

The only detailed study of the basal conglomerate is that of Rice (1987), who examined its sedimentology and provenance in the western Lake Mead area, from Rainbow Gardens to Overton Ridge. His paleocurrent data suggested that there was north-northwest paleoflow in southern exposures and south or southeast paleoflow in northern exposures. The basal conglomerate is dominated by carbonate clasts, leading Rice to suggest a source to the west or northwest off the upper plate of the Muddy Mountain thrust. However, Bohannon (1984) states that most of the clasts are gray cherty limestone, similar to the Permian Kaibab Limestone and Toroweap Formation, and red-brown sandstone of probable Mesozoic age, both of which are more likely eroded from autochthonous rocks on the north slope of the Kingman high. Rice found only minor traces of feldspar and volcanic grains in the Rainbow Gardens area, which increase in abundance southward, as well as biotite granite and gneissic clasts (3% of total clast content) in the southernmost exposure in the Rainbow Gardens area. These data indicate that a distal crystalline source lay to the south of the Rainbow Gardens area as was described above for the Virgin Mountains area.

Based on these data for the Muddy Mountains and the distribution of clasts and clast types in the Virgin Mountains, regional paleoflow in the Virgins was probably north and east (Fig. 10A). Exceptions are the north end of Wechech Basin and Tom and Cull Wash, where coarse debris was derived from the north off of an emergent area, here called the Virgin positive area. The large northeast-flowing channels at Wechech Basin may represent an axial stream system flowing on soft Cretaceous rocks at the base of the Virgin positive area; more work is needed to support this hypothesis, however.

Lacustrine facies rocks of the middle unit of the Rainbow Gardens Member contain preserved gastropods, bivalves, and ostracodes(?), minor clastic material, and bioturbation structures. The Rainbow Gardens lake was apparently shallow and subsidence was probably relatively slow given that the maximum thickness for the Rainbow Gardens is 240 m in about a 6 m.y. time span (0.04 m/1,000 yr). Using criteria from Platt and Wright (1991), the lake environment was probably a shallow, unstratified, freshwater carbonate lake with a low-gradient, low-energy margin. The general lack of evaporites and dessication features suggests that the lake was perennial. Deposition

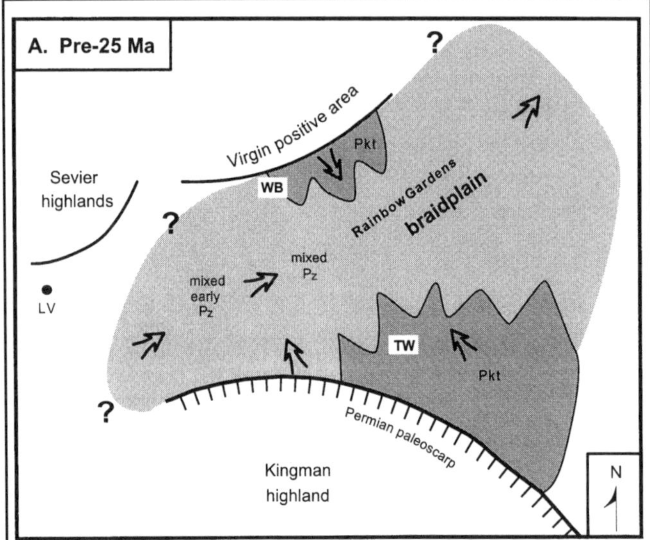

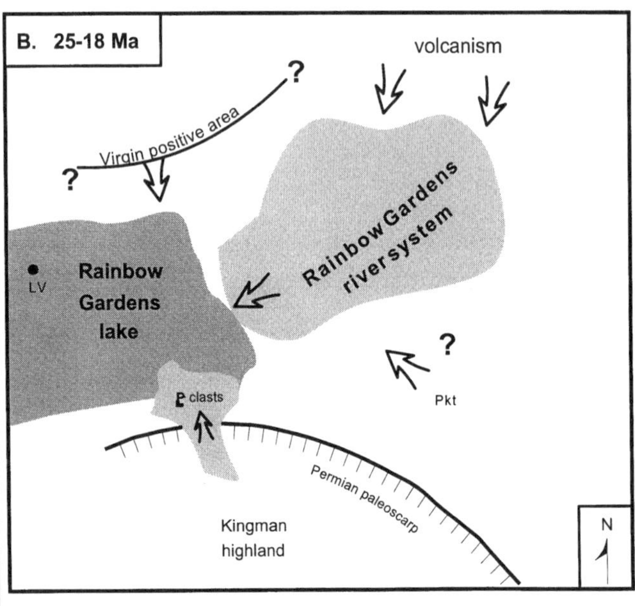

Figure 10. Generalized paleogeographic sketches of Rainbow Gardens depositional system. (A) Pre-25 Ma: basal conglomerate interpreted as braidplain or pediment deposit on the north flank of the Kingman highland and east (and south?) of the Sevier thrust system. Deposit, which rests on Permian through Cretaceous strata, composed mostly of mixed Paleozoic strata, probably derived from the Sevier highlands to the west and the Kingman highland to the south. Distinctive provenances include Permian rocks (Pkt—Kaibab Limestone and Toroweap Formation) brought into the braidplain from (1) the southwest, where eroded from a low-relief surface above the Permian paleoscarp, which has retreated northeastward from the Kingman highland; and (2) from the north, eroded from a moderate to abrupt relief area herein called the Virgin positive area. LV marks inferred position of Las Vegas, Nevada.

(B) 25–18 Ma: Active volcanism to north of Rainbow Gardens depositional system provided source for abundant volcanic material in fluvial system. Fluvial system drains to south or southwest into shallow broad ponded area between the Virgin positive area to the north or northwest and the Permian paleoscarp to the south. Lake progressively enlarges and progrades northeastward over fluvial system. A thin, distinctive Proterozoic (P) clast conglomerate within lacustrine deposits at the southern margin of the lake suggests some breaching of paleoscarp late in depositional period (see Fig. 6B).

of abundant tuff horizons, the result of fallout and possibly unwelded distal pyroclastic flows from extensive silicic volcanism to the north, must have nearly filled the lake at times, based on the abundance of volcanic material. The lacustrine deposits were near enough to the shore to be within reach of terrigenous sources that provided clastic debris, although outcrop-scale intertonguing of lacustrine and fluvial deposits is limited. The lacustrine facies deposits eventually prograded northeast across the fluvial system, interfingering with the upper (carbonate rocks) unit at Tom and Cull Wash.

The source of volcanic debris was to the north because tuffaceous fluvial facies dominate northern outcrops. Deposition of the middle unit must have been initiated by a drainage reversal from the earlier north or northeast paleoflow that deposited the basal conglomerate. Processes associated with volcanism to the north, such as regional uplift, damming of drainages, or tectonism may have been responsible for the drainage reversal.

The volcanic source is most likely the Caliente caldera complex and the Bull Valley Mountain area, the location of the southernmost and youngest magmatism associated with the "ignimbrite flareup" in the Great Basin. The main phase of volcanism occurred from about 25–18 Ma, when large volumes of ash-flow tuff were erupted; lesser volumes of rhyolite were erupted from about 16–12 Ma (R. B. Scott, 1993, personal communication). At about the time lacustrine deposition ended in the Oligocene Claron basin to the north because of volcanism (Taylor, 1993; Hintze, 1986; Anderson, 1987), tuffaceous fluvial material was transported from the north into the incipient Rainbow Gardens lake (Fig. 10B).

The upper (carbonate rocks) unit of the Rainbow Gardens Member was deposited in a marginal lacustrine or palustrine environment, using criteria set forth by Freytet and Plaziat (1982) (Fig. 11). Palustrine limestones of the upper (carbonate rocks) unit show strong evidence of subaerial exposure (Platt, 1993), including dessication (brecciation), periodic influx of sand (perhaps by eolian processes), and pedogenic overprinting of both carbonate and sandstone beds. Repeated subaerial exposure and groundwater fluctuation during slow deposition created a stack of calcrete-like beds, similar to features described for the Claron Formation (Mullett, 1989, 1990). Much of the original texture of the carbonate deposits seems to be obliterated, but they were probably spring-fed tufa deposits.

The widespread nature of this unit, overlying fluvial-facies rocks with only limited interfingering with lacustrine-facies rocks of the middle unit, indicates that the Rainbow Gardens lake shallowed and widened during deposition of the upper unit. The lake probably had a low-gradient ramp-type shoreline (Fig. 11) based on the preponderance of marginal lacustrine

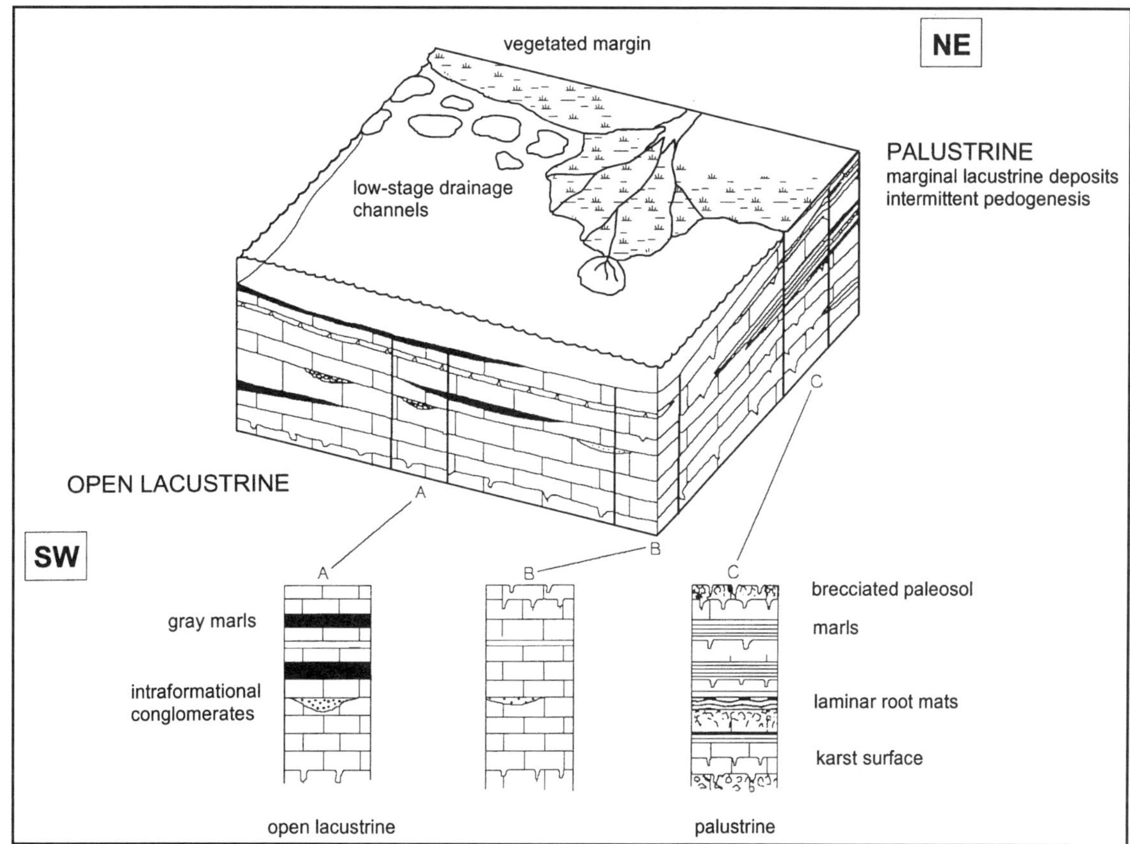

Figure 11. Figure shows low-gradient, ramp-type shoreline inferred for lacustrine facies rocks of Rainbow Gardens Member of the Horse Spring Formation, based on the abundance of marginal lacustrine and palustrine deposits. Upper (carbonate rocks) unit deposits at Lime Wash represented by vertical section A; those at Horse Spring section represented by vertical section B; and palustrine deposits at Wechech Basin section represented by vertical section C. Figure slightly modified from figure 5 in Platt and Wright (1991).

facies and the lack of cross-bedded shoreline facies (Platt and Wright, 1991). This shallowing and widening of the lake is coeval with a waning of volcanism to the north at about 18.6 Ma (R. B. Scott, personal communication, 1993).

Rainbow Gardens–Thumb unconformity

The contact of the upper (carbonate rocks) unit of the Rainbow Gardens Member with rocks of the overlying Thumb Member is a previously unrecognised erosional unconformity. The unconformity occurs throughout the Virgin Mountains area and at least locally can be demonstrably related to faulting of underlying rocks, but not tilting, as there is no angular discordance between the two members. An excellent exposure of the unconformity is found at the north end of the Wechech Basin area where two stratigraphic sections about 3 km apart were measured across the unconformity. The upper surface of the carbonate rocks at the two sections is manganese-coated white sparry limestone breccia that contains local infillings of reddish conglomeratic sandstone containing clasts of the sparry limestone breccia. The reddish conglomerate sandstone penetrates as much as 30 cm down into the Rainbow Gardens along karsted cracks. The conglomeratic sandstone loses the white sparry clasts upward, grading to coarse sandstone and siltstone which intertongue laterally and vertically with thin algal-laminated limestone beds of the Thumb Member. The conglomeratic sandstone is discontinuously exposed at the unconformity southward toward Mud Wash. The unconformity is easily located between Wechech Basin and Mud Wash, even without the conglomeratic sandstone, because of the bright white, brecciated upper surface of the Rainbow Gardens Member (Fig. 12A).

The karsted upper surface of the Rainbow Gardens Member can be traced southward to the Garden Wash section, but is more subdued in character there than at Wechech Basin (Fig. 12B). The sandstone is intermittently present and the sparry white limestone clasts gradually decrease to sand- to granule-size. A poorly exposed, undated biotite tuff is found within the lowermost meter of rocks lying above the unconformity at the MW, LRF, MH, and HS sections.

At Pakoon Ridge, the unconformity is spectacularly exposed on the south side of a northwest-striking reverse fault, where several conglomerate beds, containing carbonate pebbles to boulders

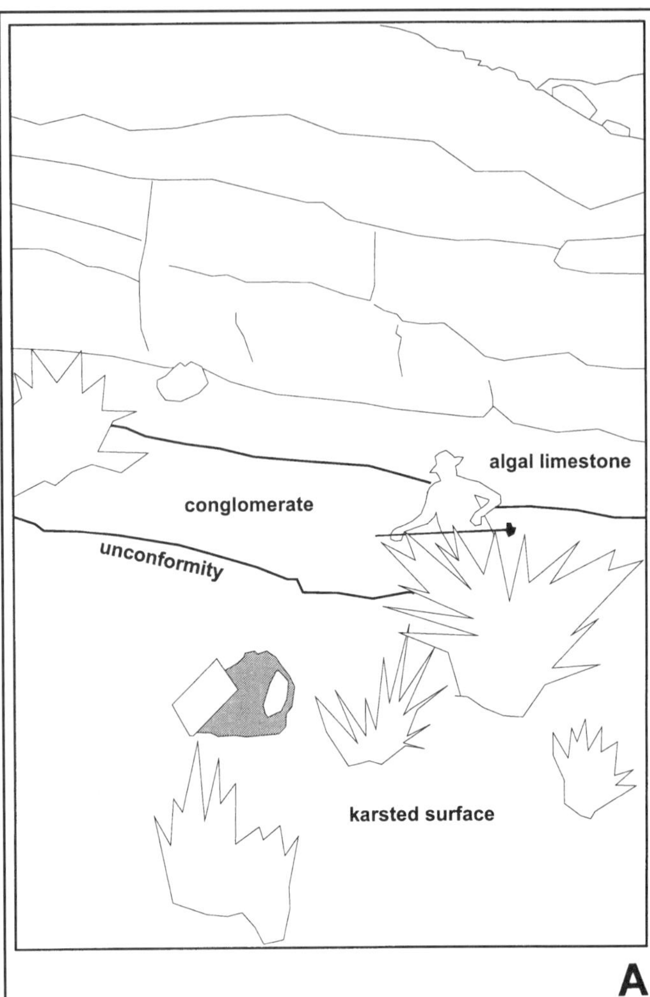

Figure 12 (on this and facing page). Nature of unconformity between Rainbow Gardens and Thumb Members of Horse Spring Formation. (A) Photograph and sketch of unconformity exposed south of Wechech Basin section. Light-colored foreground is karsted upper surface of upper (carbonate rocks) unit of Rainbow Gardens Member. Midground, behind person, is reddish conglomerate composed of angular clasts of white sparry limestone eroded from karsted surface. Above conglomerate is typical outcrop of algal-laminated limestone that contains large hemispheric algal mounds (not visible in photograph).

(as much as 2 m in diameter) of the upper unit of the Rainbow Gardens Member, intertongue laterally and vertically with limestone of the Thumb Member (Fig. 12C). The upper (carbonate rocks) unit, and perhaps part of the middle unit, of the Rainbow Gardens Member are eroded. Elsewhere in the Pakoon Ridge area, at the south end of Whitney Ridge, and in the St. Thomas Gap area, the unconformity is marked by large megabreccia blocks of Rainbow Gardens at the base of Thumb strata.

Thumb Member

Bohannon (1984) designated a reference section for the Thumb Member at Rainbow Gardens in the western Lake Mead region and defined it as rocks stratigraphically above carbonate rocks of the Rainbow Gardens Member and below carbonate rocks of the Bitter Ridge Limestone Member of the Horse Spring Formation. Stratigraphy of the Thumb Member is complicated because of rapid facies changes and structural disruptions, but in general it includes lacustrine and alluvial facies (Bohannon, 1984). Deposition of the Thumb appears to have initiated within a single basin within the south Virgin Mountains, cospacial with the earlier Rainbow Gardens basin. However, complex facies relations suggest that the basin was progressively disrupted by faulting and concomitant local uplift. It is not known if Thumb deposits in the western Lake Mead area, described by Bohannon (1984), were deposited in the same basin as those in the Virgins.

The Thumb Member is present throughout the south Virgin Mountains, but is only locally well exposed. Stratal dips everywhere are highly variable both in direction and degree because

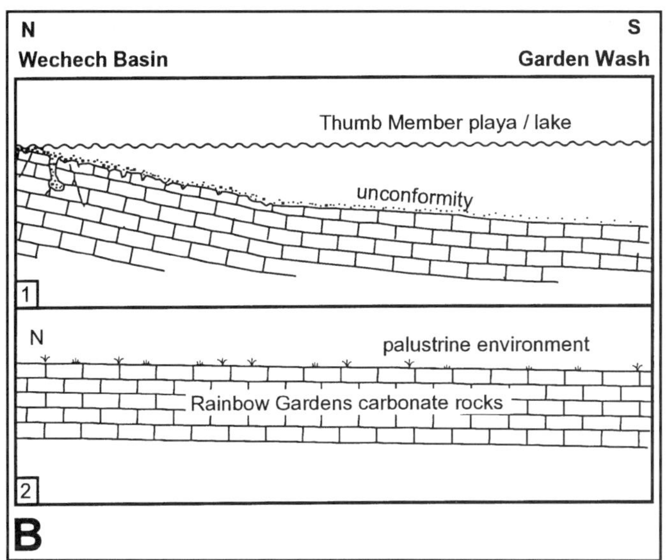

Figure 12B. Sketch showing development of unconformity. (1) Warping or uplift, probably caused by faulting, at Wechech Basin allows more pronounced development of unconformity than to south at Garden Wash. (2) Palustrine environment at end of deposition of Rainbow Gardens Member and beginning of karsting of upper surface. Modified from figure 17.12 in Vera et al. (1988).

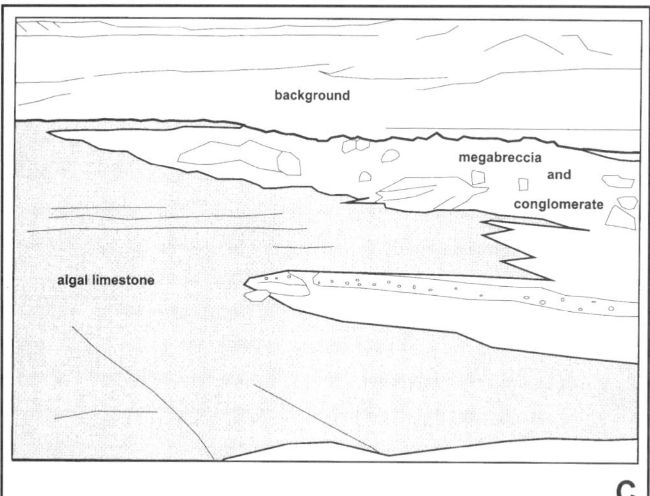

Figure 12C. Photograph and sketch of megabreccia and conglomerate above unconformity at Pakoon Ridge. Algal-laminated limestone of the Thumb Member (at left of photograph) interfingers abruptly northward (to the right) with megabreccia and conglomerate composed entirely of upper (carbonate rocks) unit of the Rainbow Gardens Member. Underlying middle unit of Rainbow Gardens is exposed just below bottom of photograph.

of complex faulting and folding (Beard and Campagna, 1991; Beard 1992, 1993). No sections of the Thumb Member were measured because of poor exposure and structural and stratigraphic complexity. The only continuous vertical section of Thumb rocks, at Wechech Basin within the Lime Ridge block, was measured by Bohannon (1984), and serves as a reference section for this study. A generalized down-plunge view of Thumb stratigraphy in the south Virgin Mountains, shown in Figure 13A, and the following descriptions of various Thumb strata are pieced together from observations made during geologic mapping (Beard and Campagna, 1991; Beard 1992, 1993).

Stratigraphy of the Thumb Member can be generalized as comprising a lower sequence of lacustrine and distal alluvial-facies rocks and an upper sequence of coarse alluvial fan-facies rocks (Fig. 13B). The lower sequence consists of lacustrine-facies carbonate rocks, gypsum, and fine-grained clastic rocks that grade laterally northward and eastward into marginal lacustrine and distal alluvial sandstone and conglomerate. Two isolated exposures of extrusive basaltic rocks occur at the basal part of the lower sequence. Megabreccia deposits occur locally both at the basal part of the Thumb and just below the upper (fanglomerate) sequence. The Gold Butte conglomerate deposit is a distinctive, poorly exposed deposit of Proterozoic-clast megabreccia and conglomerate that extends from the Gold Butte Fault northward to Horse Spring; the deposit overlies carbonate deposits of the lower sequence and underlies or intertongues with coarse alluvial fan-facies rocks of the upper sequence.

The upper (alluvial fan) sequence, which in most areas rests unconformably on the lower sequence, includes various, probably diachronous, conglomerate facies that reflect different source areas dominated by Proterozoic or Paleozoic rock. In the Wechech Basin area, a distinctive package of fluvial sandstones overlies the lacustrine and marginal lacustrine rocks of the lower sequence and is unconformably overlain by alluvial fan deposits of the upper sequence.

Lower sequence. The basal part of the lower sequence consists of laterally intertonguing gray algal-laminated limestone, vuggy gypsiferous limestone, and white limestone and dolomite; these rocks grade upward into fine-grained clastic rocks containing interbedded gypsum and fallout tuff. These carbonate rocks overlie the Rainbow Gardens Member everywhere in the south Virgins except at the south end of Wechech Basin, at

St. Thomas Gap, and locally at Pakoon Ridge. Areal distribution of the basal carbonate strata suggests that they were deposited in one basin, whereas the distribution of the overlying gypsum and fine-grained clastic rocks shows the influence of local faulting. The carbonate rocks were originally included in the Rainbow Gardens Member by Bohannon (1984) but overlie the unconformity described above and interfinger with strata characteristic of the Thumb Member.

Carbonate rocks. Algal-laminated limestone (Fig. 13) directly overlies the Rainbow Gardens Member from Wechech Basin to Mud Wash, at Pakoon Ridge, from Horse Spring to Garden Wash, and at Lime Wash. The well-bedded, wavy-laminated algal limestone locally contains hemispheroidal algal mounds. Upper surfaces of the beds are locally mud cracked. Other beds include structureless limestones with abundant algal rip-up clasts and oncoids as much as 5 cm in diameter, as well as rare cross-laminated grainstone beds that are commonly oolitic.

At Wechech Basin, the 60 m thick section of algal limestone includes 30 m of gypsum, gypsiferous limestone, and mudstone. To the south at Mud Wash, the limestone intertongues laterally with a highly visible white dolomite-limestone unit. The white dolomite-limestone unit forms the basal part of the Thumb Member south to Lime Ridge fault and across the fault to the Mud Hills. From St. Thomas Gap south to Horse Spring, thin deposits of the white dolomite-limestone unit are complexly interbedded with fine-grained clastic rocks and gypsum. At Horse Spring, the white dolomite-limestone unit intertongues abruptly with algal limestone again.

The white dolomite-limestone unit is dominantly thin-bedded dolomite interbedded with limestone, green claystone, and minor gypsum. Features suggestive of spring mounds, such as conical structures, silicified plant stems, discontinuous tufa layers, and chert, are common. Fallout tuffs are locally abundant and Brenner and Glanzman (1979) reported that many dolomite layers were probably originally volcanic ash and later dolomitized. The unit superficially resembles the white tuffaceous lacustrine deposits in the underlying Rainbow Gardens Member (Bohannon, 1984; Brenner and Glanzman, 1979), but it differs in detail and lies above the unconformity.

At Garden Wash, the carbonate rocks of the Rainbow Gardens Member are overlain by about 50 m of gypsiferous limestone and bedded structureless gypsum. The gypsiferous limestone includes low-angle cross-stratified rocks that have a distinctive vuggy texture caused by weathering of gypsum, as well as laminated rocks with local oncolitic beds. The gypsiferous horizon gives way, both upward and to the north at Horse Spring, to the algal-laminated limestone facies.

At Pakoon Ridge, facies relationships are complex. The middle unit of the Rainbow Gardens Member is unconformably overlain by tuffaceous, white dolomite-limestone rocks, which intertongue with and are overlain by algal-laminated limestone. Both units intertongue rapidly to the north with conglomerate and breccia composed of angular pebbles to boulders of Rainbow Gardens carbonate rocks (Fig. 11C). To the south the units die out quickly into fine-grained clastic rocks of the lower sequence of the Thumb Member.

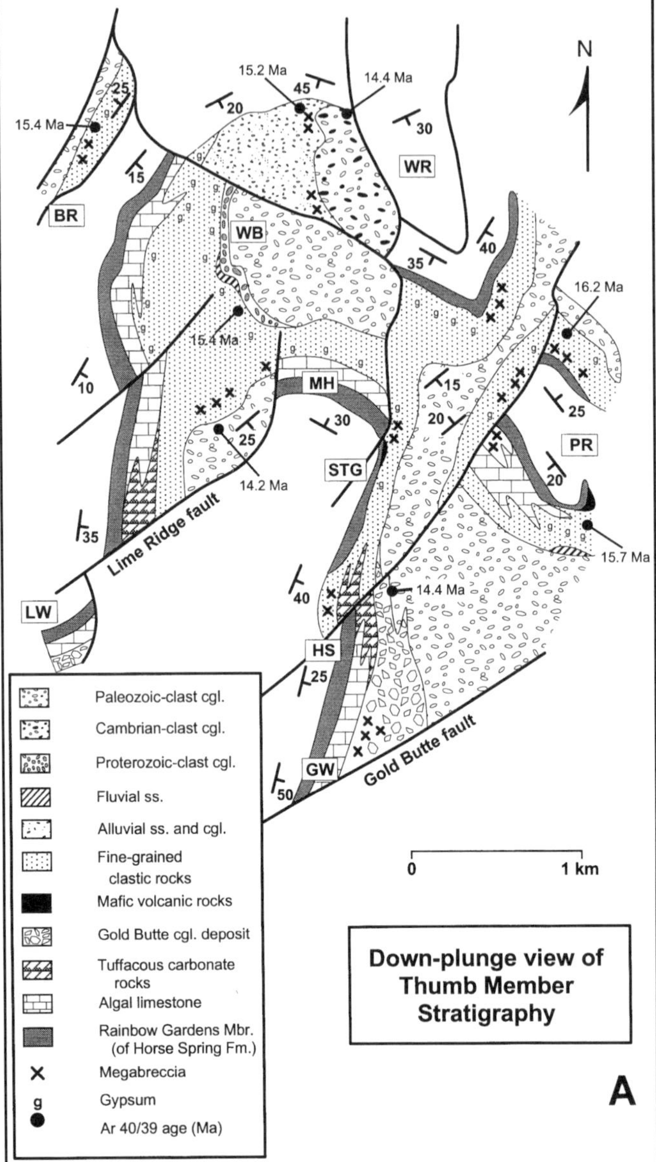

Figure 13. Stratigraphy of Thumb Member of the Horse Spring Formation.
(A) Schematic, generalized down-plunge view of stratigraphic relations between units of the Thumb Member, constructed by projecting stratigraphy down plunge along attitudes (shown on figure) of underlying strata. See text for discussion. Based on 1:24,000 scale mapping (Beard and Campagna, 1991; Beard, 1992, 1993).

Fine-grained clastic unit. The fine-grained clastic unit is best exposed at Wechech Basin, where an 800 m thick section measured by Bohannon (1984) typifies the "classic" stratigraphy of the Thumb Member. The unit is dominated by fine-grained sandstone, siltstone, and mudstone, with laterally persistent gypsum beds and fallout-tuff beds. Brown, fine-grained, well-sorted sandstone and yellowish-tan siltstone occur in thin, parallel, laterally continuous beds. Locally, coarse-grained pebbly

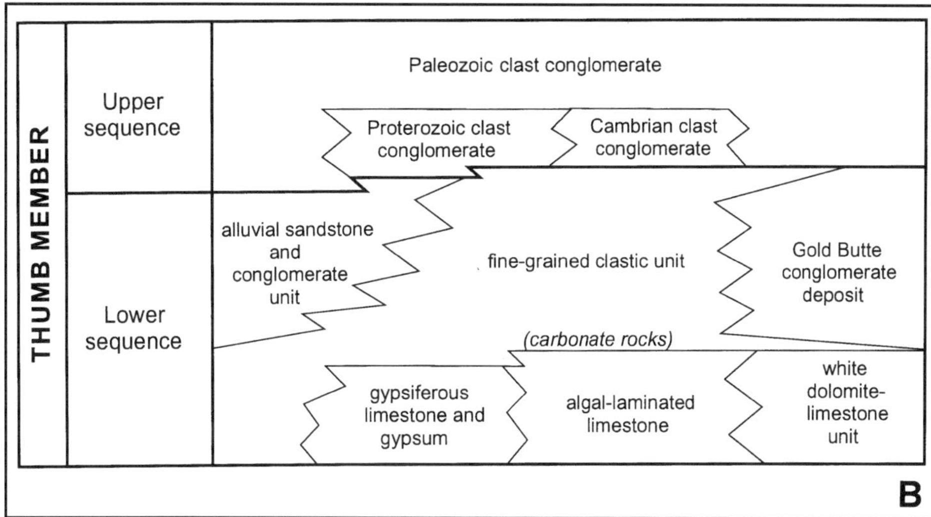

Figure 13B. Lithostratigraphic units of Thumb Member described in text. All units except gypsum-limestone shown on down-plunge view in Figure 13A.

sandstone fills broad lenticular channels, 10–30 cm thick. Interbedded minor pale brown mudstones form thin, continuous, structureless or laminated beds. In the central part of Wechech Basin, a local deposit of crinkly- to flaggy-bedded calcareous sandstones with raindrop imprints and dessication cracks interfingers laterally with gypsiferous beds.

Two to four laterally persistent, thick gypsiferous horizons occur in the middle to upper parts of the fine-grained clastic unit. Gypsum beds, from a few centimeters to as much as 0.5 m thick, are white to gray, in even- to wavy-bedded layers that are finely laminated or structureless and recrystallized. These distinctive marker beds can be traced laterally several kilometers through a series of folds.

Airfall-tuff beds, a few centimeters to 3 m in thickness, are common throughout the section. Isolated vitric fallout tuff beds are pale green to white or gray, with rare to common phenocrysts of biotite, hornblende, sanidine, and plagioclase; the tuffs are locally cross bedded in the uppermost few centimeters. Lithic grains are commonly incorporated into the tuffs.

South of Wechech Basin the fine-grained clastic unit is intermittently exposed beneath extensive Quaternary cover. From Mud Wash south to the Lime Ridge fault, the fine-grained clastic unit exposed on the east side of Tertiary Ridge is only about 200 m thick. At the south end of Whitney Ridge and in exposures on the southwest side of Pakoon Ridge the basal part of the unit truncates faults cutting the underlying Rainbow Gardens Member. The basal 10–30 m is commonly marked by rubble zones or conglomerate layers composed mostly of underlying Rainbow Gardens carbonate rocks, but locally containing boulders to house-size blocks of Mesozoic rocks (described in a following section).

Basaltic rocks. Two unusual outcrops of mafic volcanic rock occur in the basal part of the Thumb Member in the Pakoon Ridge and St. Thomas Gap areas. The first outcrop, at Pakoon Ridge, contains dark green to black, highly vesicular, altered basalt or andesite composed of plagioclase, clinopyroxene, and olivine crystals in a coarse-grained equigranular matrix. Abundant large volcanic bombs in a cindery matrix are common. The outcrop, about 200 m wide, is nearly surrounded by highly faulted and brecciated rocks of the Rainbow Gardens Member. Clasts of the mafic rocks occur in superjacent fine-grained clastic and gypsiferous sediments of the Thumb Member. An overlying tuff within 10 m of the base of the Thumb yielded an age of 15.7 ± 0.2 Ma. Field relations indicate that the rocks are proximal vent deposits, concurrent with or postdating faulting of the Rainbow Gardens.

Just south of Mud Hills and along the St. Thomas Gap road is a poorly exposed outcrop of greenish-weathering, unsorted, massive, volcanic ash in a calcite-rich matrix. The ash contains exploded fragments, from 0.5–4 cm in diameter, of altered mafic volcanic rocks of two compositions, olivine-bearing and pyroxene-bearing. The only nonvolcanic clastic material in the deposit is a few subangular carbonate clasts of the Rainbow Gardens Member. The deposit is interpreted as a phreatomagmatic deposit in the Thumb lake.

Alluvial sandstone and conglomerate unit. The alluvial sandstone and conglomerate unit is mostly exposed north and east of Wechech Basin. The unit, which coarsens to the north and east, overlies and interfingers laterally with the upper part of the fine-grained clastic unit. A tuff dated at 15.2 ± 0.2 Ma lies near the base of the alluvial sandstone and conglomerate unit, where it rests on rocks of the Late Triassic (?) and Jurassic Navajo Sandstone at the northeasternmost outcrop in Wechech Basin. At this outcrop, even, parallel beds of reddish orange sandstone and siltstone contain local channels of conglomerate. Sandstone beds are commonly cross-bedded and locally contain floating granule- to pebble-size clasts. The section coarsens upward, containing abundant angular to subangular carbonate and sandstone clasts of middle to late Paleozoic age, ranging from 1–20 cm in diameter. Locally, cobbles and boulders as much as 1 m in diameter of red sandstone, most likely either the Permian Esplanade Sandstone or Hermit Shale, are found. The conglomeratic beds exhibit both stream flow and debris flow characteristics.

Southward from the above outcrop, the alluvial sandstone and conglomerate unit surrounds a rock avalanche deposit containing brecciated middle Paleozoic carbonate rocks. Just west of this megabreccia and presumably overlying it are deposits of dark red sandstone containing rounded clasts of quartzite, carbonate rocks, and chert interbedded with layers containing large blocks of white vuggy limestone, probably derived from the Rainbow Gardens Member, as well as Mesozoic sandstone. These dark red sandstone beds may be equivalent to the fluvial sandstone unit discussed in a following section.

Megabreccia deposits. The term "megabreccia" is used here for various unstratified deposits characterized by large brecciated clasts or blocks emplaced by a variety of mechanisms, including rock avalanche, talus creep, rock slide, and debris flow. Megabreccia deposits within the Thumb Member are mostly found at two horizons—at or near the base and within the uppermost part of lower sequence. The megabreccia deposits within the Thumb are typically monolithologic, containing angular clasts, from one centimeter to tens of meters in diameter, embedded in a matrix usually composed of comminuted rock fragments of the same rock type. The lower package of megabreccia deposits is typically composed of carbonate rocks of the Rainbow Gardens Member and occurs at or near the base of the Thumb at the south end of Whitney Ridge, at St. Thomas Gap, and at Pakoon Ridge. Outcrops are mostly poorly bedded, unstratified, and unsorted rubble ranging in size from cobbles to boulders and are of probable debris-flow and rock-fall or rock-slide origin.

The upper group of megabreccias deposits is more variable in lithology and size. The two largest deposits, each covering about a square kilometer, are found west of Whitney Ridge and are interstratified with the alluvial sandstone and conglomerate unit of the lower sequence. These deposits are interpreted as large rock avalanches, using criteria outlined in Yarnold and Lombard (1989). Both avalanche deposits were probably emplaced as a result of down-to-the-west faulting along the west side of Whitney Ridge. The northern deposit, composed of Pennsylvanian carbonate rocks, is east of Wechech Basin and covers about two square kilometers. The deposit is as much as 100 m thick and consists of a stack of laterally continuous carbonate breccia sheets, each 1–2 m thick. At the northeast end of the outcrop, relict stratigraphy in individual breccia sheets is quite coherent, although fractured and faulted. At distal exposures to the southwest, the megabreccia sheets are still monolithologic, but are internally brecciated. These distal well-cemented breccia sheets have a sandy calcareous matrix.

The southern deposit, composed of middle Paleozoic carbonate rocks, is within the alluvial sandstone and conglomerate unit of the lower sequence, about 30 m below its contact with the upper sequence of coarse alluvial fan–facies rocks. The deposit, about 30 m thick and one square kilometer in areal extent, is composed of very angular, crackle-brecciated blocks, from a few centimeters to a few tens of meters in size, in a matrix of granulated Paleozoic carbonate rocks.

Smaller megabreccia deposits occur within the uppermost part of the fine-grained clastic unit, just north of the Lime Ridge fault. These deposits include large, relatively coherent blocks, several meters thick and up to 20 m long, of Mesozoic sandstone and shale, basal conglomerate of the Rainbow Gardens Member, and Rainbow Gardens carbonate rocks. I infer that the deposits were formed by rock-fall or rock-slide processes, originating to the south in the footwall of the Lime Ridge fault. The blocks are about 20 m stratigraphically below a tuff dated at 14.2 ± 0.5 Ma.

Another outcrop of megabreccia, about 40 m in diameter and 10 m thick, occurs just below the upper (alluvial fan) sequence in the Bitter Ridge area, within 80 m west of the Bitter Ridge fault. The megabreccia, which overlies gypsum beds, is a crackle-brecciated carbonate deposit that is derived either from the Rainbow Gardens Member or possibly from the Permian Pakoon Limestone. Strata overlying the megabreccia are eroded, but the deposit projects to just below the basal contact of the upper (alluvial fan) sequence, which includes clasts of similar carbonate rocks. This megabreccia deposit was probably derived from the east side of the Bitter Ridge fault.

Gold Butte conglomerate deposit. The Gold Butte conglomerate deposit overlies carbonate rocks of the lower sequence of the Thumb Member just north of the Gold Butte fault and extends northward to Horse Spring. Although poorly exposed, a few outcrops near Horse Spring demonstrate that the conglomerate has similar dips to underlying strata, about 50° eastward. The poor exposures are a function of the coarse, poorly to unbedded nature of the deposit. Just north of the Gold Butte fault, the basal part of the unit includes brecciated deposits of conglomerate of the Rainbow Gardens Member and Triassic sandstone that are either coarse megabreccias or fault breccia. Farther north, a megabreccia deposit of Paleozoic carbonate rocks occurs (Morgan, 1968, Fryxell et al., 1992). These breccias are overlain by the Gold Butte conglomerate deposit, composed of Proterozoic crystalline rocks; next to the Gold Butte fault the deposit is megabreccia and northward away from the fault it grades to coarse conglomerate. At the northernmost exposure at Horse Spring, the Gold Butte conglomerate deposit is overlain by a thin monolithologic conglomerate containing clasts of Rainbow Gardens carbonate rocks. A tuff between the Proterozoic-clast conglomerate and the overlying monolithologic conglomerate, is dated at 14.4 ± 0.3 Ma. Northward, the monolithologic conglomerate is overlain by and probably interfingers with Paleozoic clast conglomerate of the upper (alluvial fan) sequence described below.

Upper sequence. Alluvial-fan deposits of the upper sequence are widespread in the south Virgin Mountains. The deposits are both gradational with and unconformable on older units; locally they fill channels cut as much as 20 m deep and can dip as much as 5° to 10° shallower than underlying strata. A distinctive fluvial sandstone lithofacies carrying carbonate clasts of both the Rainbow Gardens Member and Mesozoic rocks underlies coarse alluvial-fan deposits in Wechech Basin, at the south end of Whitney Ridge, and at Pakoon Ridge. However, the bulk of the upper sequence is coarse conglomerate and conglomeratic sandstone. The clast compositions vary laterally within the conglomerates, reflecting differing local source areas. Although the conglomerates are mostly derived from middle to late Paleozoic rocks, they also include distinctive Proterozoic and Cambrian clast facies.

Fluvial sandstone unit. The fluvial sandstone lithofacies is discontinuously exposed within the uppermost part of the fine-grained clastic unit of the lower sequence in the Wechech Basin area (Fig. 13A). A second, isolated outcrop is found in the Pakoon Ridge area, just below the Paleozoic clast conglomerate and overlying gypsiferous beds of the lower sequence. The rocks include brown to red siltstone and fine- to coarse-grained sandstone. The clastic rocks are typically well-bedded with parallel to lenticular bedding and are commonly trough and planar cross-stratified. Locally, the deposits include abundant thin beds of channel-filling conglomeratic sandstone and conglomerate that contains (1) subrounded clasts of quartzite and chert, probably recycled from Mesozoic rocks, (2) subrounded carbonate and sandstone clasts, and (3) angular carbonate clasts of the Rainbow Gardens Member. Northeasternmost exposures in Wechech Basin and north of Mud Hills locally contain lenticular deposits of unsorted, structureless sandstone containing large (as much as 1 m) blocks of Rainbow Gardens carbonate rocks . These deposits reflect a source area dominated by Mesozoic rocks. The distinctive Cretaceous sandstone clasts could be derived only from the north or northeast. The distribution of the facies prohibits derivation from the west; furthermore, Cretaceous rocks are not present to the south because they were eroded below the sub-Tertiary unconformity.

Proterozoic clast conglomerate. The coarsest and thickest deposits of Proterozoic clast conglomerate are found in the Wechech Basin area (Fig. 13B). At the north end of Wechech Basin, the Proterozoic clast conglomerate forms a marker bed at the base of the upper sequence and includes clasts of amphibolite, garnet gneiss, and a distinctive mylonitic megacrystic granite. The probable source area is at Whitney Ridge, just a few kilometers to the northeast, where strikingly similar crystalline rocks crop out. To the south and southeast, the Proterozoic clast conglomerate interfingers with Paleozoic-clast conglomerate and is confined to wide channels. Clasts in the Proterozoic-clast conglomerate are quite angular and widely variable in size, with the largest clasts as much as 1 m in diameter. The Proterozoic clast conglomerate must be younger than 15.2 Ma, because it overlies the alluvial sandstone and conglomerate unit of the lower sequence. Based on geochemical similarities (E. I. Smith, 1989, personal commununication), the tuff at the base of the conglomerate at the north end of Wechech Basin is probably correlative with the 14.2 ± 0.5 Ma tuff just north of the Lime Ridge fault.

Paleozoic-clast conglomerate. From Mud Wash south to Horse Spring and along the south end of Whitney Ridge, the bulk of the upper (alluvial fan) sequence contains conglomerates carrying angular to subangular Paleozoic carbonate clasts, as much as 30 cm in diameter. Proterozoic clasts are rare in the basal part, but they occur sporadically higher in the section. West of the Mud Hills and just north of the Lime Ridge fault, a tuff in the basal part of the conglomerate is dated at 14.2 ± 0.5 Ma. Similar tuffs are found discontinuously throughout the Wechech Basin at the same horizon and may be correlative.

Just north of the Lime Ridge fault at the Mud Hills, the basal 1–2 m of the Paleozoic-clast conglomerate is composed of angular carbonate clasts, 2–6 cm in diameter, most likely derived from the upper part of the Rainbow Gardens Member. Similar carbonate clasts also occur within the basal part of the Paleozoic-clast conglomerate, just northeast at south Whitney Ridge, where distinctive clasts of red silicified Navajo Sandstone occur at the basal part of the unit on the south side of Whitney Ridge. The red clasts are eroded from silicified fault zones that cut the Navajo Sandstone and the Rainbow Gardens Member in the Mud Hills and at the south end of Whitney Ridge. Clasts of Rainbow Gardens carbonate rocks also occur within the basal part of the Paleozoic-clast conglomerate at Horse Spring, immediately overlying the Gold Butte conglomerate deposit.

Cambrian-clast conglomerate. A conspicuous light gray conglomerate composed mostly of clasts derived from Cambrian Undivided Dolomites forms an elongate north-south ridge east of Wechech Basin and directly west of the Whitney Ridge fault. At the west edge of the outcrop, the Cambrian-clast conglomerate overlies the alluvial sandstone and conglomerate unit of the lower sequence; at its east edge the Cambrian-clast conglomerate depositionally overlies the Cambrian Muav Limestone and Undivided Dolomites in the hanging wall of the Whitney Ridge fault. An age of 14.4 ± 0.5 Ma was obtained from a tuff near the base of the Cambrian clast conglomerate.

Interpretation of Thumb deposits. Strata of the lower sequence of the Thumb Member indicate deposition in a shallow, widespread playa-lake environment. Interfingering stromatolitic carbonate mud flats, evaporitic playas, and local hot-spring deposits grade laterally into distal alluvial fan deposits. The distribution of these facies suggests the presence of the two subbasins, one centered in the Wechech Basin area and the other from Horse Spring to Garden Wash. The intervening area, from St. Thomas Gap south to Horse Spring, was probably intermittently emergent, based on the thinness of the lower sequence, the lack of lacustrine limestone in the lower sequence, and the ubiquitous presence of megabreccia of the lower sequence at the base. At Wechech Basin, the algal-limestone facies thins northward and marginal-lacustrine facies intertongue with distal alluvial-fan facies northeastward, suggesting that the northern and eastern margins of the basin were nearby. The lack of coarse deposits implies a fairly low gradient to the basin margin. The relatively thin stack of fine-grained clastic deposits overlying the algal-limestone facies at Pakoon

Ridge and the presence of a few megabreccia deposits at the base of the lower sequence at the south end of Whitney Ridge suggest the edge of the basin was probably not far east and north. To the south, the thinning of lower sequence carbonate rock from Horse Spring south to Garden Wash suggests that the southern basin margin was nearby, as is also the case for the underlying deposits of the Rainbow Gardens Member. The western limit of the Thumb basin is not known.

The widespread presence of megabreccia deposits near the top of the lower sequence in the Wechech Basin subbasin indicates syndepositional faulting on the south side of the Virgin Peak block, the west and south sides of Whitney Ridge, the west side of Pakoon Ridge, and along the Lime Ridge and Bitter Ridge faults. These megabreccias deposits are bracketed between about 15.2 Ma and 14.2 Ma in age. The ridge separating the two subbasins probably reflects footwall uplift along the Lime Ridge fault (south side up). To the south in the Horse Spring subbasin, megabreccias of the Gold Butte conglomerate deposit were shed northward across the Gold Butte fault from the uplifted footwall on the south side. The Gold Butte conglomerate deposit immediately overlies the lower sequence carbonate rocks, in contrast to the Wechech Basin area where the carbonate rocks are overlain by as much as 1 km of fine-grained clastic lacustrine deposits. Therefore, the Gold Butte conglomerate deposit could be laterally equivalent to the fine-grained clastic rocks, in which case motion on the Gold Butte fault could be as old at 16.2 Ma (oldest fission-track age obtained from fine-grained clastic rocks of the Thumb Member in Wechech Basin, from Bohannon, 1984). Alternatively, the Gold Butte conglomerate deposit could be correlative with megabreccia horizons in the uppermost part of the lower sequence in the Wechech Basin subbasin, which would suggest motion at between 15.2 Ma and 14.4 Ma. Fission-track data from Fitzgerald et al. (1991) indicate that Proterozoic rocks just across the Gold Butte fault to the south were rapidly uplifted and unroofed at about 15 Ma. However, the data are not sufficiently precise to distinguish the probable age of the Gold Butte conglomerate deposit.

In either case, the Horse Spring subbasin must have been uplifted relative to the Wechech subbasin to the north, causing nondeposition and/or erosion of lower sequence rocks. Nondeposition could be a response to the footwall uplift along the Lime Ridge fault, suggesting that movement on the Lime Ridge fault was later than that on the Gold Butte fault. Another possibility is that uplift was caused by normal faulting within the Tramp Ridge domain concurrent with motion on the Gold Butte fault.

Lacustrine deposition in the Thumb basins ended by 14.2–14.4 Ma when alluvial fans from multiple source areas prograded out over the lake. Fluvial-facies deposits at the top of lacustrine deposits from Wechech Basin to Pakoon Ridge suggest that a northwest-southeast axial stream system developed during the late stages of filling of the lake. Alluvial fans derived from northeast of the fluvial system first eroded Mesozoic strata, and later eroded Proterozoic and Cambrian rocks exposed by normal faulting along the west side of Whitney Ridge. Fans carrying debris eroded from various Paleozoic strata also prograded from the north off of the Virgin Peak block and from the west off of the rising core of the Lime Ridge and Tramp Ridge domains. Angular discordance of up to 10° between the alluvial-fan deposits and underlying strata, as well as a decrease in dip upward, indicates that horizontal-axis tilting began at about 14 Ma.

In summary, facies distributions within the Thumb Member suggest deposition in mixed-mode type basins, influenced by both strike-slip and normal faults (Fig. 14A). The north margin of the Thumb basins, the ridge separating the two subbasins, and the south margin correspond to the location of the reverse fault bounding the south end of the Virgin Peak domain, the Lime Ridge fault, and the Gold Butte fault, respectively. The eastern margin of the Thumb depositional system probably had relatively low relief during deposition of the marginal lacustrine and distal alluvial-fan facies of the lower sequence, but was subsequently disrupted by faulting and uplift of the Whitney Ridge block. Normal faulting within the Lime Ridge block may have begun by 16.2 Ma, although tilting of the normal fault blocks is not reflected in the sedimentary record until 2 m.y. later.

DEPOSITS POSTDATING THE HORSE SPRING FORMATION

No widespread Tertiary deposits younger than the Thumb Member of the Horse Spring Formation are present in the south Virgin Mountains, although previous mapping had shown otherwise (Longwell et al., 1965, Morgan, 1968). After about 14–13 Ma, deposition ceased in the south Virgin Mountains area and the entire Virgin Mountains region became a source area for younger strata deposited to the east in the Grand Wash trough, to the west in Overton Arm, and to the north in the Virgin River depression (Fig. 14B).

Rocks of the Grand Wash trough

Red, well-bedded sandstone and minor channel-filling conglomerate gradationally overlie tilted alluvial fan deposits of the upper part of the Thumb Member southwest of Pakoon Ridge on the east side of the Virgin Mountains. Stratigraphically upward, the strata become finer grained and decrease in tilt to nearly flat lying. These flat lying strata are laterally continuous eastward with strata exposed in the Grand Wash trough.

The rocks of the Grand Wash trough were deposited in an asymmetric basin formed by motion on the Grand Wash fault. Fine-grained strata were deposited in playas and lakes in the axis of the basin, which was on the east side near the fault. These fine-grained rocks interfinger to the east with conglomerates derived from Paleozoic rocks exposed in the Grand Wash Cliffs and to the west with Proterozoic clast-bearing conglom-

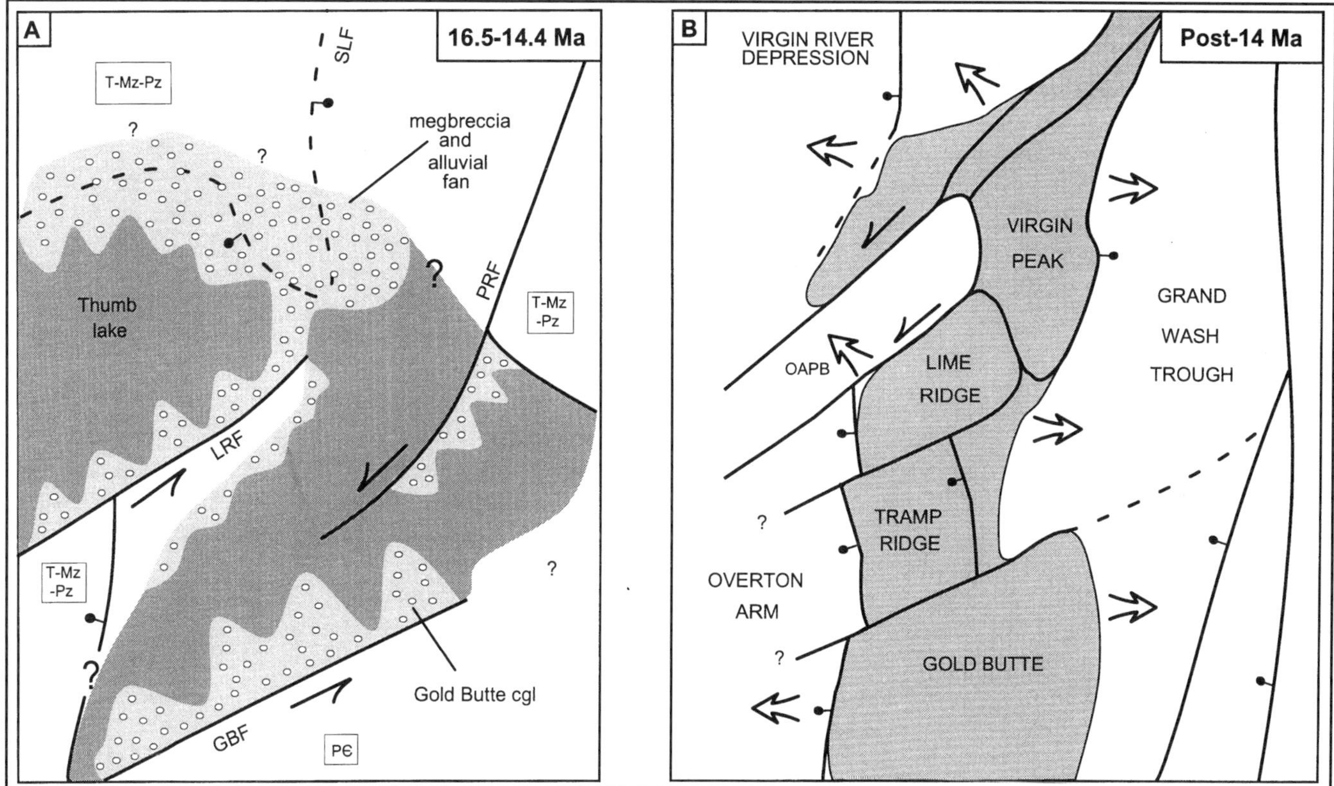

Figure 14. Paleogeographic sketch maps of depositional setting of Thumb Member of Horse Spring Formation in Virgin Mountain region. (A) 16.5–14.4 Ma: Paleoenvironment for lower sequence of Thumb Member is extensive lake with marginal lake and fluvial input mostly from north. As Lime Ridge (LRF), Pakoon Ridge (PRF), and Gold Butte (GBF) faults move, locally derived megabreccia and conglomerate deposited into lake. Dashed fault shows location of margin of uplifting Virgin Peak block, which eventually disrupts northerly derived alluvial fan (alluvial sandstone and conglomerate unit) and allows local megabreccia emplacement. Dominant lithology of source area shown by TMzPz—Tertiary, Mesozoic, and Paleozoic; P—Proterozoic rocks.

(B) Post–14 Ma: Regional uplift of Virgin Mountains region disrupts Thumb basin, so that conglomerates of upper sequence of Thumb Member prograde across Thumb lake, graded to depositional systems developing to east in Grand Wash trough, to west in Overton Arm, and north in Virgin River depression.

erates derived from the Gold Butte area (Lucchitta, 1966, 1972). Exposed strata are mostly flat-lying or gently folded. At Tassai Wash, gently dipping Proterozoic clast conglomerate unconformably overlies steeply dipping strata of the Rainbow Gardens Member.

The rocks of the Grand Wash trough are younger than the tilted upper sequence of the Thumb Member southwest of Pakoon Ridge. These uppermost Thumb rocks include an undated tuff that is likely in the 13.9–14.4 Ma range, based on correlation to dated tuffs in similar stratigraphic positions elsewhere. Fallout tuffs in the upper part of the Grand Wash trough section, below the Hualapai Limestone, yielded fission-track ages of 10.8 ± 0.8, 11.1 ± 1.3, and 11.6 ± 1.2 Ma (Bohannon, 1984). The Snap Point flow, which is interbedded on the east side of the trough with the uppermost part of the easterly derived conglomerate, has a K-Ar age of 9.0 ± 0.8 Ma (Reynolds et al., 1986). Therefore, the interior basin deposits in the Grand Wash trough are bracketed at younger than 14 Ma and older than 9 Ma in age.

Overton Arm

Gently dipping strata are widely exposed in Overton Arm west of the south Virgin Mountains. These strata unconformably overlie strongly tilted rocks of the Thumb Member exposed between the Bitter Ridge and Black Ridge faults (Beard, 1993) at the north end of Overton Arm, and farther south on the east shore of Overton Arm (Bales and Laney, 1991). The northern part of Overton Arm is underlain by the small, northeast-elongate pull-apart basin (Campagna, 1990, Campagna and Levandowski, 1991). The pull-apart basin is bounded by segments of the Lake Mead fault system—on the northwest by the Hen Springs–Rodgers Spring fault, and on the southeast by the Bitter Ridge–Hamblin Bay fault. Strata filling the basin are probably equivalent to Muddy Creek Formation in age, but rocks outside the pull-apart basin to the south are somewhat older and are probably equivalent in age to the rocks of the Grand Wash trough.

Exposures near Bitter Ridge, on the east edge of the Overton Arm basin, underlie a mesa capped by a thick calcrete soil that is probably correlative with the Mormon Mesa calcrete, a thick widespread calcrete soil that formed on the upper surface of rocks of the Muddy Creek Formation in the Virgin River depression (Gardener, 1972). The rocks are eroded and buried beneath Quaternary deposits to the west and are cut by the Bitter Ridge fault to the east. The bulk of the strata are fine-grained siltstone and mudstone, with thin discontinuous lenses of sandstone of probable lacustrine origin. These rocks grade laterally eastward, toward the Bitter Ridge fault, into sandstone and conglomerate of alluvial fan origin; clasts in the conglomerate include Paleozoic carbonate rocks and Proterozoic crystalline rocks similar to those exposed to the north at Bunkerville Ridge. Both the lacustrine and alluvial-fan deposits are overlain by a thick section of subrounded, well-imbricated, and cross-stratified conglomerate that was deposited in a large westerly flowing, axial, fluvial channel in the uppermost part of the Muddy Creek. The clasts are predominantly mixed Paleozoic carbonate rock. These rocks are tilted to the same degree as the underlying Muddy Creek strata. Similar fluvial deposits occur near the top of the Muddy Creek to the north in the Virgin River depression (Kowallis and Everett, 1986).

East of the Bitter Ridge fault no axial fluvial deposits are found; equivalent flat-lying strata are thin alluvial fan deposits, derived from the west side of Bitter Ridge, that overlie strongly tilted and deformed strata of the Thumb Member. These strata include fine-grained clastic rocks and gypsum of the lower sequence of the Thumb, overlain by the upper conglomerate of the upper sequence. A tuff near the top of the lower sequence is dated at 15.4 Ma. The deposits of Muddy Creek Formation in the pull-apart basin, then, are younger than about 14 Ma and older than the overlying Mormon Mesa calcrete, which is undated but is no older than 5.5 Ma (D. Schmidt, personal communication, 1991).

South of the Overton Arm pull-apart basin, Tertiary strata are mostly covered by Quaternary alluvial fan deposits. However, a thick sequence of conglomerate that includes large blocks (up to several meters in diameter) of rapakivi granite derived from Gold Butte is exposed just west of the Lime Ridge and Gold Butte domains on the east side of Overton Arm. These conglomerates unconformably overlie tilted and faulted rocks of the Horse Spring Formation and Mesozoic through Proterozoic strata. The conglomerate is overlain by the informally named Gold Butte basalt, dated by Cole (1989) at 9.5–9.2 Ma. Tertiary strata south of the Overton Arm basin, therefore, are similar in age to the rocks of the Grand Wash trough to the east.

Virgin River depression

A detailed study of the subsurface geometry and stratigraphy of the Virgin River depression was presented by Bohannon et al. (1993). They interpreted the Muddy Creek Formation, which is widely exposed in the Virgin River depression, to be about 1–2 km thick. The Muddy Creek buries two older, less extensive subbasins interpreted to be filled with the red sandstone unit and the Lovell Wash Member of the Horse Spring Formation. According to Bohannon et al. (1993), the western basin also contains the Rainbow Gardens and Thumb Members of the Horse Spring. A sandstone section that included grains of Proterozoic crystalline rocks, encountered by the Virgin 1A test well drilled by Mobil, was interpreted by Bohannon et al. (1993) as the red sandstone unit. The crystalline rock grains suggest that Proterozoic rocks were exposed at the only likely source areas, Bunkerville and Black Ridges, when the red sandstone unit was deposited.

Bohannon et al. (1993) define three stages of evolution of the Virgin River depression beginning with an early episode of deposition of the Rainbow Gardens and Thumb Members from about 26–13 Ma. They recognize a period of locally significant deformation at about 13 Ma, in which the upper crust extended about 60% in the Virgin River depression, forming fault-controlled basins that accumulated significant amounts of sediment in the two subbasins. The final stage, post–10 Ma, was the widespread filling of the Virgin River Basin by deposits of the Muddy Creek Formation. The beginning of the middle stage, 13–10 Ma, roughly corresponds to the uplift and tilting of the Virgin and south Virgin Mountain region, and the filling of the Grand Wash trough and Overton Arm depressions.

PALEOGEOGRAPHY AND TECTONICS

Rainbow Gardens Member

The southern margin of the Rainbow Gardens basin was a low-relief, north- to northeast-sloping, beveled surface formed on the north-plunging nose of the Kingman highland. Projection of the well-documented paleogeography of the western Colorado Plateau westward to the Lake Mead area suggests that the beveled surface below the Permian paleoscarp lay not far to the south and west. In the Virgin Mountains area, clasts in the basal conglomerate of the Rainbow Gardens Member are mostly derived from middle and late Paleozoic strata except in westernmost exposures where they were eroded from exclusively early Paleozoic strata. In contrast to the low relief southern margin, monolithologic deposits of coarse, angular Permian limestone clasts at northern exposures of the basal conglomerate at Wechech Basin and Tom and Cull Wash indicate that the Virgin positive area to the north had relatively high relief. The Virgin positive area disrupted the northerly paleoflow direction, forcing a more northeasterly flow along the base of the positive area and controlling the location of a northeast-flowing axial stream system. I infer that the sediments transported during deposition of the basal conglomerate were deposited in the Oligocene Claron basin to the north.

The Rainbow Gardens basin, represented by the middle and upper units of the Rainbow Gardens Member, has commonly been assumed to have formed in response to early extension (e.g., Bohannon et al., 1993, Carpenter, 1988). However, an average sedimentation rate of 0.04 m /1,000 yr is more consistent with cratonic basin sedimentation than extensional basin subsidence (Schwab, 1976). In addition, peak extensional tectonism in both the Virgins and areas to the north occurred at 16–14 Ma (Anderson, 1987) postdating the deposition of the Rainbow Gardens Member. Rather than reflecting the onset of extension, the Rainbow Gardens basin probably formed as a result of blockage of northeast drainage into the Oligocene Claron basin. I infer that blockage was caused by damming and uplift to the north related to intrusion of shallow plutons and voluminous eruptions of ash-flow tuff that are interbedded with the upper part of the Claron Formation. This extensive volcanism occurring from about 25–18.7 Ma ended deposition within the Claron basin (Siders and Shubert, 1986, Anderson, 1987).

The deposition of the tuffaceous part of the Rainbow Gardens Member corresponds in age to the main eruptive phase of the Caliente caldera complex. The last extensive eruptions in the Caliente caldera complex are the 18.7 Ma Racer Canyon Tuff and the 18.2 Ma Hiko Tuff (L. W. Snee, personal communication). These compare closely to the 18.8 Ma date from a tuff within a few meters of the top of the tuffaceous lacustrine sequence of the Rainbow Gardens Member. Palustrine deposits in the uppermost part of the Rainbow Gardens Member correspond to a waning of volcanism.

Thumb Member

In contrast to the Rainbow Gardens Member, deposition of the Thumb Member was strongly controlled by extensional deformation that was synchronous with renewed regional volcanism. Extension was manifested by both dip-slip and strike-slip faulting and development of local relief, resulting in complex mixed-mode depositional patterns (Gibbs, 1987). A sediment accumulation rate of about 0.5 m/1,000 yr for Thumb strata is consistent with its deposition in a rift basin (Schwab, 1976). The unconformity developed between the Thumb and Rainbow Gardens Members records the first evidence of faulting. The unconformity is more pronounced in northern exposures (Wechech Basin and Pakoon Ridge) of basal Thumb strata than it is to the south, suggesting faulting and uplift along a roughly northwest-southeast trend.

This mixed-mode deposition, strongly controlled by strike-slip faults of the Lake Mead fault system, is in contrast to coeval extension to the southwest of the Virgins, in the Black and El Dorado Mountains, where simple half-graben deposition is controlled by large-scale normal faults (Faulds, 1992). However, in both the Virgins and the Black–El Dorado Mountains area, large-scale horizontal-axis rotation of strata along normal faults after about 14 Ma is at the end of, rather than synchronous with, extensional tectonism. In the Virgins, horizontal-axis rotation is partitioned into domains between the strike-slip faults, whereas in the Black and El Dorado Mountains large linear fault blocks are rotated. This late rotation is evidenced by the lack of incremental fanning dips of synextensional strata, as would be expected if horizontal-axis rotation kept pace with extension. The pattern of late-stage horizontal axis rotation is not uncommon; similar relations are reported for the red sandstone unit in the western Lake Mead area (Duebendorfer and Wallin, 1991) and in the Sacramento Mountains (Fedo and Miller, 1992) in southeastern California. In these areas, thick synextensional sequences have conformable dips and are disrupted only late in the extensional episode by block faulting and tilting, which effectively ends deposition. Explanations for such relations include very rapid sedimentation relative to block rotation and diachroneity of horizontal-axis rotation across extension-parallel transfer faults.

In the Virgin Mountains area, deposition into the mixed-mode basins of the Thumb Member shifted into half-grabens to the east in the Grand Wash trough and to the west into Overton Arm after 14 Ma. At roughly the same time, about 13 Ma, half-graben deposition occurred to the north in the Virgin River depression. The Lime Ridge and Gold Butte faults were mostly inactive after 14 Ma. However, movement along the Bitter Ridge–Hen Spring faults system formed the Overton Arm pull-apart basin (Campagna and Aydin, 1994), probably sometime after 10 Ma. Wholesale uplift of the Virgin Mountains area, starting at about 14 Ma, may have been responsible for the east, west, and northward shift of extensional deformation, including both normal and strike-slip faulting. The coincidence of large-scale horizontal-axis rotation and wholesale uplift of the Virgin Mountains area suggests a genetic relationship between the two processes that is at present poorly understood.

Implications for palinspastic reconstructions

Bohannon (1984) described a distinctive facies in the Rainbow Gardens Member that occurs at Frenchman Mountain, in slivers along the Hamblin Bay fault, and in the Wechech Basin in the Virgin Mountains. He suggested that these strata must have been deposited in close proximity and that restoration of these widely separated areas would require about 65 km of southwestward transport. The kinematic sequence by which the transport occurred is poorly understood, but a significant amount of that transport most likely occurred along the Bitter Ridge–Hen Spring fault system and its western counterpart, the Hamblin Bay fault. In the Wechech Basin, this distinctive facies corresponds to a fairly thin section of dominantly palustrine facies deposited near the edge of the basin south of the Virgin positive area. If Bohannon's restoration is correct, the Muddy Mountain thrust, which lay to the north and west of the Frenchman Mountain and Hamblin Bay sections, would be placed just west of the south Virgin Mountains. This suggests that the Virgin positive area may have been part of the thrust terrane.

In contrast to potentially large amounts of transport along the Bitter Ridge–Hen Spring segment of the eastern Lake Mead fault system, the amount of slip along the Lime Ridge fault was probably less than 10 km. The Rainbow Gardens Member at the Lime Ridge fault section, just north of the fault, is strikingly similar to the Mud Hills section just south of the fault, and they are 10 km apart. Restoration of these two sections would allow no more than about 10 km left-lateral separation along the eastern part of the fault. The amount of slip could increase to the west because of down-to-the-west displacement along north-south normal faults that fed into the Lime Ridge fault.

Displacement on the Gold Butte fault is not well constrained, but it exceeds that along the Lime Ridge fault. Left-lateral strike separation of Paleozoic strata across the fault suggests about 14 km of displacement, but there are no known piercing points. The section of the Rainbow Gardens at Garden Wash, just north of the Gold Butte fault may have originated near, but slightly east of the Tassai Wash section. This assumption is suggested by the beveled surface beneath the sub-Tertiary unconformity and the similarity between the two sections. The Tassai Wash section rests on Permian rocks near the beveled edge of the northwest-striking Kaibab paleoscarp and the Garden Wash section overlies Early Triassic strata. Thus, Garden Wash would not restore much farther west of Tassai Wash. A range of permissible slip between 14–25 km is suggested. However, to the west along the fault, displacement could be much greater, much like the increase in slip seen westward along the Lime Ridge fault.

The Gold Butte block was interpreted by Wernicke and Axen (1988) and Fryxell et al. (1992) as a large, east-tilted block exposing as much as 15–17 km of crust. An alternative interpretation is that the bulk of the block is flat lying, with east-tilting on the east end of the block caused by a rollover anticline as a result of slip on the Grand Wash fault. Fission–track data from Fitzgerald et al. (1991) indicates rapid cooling of the Gold Butte block at about 15 Ma. This corresponds closely in age to the Gold Butte conglomerate deposit, which was shed northward into the Thumb basin from the Gold Butte block during left-lateral and down-to-the-north motion on the Gold Butte fault, but before large-magnitude horizontal-axis rotation occurred within the Tramp Ridge domain. Because the Gold Butte conglomerate deposit contains little or no material derived from Paleozoic strata, it was most likely not transported in a left-lateral sense very far from its source, because Paleozoic rocks are exposed just 6 km to the east on the south side of the fault. Although Fryxell et al. (1992) suggest that top-to-the-west tectonic denudation of the Gold Butte block is responsible for the vast exposures of Proterozoic rocks within the Gold Butte block, it is probable that Proterozoic rocks were already near the surface because of the pre-Miocene paleogeography (Fig. 4). The Gold Butte conglomerate deposit is conspicuously lacking in clasts of Paleozoic strata, which would be expected to occur within the basal part of the section during progressive unroofing.

The fission-track data of Fitzgerald et al. (1991) is equivocal. The data that show Miocene cooling ages are within statistical error of each other from east to west across the northern end of the block (14–17 Ma), such that they could also be interpreted as reflecting wholesale vertical uplift rather than progressive east tilting. In addition, the oldest age, 77 Ma, was obtained from the southwesternmost sample. Fitzgerald et al. (1991) interpret this as an upper plate sliver of crystalline rocks from just below the Paleozoic unconformity that was stranded on the lower plate during low-angle detachment faulting. However, the sample could be from in situ rocks that contain an older cooling age because of Late Cretaceous erosion of overlying Paleozoic strata. Fission-track data from Proterozoic rocks at the sub-Tertiary unconformity farther south in the Gold Butte block would help resolve the two interpretations.

Lastly, the paleogeography of the Thumb Member within the Virgins allows speculation regarding the original position of Frenchman Mountain. Most workers agree that Frenchman Mountain, now east of Las Vegas, originated near the Virgin Mountains and that Gold Butte was the source of the megabreccia deposits within the Thumb. However, the exact location is controversial. The megabreccia deposits of the Thumb at Frenchman Mountain are mostly Proterozoic crystalline clasts with early Paleozoic clasts only locally present (Parolini, 1986). This suggests that the Proterozoic rocks were exposed in the Gold Butte block by 16 Ma.

Bohannon (1984), placed Frenchman Mountain north of Overton, Nevada and west of the Bunkerville and Black Ridge areas. However, Schmitt and Rice (1988) thought that the megabreccia deposits in the Thumb Member were too proximal to be derived from Gold Butte and then transported 30 km north to the Overton area. They suggested instead that the source was in the northern Black Mountains. Rowland et al. (1990), using several lines of evidence, concluded that Frenchman Mountain originated just north of the Gold Butte block. Fryxell and Duebendorfer (1990) correlated mylonitic and chloritic rocks in the megabreccia deposits at Frenchman Mountain with similar rocks that crop out on the west and northwest sides of the Gold Butte block. They proposed that the Frenchman Mountain block formed the hanging wall to the Gold Butte block and was transported westward along a major low-angle fault system. Parolini (1986) demonstrated that the megabreccia deposits were derived from the south rather than the east. Assuming no vertical-axis rotation of the Frenchman Mountain block (Rowland et al., 1990), the southerly derivation of megabreccia deposits supports a position north of, rather than west of, the Gold Butte fault.

I suggest that Frenchman Mountain was located west of the Tramp Ridge block and north of the Gold Butte fault and was transported westward along the Gold Butte fault to a position north of the Wilson Ridge pluton. At this location, flows of dacite from the River Mountains andesitic stratovolcano, which is comagmatic with the Wilson Ridge pluton (Weber and Smith, 1987), interfinger with the upper part of the Thumb Member at the south end of the Frenchman Mountain block. A prevolcanic fault that has been inferred to separate the subvolcanic basement rocks of the River Mountains from Frenchman Mountain (Rowland et al., 1990) may be the Gold Butte fault. This prevolcanic fault was subsequently translated northwest to its present position, along with the now-joined Frenchman Mountain–River Mountains block, by motion on the Las Vegas Valley Shear zone (Duebendorfer and Wallin, 1991).

SUMMARY

The areal distribution of the depositional system responsible for the formation of the Rainbow Gardens Member of the Horse Spring Formation was most likely controlled by low-relief emergent areas to the south, west, and probably northwest, whereas the mixed-mode style deposition responsible for the genesis of the Thumb Member was controlled by local relief generated by slip on several splays of the Lake Mead fault system. The mostly synvolcanic Rainbow Gardens Member was deposited in a single shallow sag basin and as a result its stratigraphy is fairly simple and predictable. In contrast, the Thumb Member was deposited in a complex system of partially connected basins and lateral facies changes are rapid. The overall sequence comprising the Horse Springs Formation in the Virgin Mountains area, from base to top, is that of a fluvial to lacustrine transition in the Rainbow Gardens Member, accompanied by major volcanism to the north. The waning of this volcanism corresponds to intermittent emergence of lacustrine strata and pedogenic alteration. A previously unrecognized unconformity at the base of the Thumb Member marks structural disruption by faults of the Lake Mead fault system, followed by renewed and rapid deposition in complex lacustrine and marginal lacustrine environments. The deposition of the Thumb ends with final coarsening upward of strata because of progradation of alluvial fans over the lake deposits. By 14 Ma, sedimentation within the south Virgins waned and depocenters shifted to the east and west, accompanied by large-scale horizontal-axis tilting and uplift of the Virgin Mountains area relative to the surrounding regions.

In summary, the Horse Spring Formation records the transition from pre- to synextensional deposition within a mixed strike-slip and normal-faulting regime. Segmentation and rotation of the Horse Spring by 14 Ma is attributed to the transfer of strike-slip and extensional deformation to the west and north, which may have been either a result of or driving force for wholesale uplift of the Virgin Mountains region.

ACKNOWLEDGMENTS

Discussions with Ernie Anderson, Bob Bohannon, Gary Calderone, David Campagna, Ernie Duebendorfer, Joan Fryxell, Eugene Smith, Dwight Schmidt, Wanda Taylor, and Van Williams vastly improved my understanding of the Tertiary stratigraphy and structure. I thank Larry Snee, Steve Harlan, and Ross Yeoman for their invaluable assistance during use of the U.S. Geological Survey $^{40}Ar/^{39}Ar$ lab in Denver. Sonja Ward assisted in both field and office during study of the Rainbow Gardens Member. Debra Block provided editorial and drafting assistance. I thank Bob Bohannon, Bill Dickinson, and Tim Wallin for thorough reviews. Lastly, these studies were greatly helped by the generosity of the Bureau of Land Management in St. George, Utah, and especially the Nay family at Nay Ranch on the Arizona-Nevada border.

REFERENCES CITED

Anderson, J., and Kurlich, R., III, 1989, Post-Claron Formation, pre-regional ash-flow tuff early Tertiary stratigraphy of the southern High Plateaus of Utah: Geological Society of America Abstracts with Programs, v. 21, no. 5, p. 50.

Anderson, R. E., 1973, Large-magnitude late Tertiary strike-slip faulting north of Lake Mead, Nevada: U.S. Geological Survey Professional Paper 794, p. 18.

Anderson, R. E., 1987, Neogene geologic history of the Nevada-Utah border area at and near latitude 37°30′ N: Geological Society of America Abstracts with Programs, v. 19, no. 7, p. 572.

Anderson, R. E., 1990, West-directed tectonic escape between the northern and southern sectors of the Basin and Range Province: Geological Society of America Abstracts with Programs, v. 22, no. 3, p. 2.

Anderson, R. E., and Barnhard, T., 1993a, Aspects of three-dimensional strain at the margin of the extensional orogen, Virgin River depression area, Nevada, Utah, and Arizona: Geological Society of America Bulletin, v. 105, p. 1019–1052.

Anderson, R. E., and Barnhard, T., 1993b, Heterogeneous Neogene strain and its bearing on horizontal extension and horizontal and vertical contraction at the margin of the extensional orogen, Mormon Mountains area, Nevada and Utah: U.S. Geological Survey Bulletin 2011, 43 p.

Anderson, R. E., and Bohannon R. G., 1993, Three-dimensional aspects of the Neogene strain field, Nevada-Utah-Arizona tricorner area, *in* Lahren, M. M., Trexler, J. H., Jr., and Spinosa, Claude, eds., Crustal evolution of the Great Basin and the Sierra Nevada: Geological Society of America Cordilleran/Rocky Mountain Section Meeting Guidebook: Reno, University of Nevada, p. 167–196.

Anderson, R. E., Longwell, C., Armstrong, R., and Marvin, R., 1972, Significance of K-Ar ages of Tertiary rocks from the Lake Mead region, Nevada-Arizona: Geological Society of America Bulletin, v. 83, p. 273–288.

Axen, G., Taylor, W., and Bartley, J. M., 1993, Space-time patterns and tectonic controls of Tertiary extension and magmatism in the Great Basin of the western United States: Geological Society of America Bulletin, v. 105, p. 56–76.

Bales, J. T., and Laney, R. L., 1991, Geohydrologic reconnaissance of Lake Mead National Recreation Area—Virgin River, Nevada, to Grand Wash Cliffs, Arizona: U.S. Geological Survey Water-Resources Investigations Report 91-4185, 29 p.

Beal, L., 1965, Geology and mineral deposits of the Bunkerville mining district, Clark County, Nevada: Nevada Bureau of Mines Bulletin 63, 96 p.

Beard, L. S., 1991, Contrasting structural geometries along the Lake Mead shear zone in the Virgin and South Virgin Mountains: Geological Society of America Abstracts with Programs, v. 23, no. 5, p. 234.

Beard, L. S., 1992, Preliminary geologic map of the St. Thomas Gap 7.5-minute quadrangle, Clark County, Nevada and Mohave County, Arizona: U.S. Geological Survey Open-File Report 92-326, scale 1:24,000.

Beard, L. S., 1993a, Preliminary geologic map of the Whitney Pocket 7.5 minute quadrangle, Clark County, Nevada, U.S. Geological Survey Open-File Report OF 93-716.

Beard, L. S., 1993b, Tertiary stratigraphy of the South Virgin Mountains, southeast Nevada and the Grand Wash trough, northwest Arizona, *in* Sherrod, D. R., and Nielson, J. E., Tertiary stratigraphy of the highly extended terranes, California, Arizona, and Nevada: U.S. Geological Survey Bulletin 2053, p. 29–32.

Beard, L. S., 1994, Kinematics and timing of left-slip faulting on the eastern Lake Mead fault system: Geological Society of America Abstracts with Programs, Cordilleran Meeting, San Bernadino, California, v. 26, p. 37.

Beard, L. S., 1995, Geologic map of the Virginia Peak 7.5-minute quadrangle, Clark County, Arizona, and Mohave County, Nevada: U.S. Geological Survey Open-File Report, in press.

Beard, L. S., and Campagna, D., 1991, Preliminary geologic map of the Devil's Throat 7.5-minute quadrangle, Clark County, Nevada, U.S. Geological Survey Open-File Report OF 91-132, scale 1:24,000.

Beratan, K., 1991, Miocene synextensional sedimentation patterns, Whipple Mountains, southeastern California: Implications for the geometry of the Whipple detachment system: Journal of Geophysical Research, v. 96, p. 12425–12442.

Best, M., and 10 others, 1993, Oligocene-Miocene caldera complexes, ash-flow sheets, and tectonism in the central and southeastern Great Basin, *in* Lahren, M. M., Trexler, J. H., and Spinosa, C., Crustal evolution of the Great Basin and the Sierra Nevada: Geological Society of America Cordilleran /Rocky Mountain Section Meeting Guidebook: Reno, University of Nevada, p. 285–311.

Blair, W. N., and Armstrong, A. K., 1979, Hualapai Limestone Member of the Muddy Creek Formation; the youngest deposit predating the Grand Canyon, southeastern Nevada and northwestern Arizona: U.S. Geological Survey Professional Paper 1111, 14 p.

Blank, H. R., Rowley, P. D., and Hacker, D. B., 1992, Miocene monzonitic intrusions and associated megabreccias of the iron axis region, southwestern Utah, *in* Field guide to geologic excursions in Utah and adjacent areas of Nevada, Idaho, and Wyoming: Salt Lake City, Utah Geological Survey Miscellaneous Publication 92-3, p. 399–420.

Bohannon, R. G., 1979, Strike-slip faults of the Lake Mead region of southern Nevada, *in* Armentrout, J. M., Cole, M. R., and TerBest, H., Jr., Cenozoic paleogeography of the Western United States: Pacific Coast Paleogeography Symposium: Anaheim, California, Society of Economic Paleontologists and Mineralogists, Pacific Section, p. 129–139.

Bohannon, R. G., 1983, Mesozoic and Cenozoic tectonic development of the Muddy, North Muddy, and northern Black Mountains, Clark County, Nevada, *in* Miller, D. M., Todd, V. R., and Howard, K. A., Tectonic and stratigraphic studies in the Eastern Great Basin: Boulder, Colorado, Geological Society of America Memoir 157, p. 125–148.

Bohannon, R. G., 1984, Nonmarine sedimentary rocks of Tertiary age in the Lake Mead region, southeastern Nevada and northwestern Arizona: U.S. Geological Survey Professional Paper 1259, 72 p.

Bohannon, R. G., 1992, Geologic map of the Red Pockets quadrangle, Mohave County, Arizona: U.S. Geological Survey Miscellaneous Investigations Series I-2288, scale 1:24,000.

Bohannon R. G., Grow, J., Miller, J., and Blank, R., 1993, Seismic stratigraphy and tectonic development of Virgin River depression and associated basins, southeastern Nevada and northwestern Arizona: Geological Society of American Bulletin, v. 105, p. 501–520.

Brenner, E. F., and Glanzman, R., 1979, Tertiary sediments in the Lake Mead area, Nevada, *in* Newman, G. W., and Goode, H. D., Basin and Range Symposium: Rocky Mountain Association of Geologists and Utah Geological Association, p. 313–323.

Brenner-Tourtelot, E., and Glanzman, R., 1977, Lithium-bearing rocks of the Horse Spring Formation, Clark County, Nevada: Energy, v. 3, p. 255–262.

Campagna, D., 1990, The Lake Mead fault system and the Las Vegas Valley shear zone: strike-slip faulting and associated deformation in the Basin and Range, southeastern Nevada [Ph.D. dissert.]: West Lafayette, Indiana, Purdue University, 80 p.

Campagna, D., and Aydin, A., 1991, Tertiary uplift and shortening in the Basin and Range: The Echo Hills, southeastern Nevada: Geology, v. 19, p. 485–488.

Campagna, D., and Aydin, A., 1994, Basin genesis associated with strike-slip faulting in the Basin and Range, southeastern Nevada: Tectonics, v. 13, p. 327–341.

Campagna, D., and Levandowski, D., 1991, The recognition of strike-slip fault systems using imagery, gravity, and topographic data sets: Photogrammetric Engineering and Remote Sensing, v. 57, p. 1195–1201.

Carpenter, D. G., Carpenter, J. A., Bradley, M. D., Franz, U. A., and Reber, S. J., 1989, On the role of isostasy in the evolution of normal fault systems: Comment: Geology, v. 17, p. 774–775.

Carpenter, J., 1988, Reinterpretation of Mormon Peak detachment in Mormon Mountains, southern Nevada: American Association of Petroleum Geologists Bulletin, v. 72, no. 2, p. 168–169.

Cole, E. D., 1989, Petrogenesis of Late Cenozoic alkalic basalt near the eastern boundary of the Basin-and-Range, Upper Grand Wash Trough, Arizona and Gold Butte, Nevada: [M.S. thesis]: Las Vegas, University of Nevada, 68 p.

Damon, P. E., Shafiqullah, M., and Scarborough, R. B., 1978, Revised chronology for critical stages in the evolution of the lower Colorado River: Geological Society of America Abstracts with Programs, v. 10, p. 101–102.

Dickinson, W., Klute, M., Hayes, M., Janecke, S., Lundin, E., McKittrick, M., and Olivares, M., 1988, Paleogeographic and paleotectonic setting of Laramide sedimentary basins in the central Rocky Mountain region: Geological Society of American Bulletin, v. 100, p. 1023–1039.

Duebendorfer, E., and Black, R., 1992, Kinematic role of transverse structures in continental extension: An example from the Las Vegas Valley shear zone, Nevada: Geology, v. 20, p. 1107–1110.

Duebendorfer, E., and Simpson, D. A., 1994, Kinematics and timing of Tertiary extension in the western Lake Mead region, Nevada: GSA Bulletin, v. 106, p. 1057–1073.

Duebendorfer, E., and Wallin, E. T., 1991, Basin development and syntectonic sedimentation associated with kinematically coupled strike-slip and detachment faulting, southern Nevada: Geology, v. 19, p. 87–90.

Eberly, L. D., and Stanley, T. B. J., 1978, Cenozoic stratigraphy and geologic history of southwestern Arizona: Geological Society of America Bulletin, v. 89, p. 921–940.

Evans, J. P., and Neves, D. S., 1992, Footwall deformation along Willard thrust, Sevier orogenic belt: Implications for mechanisms, timing, and kinematics: Geological Society of America Bulletin, v. 104, p. 516–527.

Faulds, J., 1992, Miocene stratigraphy of the central Black Mountains, northwestern Arizona: Variations across a major accommodation zone, *in* Sherrod, D. R., and Nielson, J. E., Tertiary stratigraphy of the highly extended terranes, California, Arizona, and Nevada: U.S. Geological Survey Bulletin 2053, 46 p.

Faulds, J., and Geissman, J., 1992, Implications of paleomagnetic data on Miocene extension near a major accommodation zone in the Basin and Range province, northwestern Arizona and southern Nevada: Tectonics, v. 11, p. 204–227.

Fedo, C., and Miller, J. G., 1992, Evolution of a Miocene half-graben basin, Colorado River extensional corridor, southeastern California: Geological Society of America Bulletin, v. 104, p. 481–493.

Fillmore, R., 1993, Sedimentation and extensional basin evolution in a Miocene metamorphic core complex setting, Alvord Mountain, central Mojave Desert, California: Sedimentology, v. 40, p. 721–742.

Fitzgerald, P. G., Fryxell, J., and Wernicke, B. P., 1991, Apatite fission track constraints on the extensional evolution of the Gold Butte crustal section, South Virgin Mountains, Nevada: Geology, v. 19, p. 1013–1016.

Freytet, P., and Plaziat, J., 1982, Continental carbonate sedimentation and pedogenesis—Late Cretaceous and Early Tertiary of Southern France: Contributions to Sedimentology, v. 12, p. 213.

Fryxell, J., and Duebendorfer, E. M., 1990, Origin and trajectory of the Frenchman Mountain block, southern Nevada: Geological Society of America Abstracts with Programs, v. 22, no. 7, p. 226.

Fryxell, J., Salton, G., Selverstone, J., and Wernicke, B., 1992, Gold Butte crustal section, South Virgin Mountains, Nevada: Tectonics, v. 11, p. 1099–1120.

Garces, B., 1993, Lacustrine deposition and related volcanism in a transtensional tectonic setting: Upper Stephanian–Lower Autunian in the Aragon-Bearn Basin, western Pyrenees (Spain-France): Sedimentary Geology, v. 83, p. 133–160.

Gardener, L. R., 1972, Origin of the Mormon Mesa caliche, Clark County, Nevada: Geological Society of America Bulletin, v. 83, p. 143–156.

Gibbs, A., 1987, Development of extension and mixed-mode sedimentary basins, *in* Coward, M. P., Dewey, J. F., and Hancock, P. L., Continental extensional tectonics: London, Geological Society of London Special Publication No. 28, p. 19–33.

Goetz, A. F. H., Billingsley, F. C., Gillespie, A. R., Abrams, M. J., Squires, R. L., Shoemaker, E. M., Lucchitta, I., and Elston, D. P., 1975, Application of ERTS images and image processing to regional geologic problems and geologic mapping in northern Arizona: Jet Propulsion Laboratory Technical Report 32-1597, 188 p.

Goldstrand, P. M., 1990a, Stratigraphy and ages of the basal Claron, Pine Hollow, Canaan Peak, and Grapevine Wash Formations, southwest Utah: Utah Geological and Mineral Survey Open-File Report 193, 186 p.

Goldstrand, P. M., 1990b, Stratigraphy and paleogeography of late Cretaceous and Early Tertiary rocks of southwest Utah: Salt Lake City, Utah Geological and Mineralogical Survey Miscellaneous Publication MP90-2, 58 p.

Goldstrand, P. M., 1992, Evolution of Late Cretaceous and Early Tertiary basins of southwest Utah based on clastic petrology: Journal of Sedimentary Petrology, v. 62, p. 495–507.

Graf, W., Hereford, R., Laity, J., and Young, R. A., 1987, Colorado Plateau, *in* Graf, W., ed., Geomorphic systems of North America: Boulder, Colorado, Geological Society of America, Centennial Special Volume 2, p. 259–302.

Gray, R. S., 1964, Late Cenozoic geology of Hindu Canyon, Arizona: Arizona Academy of Science Journal, v. 3, p. 39–42.

Hintze, L. F., 1986, Stratigraphy and structure of the Beaver Dam Mountains, southwestern Utah, *in* Griffen, D. T., and Phillips, W. R., eds., Thrusting and extensional structures and mineralization in the Beaver Dam Mountains, southwestern Utah: Utah Geological Association Publication 15, p. 1–36.

Kowallis, B., and Everett, B., 1986, Sedimentary environments of the Muddy Creek Formation near Mesquite, Nevada, *in* Griffen, D. T., and Phillips, W. R., eds., Thrusting and extensional structures and mineralization in the Beaver Dam Mountains, southwestern Utah: Utah Geological Association Publication 15, p. 69–76.

Longwell, C. R., 1921, Geology of the Muddy Mountains, Nevada, with a section to the Grand Wash Cliffs in western Arizona: American Journal of Science, v. 201, p. 39–62.

Longwell, C. R., 1922, The Muddy Mountain overthrust in southeastern Nevada: Journal of Geology, v. 30, p. 63–72.

Longwell, C. R., 1936, Geology of the Boulder reservoir floor, Arizona-Nevada: Geological Society of America Bulletin, v. 47, p. 1393–1476.

Longwell, C. R., 1971, Measure of lateral movement on Las Vegas shear zone, Nevada: Geological Society of America Abstracts with Programs, v. 3, p. 152.

Longwell, C. R., 1974, Measure and date of movement on Las Vegas Valley shear zone, Clark County, Nevada: Geological Society of American Bulletin, v. 85, p. 985–990.

Longwell, C. R., Pampeyan, E. H., Bower, B., and Roberts, R. J., 1965, Geology and mineral deposits of Clark County, Nevada: Nevada Bureau of Mines Bulletin 62, 218 p.

Lucchitta, I., 1966, Cenozoic geology of the upper Lake Mead area adjacent to the Grand Wash Cliffs, Arizona [Ph.D. dissert.]: University Park, Pennsylvania State University, 218 p.

Lucchitta, I., 1972, Early history of the Colorado River in the Basin and Range province: Geological Society of America Bulletin, v. 83, p. 1933–1948.

Lucchitta, I., 1979, Late Cenozoic uplift of the southwestern Colorado Plateau and adjacent lower Colorado River region: Tectonophysics, v. 61, p. 63–95.

Lucchitta, I. and Young, R. A., 1986, Structure and geomorphic character of western Colorado Plateau in the Grand Canyon–Lake Mead region, *in* Nations, J. D., Conway, C. M., and Swann, G. A., eds., Geology of central and northern Arizona: Geological Society of America, Rocky Mountain Section Meeting, Field Trip Guidebook: Flagstaff, Northern Airzona University, p. 159–176.

Machette, M., 1985, Calcic soils of the southwestern United States, *in* Weide, D. L., ed., Soils and Quaternary geology of the southwestern United States: Geological Society of America Special Paper 203, p. 1–21.

Mackin, J. H., 1960, Structural significance of Tertiary volcanic rocks in southwestern Utah: American Journal of Science, v. 358, p. 81–131.

Matthews, J., 1976, Paleozoic stratigraphy and structural geology of the Wheeler Ridge area, northwestern Mohave County, Arizona [M.S. Thesis]: Flagstaff, Northern Arizona University, 145 p.

McNair, A. H., 1951, Paleozoic stratigraphy of part of northwestern Arizona: American Association of Petroleum Geologists Bulletin, v. 35, p. 503–541.

Moore, R., 1972, Geology of the Virgin and Beaverdam Mountains, Arizona: Arizona Bureau of Mines Bulletin 186, 65 p.

Morgan, J. R., 1968, Structure and stratigraphy of the northern part of the South Virgin Mountains, Clark County, Nevada [M.S. thesis]: Albuquerque, University of New Mexico, 103 p.

Mullett, D., 1989, Interpreting the Early Tertiary Claron Formation of southern Utah: Geological Society of America Abstracts with Programs, v. 21, no. 5, p. 120.

Mullet, D., 1990, Soil fabrics and horizontal cracking in the Paleogene Claron Formation of southern Utah: Geological Society of America Abstracts with Programs, v. 22, no. 7, p. A335.

Parolini, J., 1986, Debris flows within a Miocene alluvial fan, Lake Mead region, Clark County, Nevada [M.S. thesis]: Las Vegas, University of Nevada, 119 p.

Platt, N. H., 1993, Fresh-water carbonates from the Lower Freshwater Molasse (Oligocene, western Switzerland): Sedimentology and stable isotopes: Sedimentary Geology, v. 78, p. 81–99.

Platt, N. H., and Wright, V. P., 1991, Lacustrine carbonates: Facies models, facies distributions and hydrocarbon aspects: Special Publications of the International Association of Sedimentologists No. 13, p. 57–74.

Platt, N. H., and Wright, V. P., 1992, Palustrine carbonates and the Florida Everglades: Towards an exposure index for the fresh-water environment?: Journal of Sedimentary Petrology, v. 62, p. 1058–1071.

Reynolds, S. J., Florence, F. P., Welty, J. W., Roddy, M. S., Currier, D. A., Anderson, A. V., and Keith, S. B., 1986, Compilation of radiometric age determinations in Arizona: Arizona Bureau of Geology and Mineral Technology Bulletin 197, 258 p.

Rice, J. A., 1987, Sedimentology, provenance, and tectonic significance of the basal conglomerate of the Rainbow Gardens Member of the Miocene Horse Spring formation, Lake Mead area, southeastern Nevada [M.S. thesis]: Las Vegas, University of Nevada, 171 p.

Ron, H., Aydin, A., and Nur, A., 1986, Strike-slip faulting and block rotation in the Lake Mead fault system: Geology, v. 14, p. 1020–1023.

Rowland, S., Parolini, J., Eschner, E., and McAllister, A., 1990, Sedimentologic and stratigraphic constraints on the Neogene translation and rotation of the Frenchman Mountain structural block, Clark County, Nevada, *in* Wernicke, B. P., ed., Basin and Range extensional tectonics near the latitude of Las Vegas, Nevada: Geolological Society of America Memoir 176, p. 99–122.

Schmitt, J. G. and Rice, J. A., 1988, Sedimentology and provenance of Miocene Horse Spring Formation conglomerates: Implications for estimates of motion along the Lake Mead fault system, Nevada: Geological Society of America Abstracts with Programs, v. 20, no. 3, p. 228.

Schwab, F. L., 1976, Modern and ancient sedimentary basins: Comparative accumulation rates: Geology, v. 4, p. 723–727.

Seager, W., 1966, Geology of the Bunkerville section of the Virgin Mountains, Nevada and Arizona [Ph.D. dissert.]: Tucson, University of Arizona, 124 p.

Seager, W., 1970, Low-angle gravity glide structures in the northern Virgin Mountains, Nevada: Geological Society of America Bulletin, v. 81, p. 1517–1538.

Shafiqullah, M., Damon, P. E., Lynch, D. J., Reynolds, S. J., Rehrig, W. A., and Raymond, R. H., 1980, K-Ar geochronology and geologic history of southwestern Arizona and adjacent areas, *in* Jenney, J. P., and Stone, C., Studies in western Arizona: Arizona Geological Society Digest v. 12, p. 201–260.

Sherrod, D. R., and Nielson, J. E., 1993, Tertiary stratigraphy of highly extended terranes, California, Arizona, and Nevada: U.S. Geological Survey Bulletin 2053.

Siders, M. A., and Shubert, M. A., 1986, Stratigraphy and structure of the northern Bull Valley Mountains and Antelope Range, Iron County, Utah, *in* Griffen, D. T., and Phillips, W. R., Thrusting and extensional structures and mineralization in the Beaver Dam Mountains, southwestern Utah: Utah Geological Association Publication 15, p. 87–102.

Taylor, W., 1993, Stratigraphic and lithologic analysis of the Claron Formation in southwestern Utah: Salt Lake City, Utah Geological Survey Miscellaneous Publication 93–1, 52 p.

Timm, J. J., 1985, Age and significance of Paleozoic sedimentary rocks in the southern River Mountains, Clark County, Nevada [M.S. thesis]: Las Vegas, University of Nevada, 62 p.

Vera, J. A., Ruiz-Ortiz, P. A., Garcia-Hernandez, M., and Molina, J. M., 1988, Paleokarst and related pelagic sediments in the Jurassic of the Subbetic Zone, southern Spain, *in* James, N. P., and Choquette, P. W., Paleokarst: New York, Springer-Verlag, 416 p.

Volborth, A., 1962, Rapakivi-type granites in the Precambrian complex of Gold Butte, Clark County, Nevada: Geological Society of American Bulletin, v. 73, p. 813–832.

Weber, M. E., and Smith, E. I., 1987, Structural and geochemical constraints on the reassembly of disrupted mid-Miocene volcanoes in the Lake Mead–El Dorado Valley area of southern Nevada: Geology, v. 15, p. 553–556.

Wernicke, B., Axen, G., 1988, On the role of isostasy in the evolution of normal fault systems: Geology, v. 16, p. 848–851.

Wernicke, B., Spencer, J. E., Burchfiel, B. C., and Guth, P. L., 1982, Magnitude of crustal extension in the southern Great Basin: Gelology, v. 10, p. 499–502.

Wernicke, B., Guth, P. L., and Axen, G. J., 1984, Tertiary extensional tectonics in the Sevier thrust belt of southern Nevada: Geological Society of America Annual Meeting Guidebook No. 4, Field Trip 19: Reno, Nevada, McKay School of Mines, p. 473–510.

Wernicke, B., Guth, P. L., and Axen, G., J., 1984, Tertiary extensional tectonics in the Sevier thrust belt of southern Nevada: Geological Society of America Annual Meeting Guidebook No. 4, Field Trip 19: Reno, Nevada, McKay School of M ines, p. 473–510.

Wernicke, B., Axen, G. J., and Snow, J. K., 1988, Basin and Range extensional tectonics at the latitude of Las Vegas, Nevada: Geological Society of American Bulletin, v. 100, p. 1738–1757.

Williams, V. S., Bohannon, R. G., and Hoover, D. L., 1995, Geologic Map of Riverside Quadrangle: U.S. Geological Survey Map GQ 1770, in press.

Yarnold, J. and Lombard, J., 1989, A facies model for large rock-avalanche deposits formed in dry climates, *in* Colburn, I., Abbott, P., and Minch, J., eds., Conglomerates in basin analysis: A symposium dedicated to A. O. Woodford: Society of Economic Paleontologists and Mineralogists Pacific Section, v. 62, p. 9–31.

Young, R. A., 1966, Cenozoic geology along the edge of the Colorado Plateau in northwestern Arizona, St. Louis, Washington University, 167 p.

Young, R. A., 1979, Laramide deformation, erosion and plutonism along the southwestern margin of the Colorado Plateau: Tectonophysics, v. 61, p. 25–47.

Young, R. A., 1982, Paleogeomorphic evidence for the structural history of the Colorado Plateau margin in Arizona, *in* Frost, E. G., and Martin, D. L., eds., Mesozoic-Cenozoic tectonic evolution of the Colorado River region, California, Arizona, and Nevada: San Diego, California, Cordilleran Publishers, p. 29–39.

Young, R. A., 1985, Geomorphic evolution of the Colorado Plateau margin in west-central Arizona; a tectonic model to distinguish between the causes of rapid symmetrical scarp retreat and scarp dissection, *in* Hack, J. T., and Morisawa, M., eds., Tectonic geomorphology: Binghamton Symposia in Geomorphology International Series 15: London, Allen and Unwin, p. 261–278.

Manuscript Accepted by the Society April 21, 1995

Printed in U.S.A.

Geological Society of America
Special Paper 303
1996

The Mud Hills, Mojave Desert, California: Structure, stratigraphy, and sedimentology of a rapidly extended terrane

Raymond V. Ingersoll
Department of Earth and Space Sciences, University of California, Los Angeles, California 90095-1567
Kathleen A. Devaney
4792 Sandia Drive, Los Alamos, New Mexico 87544
Jeffrey K. Geslin
U.S. Geological Survey, Suite 860, MS 509, 101 Convention Center Dr., Las Vegas, Nevada 89109
William Cavazza
Department of Mineralogical Sciences, University of Bologna, Piazza di Porta S. Donato 1, 40126 Bologna, Italy
David S. Diamond
Engineering Science, 1301 Marina Village Parkway, Suite 200, Alameda, California 94501
William A. Heins*
Department of Geology and Geography, Box 255, Vassar College, Poughkeepsie, New York 12601
Keith J. Jagiello
24314 Travis House, Katy, Texas 77493
Kathleen M. Marsaglia
Department of Geological Sciences, University of Texas at El Paso, El Paso, Texas 79968-0555
Earnest D. Paylor, II
M.S. 183-501, Tectonics and Geophysics Group, Jet Propulsion Laboratory, 4800 Oak Grove Drive, Pasadena, California 91109
Paul F. Short
James M. Montgomery Consulting, 301 North Lake Avenue, Suite 600, Pasadena, California 91101

ABSTRACT

The Mud Hills expose pre-, syn-, and postextensional strata related to Miocene development of the Waterman Hills detachment fault. Synextensional breccia (Mud Hills Formation) was deposited during the final stages of crustal extension of the upper plate (20–18 Ma). Previous workers have misinterpreted fault contacts as stratigraphic contacts, and have developed intricate pseudostratigraphy to explain their observations. Detailed mapping, combined with stratigraphic and sedimentologic data, documents that the volcaniclastic Pickhandle Formation (Miocene) is conformably overlain by the plutoniclastic Mud Hills Formation (Miocene) with no interfingering. Repetition of these south-dipping lithologic units is caused by imbricate, north-dipping listric faults. These relations are demonstrated by the systematic northward "v"ing of fault contacts and southward "v"ing of stratigraphic contacts in canyons. Stratigraphic dips decrease upsection, consistent with incremental rotation of basinal strata during deposition. Most of the Mud Hills Formation consists of rock-avalanche breccia and megabreccia derived from Mesozoic granodiorite, which is

*Present address: Department of Geology/Earth Sciences, Lewis-Clark State College, Lewiston, Idaho 83501.

Ingersoll, R. V., Devaney, K. A., Geslin, J. K., Cavazza, W., Diamond, D. S., Heins, W. A., Jagiello, K. J., Marsaglia, K. M., Paylor, E. D., II, Short, P. F., 1996, The Mud Hills, Mojave Desert, California: Structure, stratigraphy, and sedimentology of a rapidly extended terrane, *in* Beratan, K. K., ed., Reconstructing the History of Basin and Range Extension Using Sedimentology and Stratigraphy: Boulder, Colorado, Geological Society of America Special Paper 303.

identical to basement exposed beneath the Pickhandle and Jackhammer (Miocene?) Formations to the north. A distinctive rhyolite-tuff breccia with granodiorite matrix provides a marker horizon within the Mud Hills Formation; the rhyolite tuff is presumed to have been derived from the Waterman Hills area to the south. The Mud Hills Formation was derived from now-buried granodiorite of a stranded upper-plate block to the south, as demonstrated by northward paleocurrents, and the presence of fine-grained units close to the presumed master fault (as is typical of half-graben sedimentation). Unconformably overlying the Mud Hills Formation is the Owl Conglomerate (basal member of the Barstow Formation), which has mixed provenance with southward paleocurrents; the Owl Conglomerate was derived from residual highlands after extension ceased.

Our palinspastic model for development of the Mud Hills and Waterman Hills area includes the following: (1) eruption and deposition of andesitic to rhyodacitic volcaniclastic detritus (Jackhammer and Pickhandle Formations) at 24–21 Ma; (2) initiation of extension no later than 21 Ma; (3) development of a supradetachment basin that was quickly dismembered (22–21 Ma) (strata now located in Waterman Hills); (4) initiation of an upper-plate normal fault with down-to-the-north movement, beginning Mud Hills deposition, just prior to eruption of rhyolite at 20.0 Ma; (5) continued movement on the fault and subsidiary faults, which rotated half-graben strata coincident with sedimentation; (6) cessation of extension before 16 Ma, coincident with deposition of the basal Barstow Formation; (7) accumulation of the Barstow Formation, derived from residual highlands both to the north and the south; and (8) strike-slip faulting and related folding after 10 Ma, which formed the Barstow syncline.

Integration of structural, stratigraphic, and sedimentologic information is essential for correct reconstruction of highly extended terranes. Interpretations derived from cursory examination of such terranes may result in misinterpretation of complex relations.

INTRODUCTION

The Mud Hills, north of Barstow, California (Fig. 1), expose pre-, syn- and postextensional strata related to Miocene development of the Waterman Hills detachment fault. Postextensional strata (Barstow Formation) have been studied extensively, especially in the Rainbow basin area, which has been a favorite introductory mapping area for university students. The older Tertiary strata (syn- and pre-extensional) have been relatively less studied because of greater difficulty of access, stratigraphic and structural complexity, and paucity of chronostratigraphic data (faunal or radiometric).

We undertook a detailed study of a relatively small and beautifully exposed area in the Mud Hills in order to unravel the complex interplay of extensional tectonics, erosion, sedimentation, and basin development. We began the study in the winter of 1988 and completed it in 1993. During this time, we discovered that the structure was more complex than previously realized, and that lack of recognition of this complexity had led to misinterpretations of the stratigraphy and the paleogeographic and paleotectonic history of the area.

We present sedimentologic, stratigraphic, and structural data and interpretations that constrain how the Mud Hills area evolved during the Miocene. Our palinspastic reconstruction, in turn, constrains how the Waterman Hills detachment evolved. The paleogeographic and paleotectonic history is contained in the details, which can only be determined through integrated study of the stratigraphy, sedimentology, petrology, and structure of such an area (Ingersoll and Diamond, 1990; Ingersoll et al., 1993). The lessons learned through this investigation should be applicable to studies of other highly extended terranes.

PREVIOUS WORK

Dibblee (1967, 1968) summarized early work in the western Mojave Desert, including the present study area. He formalized stratigraphic terminology of the Jackhammer and Pickhandle Formations, units first described by McCulloh (1952); he also redefined the Barstow Formation (modified after Merriam, 1919), including the Owl Conglomerate Member. Durrell (1953) also measured the Pickhandle and Barstow Formations within the study area. Bowen (1954) defined the Waterman Gneiss and Dibblee (1967) described quartz monzonite and granodiorite found in the hills north of Barstow.

Dibblee (1980a, 1980b, 1980c) provided additional descriptions of stratigraphy, basement rocks, and structure of the study area; Burke et al. (1982) provided chronostratigraphic constraints, including radiometric ages. The Barstow Formation has recently been studied extensively (e.g., Woodburne et al., 1982, 1990; MacFadden et al., 1990).

There has been a surge of interest in the region during the last decade, as it has become clear that the Barstow area exemplifies an evolving low-angle detachment fault with associated sedimentary basins (e.g., Dokka, 1986, 1989a, 1989b; Dokka and Woodburne, 1986; Dokka et al., 1988; Glazner et al., 1988, 1989a, 1989b, 1994; Bartley et al., 1990; Ingersoll and Diamond, 1990; Travis et al., 1990; Walker et al., 1990; Travis, 1992; Fillmore and Walker, 1993, this volume; Ingersoll et al., 1993). The interested reader is referred to these publications for additional information.

REGIONAL SETTING

Discussion of the pre-Miocene history of the Mud Hills area is beyond the scope of the present study; many of the above references provide details. The most important aspect of this history is that extensive Mesozoic intrusive rocks (mostly granite, quartz monzonite, and granodiorite) intruded complex Paleozoic and Mesozoic metasedimentary and metavolcanic strata (e.g., Dibblee, 1967, 1968, 1980c; Miller, 1981; Walker, 1988; Walker et al., 1990; Martin and Walker, 1991, 1992).

Hewett (1954) suggested that pre-Miocene erosion of the central Mojave removed 4.5 km of crust, although this figure was developed prior to appreciation of low-angle detachment faulting (e.g., Wernicke, 1985) in the Mojave Desert. Burke et al. (1982, p. E14) stated that "welded quartz latite ash-flow sheets that are preserved only at Lane Mountain were deposited on a deeply eroded pre-Cenozoic basement surface of low relief during early Miocene time." The Jackhammer, Pickhandle, Mud Hills, and Barstow Formations were deposited sequentially on this deeply eroded pre-Cenozoic basement surface (see discussion that follows).

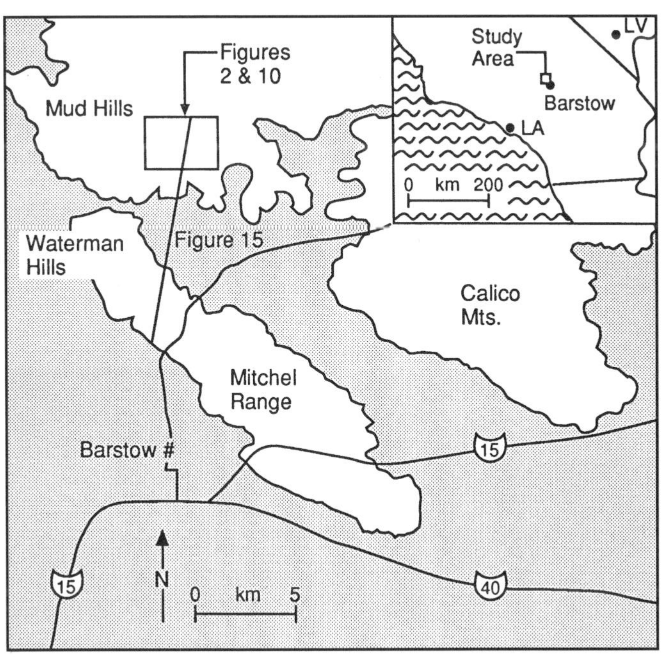

Figure 1. Regional location map, showing study area in Mud Hills, north of Barstow, California. Also shown are locations of Figures 2, 10, and 15. LA = Los Angeles; LV = Las Vegas.

The Barstow Formation is exposed in the Barstow syncline, which trends east to west in the southern Mud Hills (see Dibblee, 1967, figs. 54 and 55). The Jackhammer, Pickhandle, and Mud Hills Formations are exposed on the northern limb of the syncline, unconformably underlying the Barstow Formation.

STRATIGRAPHY OF THE MUD HILLS

Introduction

Early workers (McCulloh, 1952; Durrell, 1953) measured stratigraphic sections and drew N-S cross sections through the eastern Mud Hills (eastern part of our study area) that showed few faults within the granitic basement or overlying Jackhammer (Tj), Pickhandle (Tp) and Barstow (Tb) Formations. They showed these three formations dipping south in a simple homocline, with decreasing dip upsection, culminating in the axis of the Barstow syncline.

Dibblee (1967, 1968) utilized McCulloh's (1952) stratigraphic nomenclature in the first published maps of our study area. His mapping showed several repetitions of volcaniclastic and plutoniclastic units within McCulloh's Pickhandle Formation; he interpreted these units as stratigraphically interfingering, without substantial faulting. Subsequent workers (Dokka and Woodburne, 1986; Woodburne et al., 1990; Dokka et al., 1991; Travis, 1992) have accepted this basic interpretation, although with modifications of stratigraphic details.

We began our study by measuring stratigraphic sections in the Owl Canyon and Solomon Canyon areas (Figs. 2 and 3). The volcaniclastic section in Owl Canyon is repeated imbricately by normal faults, with one distinctive ashflow tuff bed repeated nine times on a single hill side. Our measured sections show true stratigraphic thickness (with removal of small-scale fault repetitions of strata), although some small-scale faults are difficult to map in the plutoniclastic breccia. The Solomon Canyon section has several fault repetitions of mappable stratigraphic contacts; the faulted nature of these repetitions was not recognized during the initial measuring. These faults are now indicated in the measured section of Figure 3.

Jackhammer and Pickhandle Formations

We accept previous workers' (e.g., McCulloh, 1952; Dibblee, 1967, 1968) descriptions and interpretations of the Jackhammer and Pickhandle Formations (excluding the Mud Hills Formation; see next section).

The Pickhandle Formation includes ashflow tuff that was deposited by pyroclastic flows that may have traveled tens of kilometers (primarily from the Black Canyon and Opal Moun-

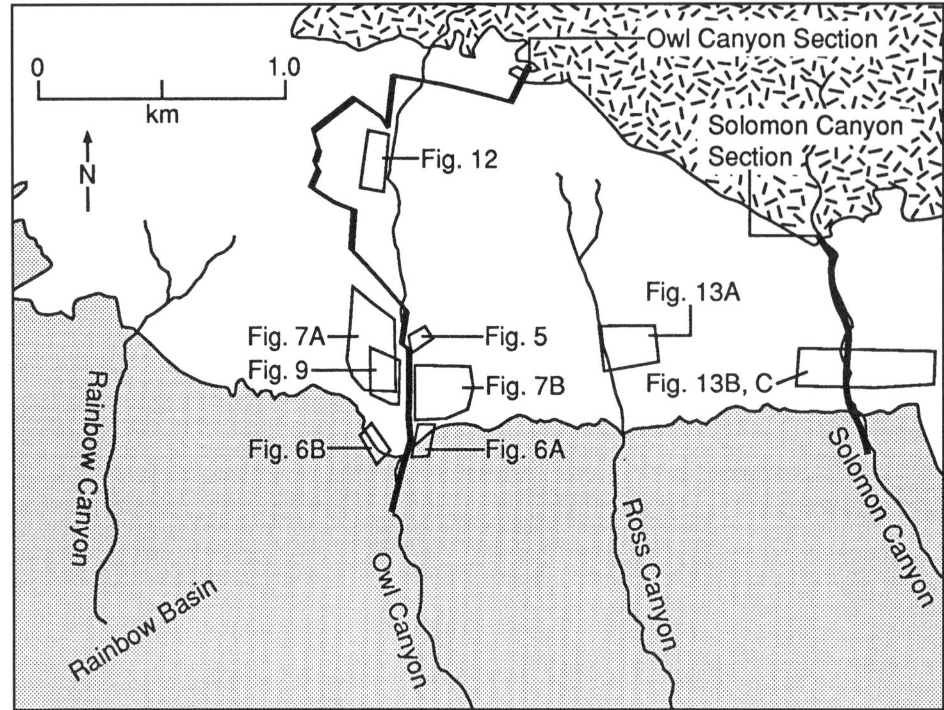

Figure 2. Outline map of study area in Mud Hills, showing place names and locations of measured sections and photo figures. Base map is from Figure 10, with granite symbol (random dashes) representing area underlain by Mesozoic granitoids; shaded area is underlain by Barstow Formation; unpatterned area is underlain by Jackhammer, Pickhandle, and Mud Hills Formations. Quaternary units not shown. Heavy lines of Owl Canyon section shown offset along faults (lighter lines). Faults that repeat Solomon Canyon section are not shown in this figure (see Figs. 3 and 10).

tain area) (Dibblee, 1967, 1968). As a result, some Pickhandle lithofacies are tabular bodies with uniform thickness at the map scale of our study area. Lahars also make up a significant proportion of the Pickhandle Formation (Fillmore and Walker, this volume).

Mud Hills Formation

Our initial work convinced us that including both volcaniclastic units (mostly ashflow tuff and lahars) and plutoniclastic units (mostly monolithologic breccia and megabreccia) within the Pickhandle Formation made little sense, from either a genetic or a practical (mapping) viewpoint. McCulloh's (1952) type sections of the Jackhammer and Pickhandle Formations are quite different from the exposures in our study area. The plutoniclastic breccia and megabreccia (including the distinctive intermixed rhyolite-tuff breccia) deserve separate recognition as map units. Working independently and concurrently, Travis et al. (1990), Dokka et al. (1991) and Travis (1992) came to the same conclusion and named the plutoniclastic unit the "Mud Hills Formation," which is thereby removed from the Pickhandle Formation. We accept this name as a useful refinement of the stratigraphy. We disagree, however, with these authors' interpretation of the stratigraphic and structural relations between the primarily volcaniclastic Pickhandle Formation and the primarily plutoniclastic Mud Hills Formation. Contrary to the complex interfingering of units they show (figs. 22 and 23 of Dokka et al., 1991; figs. 3.3, 3.4, and 3.5 of Travis, 1992), field relations indicate that *everywhere in our study area, the Mud Hills Formation stratigraphically overlies the Pickhandle Formation, with only minor reworking of the latter into the base of the former.* The complexity of stratigraphy illustrated by these authors (and all previous workers) is an artifact of the lack of recognition of structural repetition of the Pickhandle–Mud Hills contact, as illustrated in a following section.

We, hereby, formally define the Mud Hills Formation as being the interbedded plutoniclastic breccia and megabreccia, rhyolite-tuff breccia, and sandstone units described herein and exposed in Owl Canyon and adjoining areas. The measured section from Owl Canyon (Fig. 3) serves as the type section, where the Mud Hills Formation is 230 m thick. The Mud Hills Formation is underlain by the Pickhandle Formation, consisting predominantly of volcaniclastic strata, and is overlain by the Barstow Formation. All of these units are Lower to Middle Miocene. Additional description of the Mud Hills Formation may be found in Travis (1992) and our following discussion.

In contrast to the Pickhandle Formation, the plutoniclastic breccia and megabreccia of the Mud Hills Formation "are the results of large landslides or rockfalls" (McCulloh, 1952, p. 118); depositional characteristics and grain size (as many as 60 m for one possible clast in Owl Canyon) dictate that they

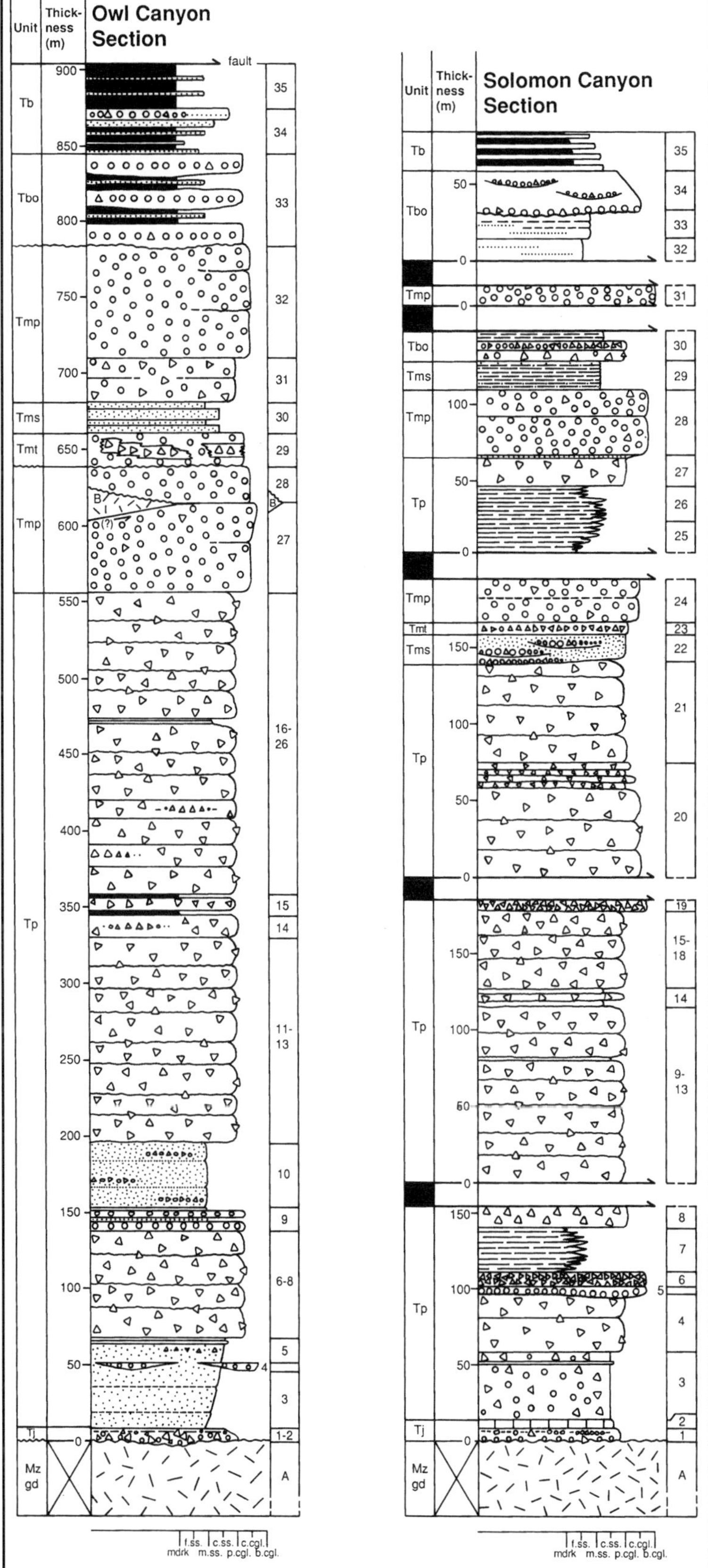

Figure 3. Measured stratigraphic sections from Owl Canyon and Solomon Canyon. Locations shown in Figure 2. Stratigraphic symbols shown in Figure 4. Triangles denote volcaniclastic conglomerate and breccia; circles denote plutoniclastic conglomerate, breccia, and megabreccia. Unit 31 of Solomon Canyon section is a fault-bounded block, and is not shown in Figure 10.

were locally derived. As a result, the Mud Hills Formation consists of wedges of detritus with extreme facies changes within the study area. The fault repetition of these facies changes is initially stratigraphically bewildering. Our schematic stratigraphic diagram of the Mud Hills Formation and adjacent units (Fig. 4) is consistent with the following descriptions of structure and sedimentology.

The Pickhandle–Mud Hills contact is planar and only slightly disconformable (Fig. 5). The contact of both of these units with the overlying basal Barstow Formation (Owl Conglomerate Member in most locations), however, is a buttress unconformity developed on significant sub-Barstow topography (Fig. 6). We suggest that the Mud Hills Formation was deposited during listric normal faulting (see following discussion) within a rotating half graben; as a result, dips flatten stratigraphically upward (Fig. 7). Basal beds of the Mud Hills Formation dip as steeply as uppermost Pickhandle beds; upper beds of the Mud Hills Formation dip only slightly less than the overlying Owl Conglomerate.

We disagree with Woodburne et al.'s (1990, p. 466) statement that "information recently developed by us shows that the base of the Barstow in part interfingers with the top of the Pickhandle (i.e., Mud Hills Fm.) on the north limb of the syncline, in

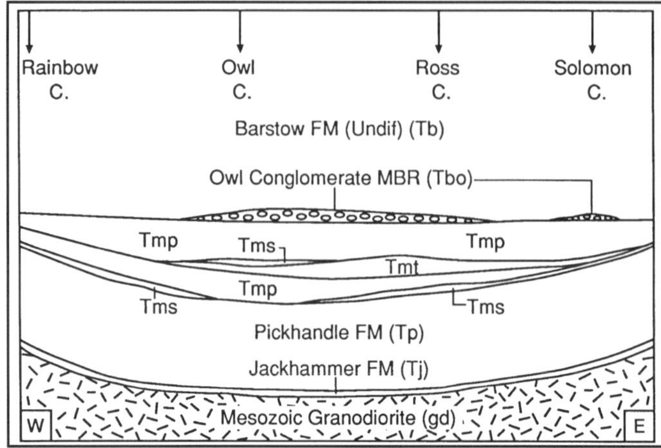

Figure 4. Schematic west-east cross section of the Mud Hills Formation and adjacent units, showing relative stratigraphic thicknesses. Curviplanar base and westward and eastward thinning of Mud Hills Formation may be controlled by synsedimentary north-dipping listric normal faults (e.g., fault-growth model of Schlische, 1991). See Figure 10 for present map pattern. Members of Mud Hills Formation: Tms = sandstone; Tmp = plutoniclastic breccia and megabreccia; and Tmt = rhyolite-tuff breccia.

Figure 5. Contact between Pickhandle (Tp) and Mud Hills (Tmp) Formations. (See Figures 2 and 10 for locations and relations). (A) View to east along contact, which dips to south at approximately 60°. W. A. Heins for scale at lower right. (B) Avalanche-groove casts (parallel to hammer handle) along base of Mud Hills Formation. Slight asymmetry along some grooves suggests south-to-north transport (see Fig. 16).

Owl Canyon (Figs. 3 and 4)." This interpretation seems to derive from a questionable correlation based on preliminary radiometric dating of their "Red Tuff" within the basal Barstow Formation (which we think may be a reworked tuffaceous sandstone).

The buttress unconformity beneath the Barstow Formation (Fig. 6) has, at the very least, 5 m of vertical relief, based on exposures in Owl and Solomon Canyons. The following contrasts are found across this unconformity: (1) There is a reversal of paleocurrent directions, with the underlying Mud Hills Formation derived from the south and the Owl Conglomerate derived from the north. (2) Mud Hills breccia is matrix supported and chaotic, whereas Owl Conglomerate consists of organized (well imbricated) alluvial deposits. (3) Mud Hills clasts are monolithologic, whereas Owl Conglomerate clasts are diverse, but derived primarily from the Pickhandle and Mud Hills Formations. (4) Basal Barstow Formation is sandstone, shale, and locally lacustrine limestone away from Owl Canyon, so that the unconformity is even more profound than illustrated in Figure 6; this variability of the basal Barstow Formation reflects topographic and depositional variability following cessation of normal faulting which produced the Mud Hills Formation. (5) Stratigraphic dip in the upper Mud Hills Formation averages near 40°, whereas basal Barstow dips are near 25°. All of these characteristics argue for a profound break in sedimentation between deposition of these two units.

The Mud Hills Formation consists of three facies (see Fig. 4): plutoniclastic breccia and megabreccia (Tmp), rhyolite-tuff breccia with plutoniclastic matrix (Tmt), and sandstone (Tms) (Fig. 8). All of these facies interfinger over short distances. The presence of any of the distinctive rhyolite-tuff breccia within a unit resulted in its being mapped as Tmt.

Tmt represents the middle of the Mud Hills Formation, separating two Tmp-Tms wedges. Exposures of the basal Tmp-Tms members on the north side of the imbricately faulted homocline

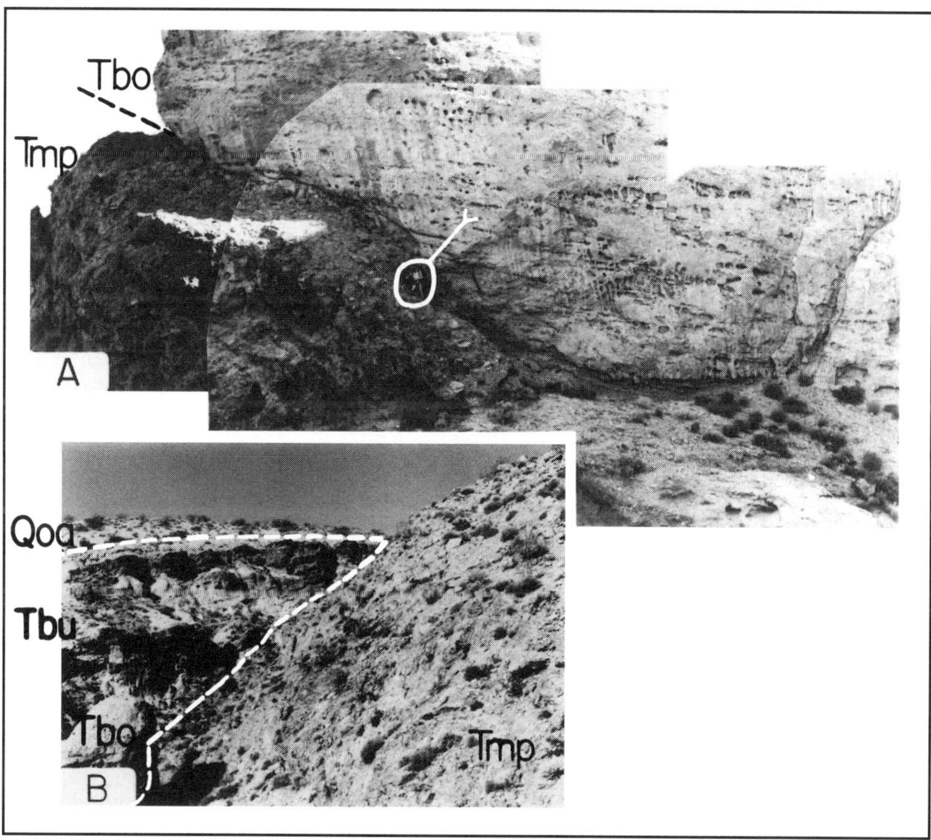

Figure 6. Buttress unconformity between Mud Hills and Barstow Formations. (See Figures 2 and 10 for locations and relations). (A) Looking east in Owl Canyon. Avalanche breccia of Mud Hills Formation (Tmp), dipping approximately 45° to south, overlain by Owl Conglomerate of Barstow Formation (Tbo), dipping approximately 25° to south. Monolithologic plutoniclastic breccia dominates Mud Hills Formation (below geologists [circled]); imbricated, well organized alluvial conglomerate dominates Owl Conglomerate (above geologists). (B) Looking west in Owl Canyon at Mud Hills breccia (Tmp) on right (dipping approximately 45° to south), overlain by Barstow Formation on left (dipping approximately 25° to south). Knob at lower left is Owl Conglomerate (Tbo), which is overlain by other facies of the Barstow Formation (Tbu), including dark limestone beds. These lacustrine limestone beds are overlain by Quaternary pediment gravel and eolian sand (Qoa). Upper limestone bed is approximately 2 m thick.

contain more distal parts of the depositional wedge, where Tms interfingers with Tmp. In fact, in the northeast corner of our map area (Solomon Canyon), Tmt lies directly on Tms (Fig. 4 and below). Our interpretation is that the Tmt avalanche flowed farther to the northeast than did the basal Tmp avalanche, whereas both upper and lower Tmp avalanches flowed farther to the northwest than did the Tmt avalanche. As a result, west of Owl Canyon, Tmt is not exposed (Fig. 4).

Exposures of the upper Tmp on the south side of the imbricated homocline contain more proximal parts of the depositional wedge; here, very coarse Tmp and Tmt surround one lens of Tms exposed in the deepest part of Owl Canyon (Fig. 9), in the position predicted by models for half-graben sedimentation (e.g., Leeder and Gawthorpe, 1987). These models predict that the finest sediment should be found closest to the fault scarp because this is where subsidence is fastest and sedimentary processes cannot keep pace with tectonic processes, except in the case of catastrophic rock avalanches (e.g., Yarnold and Lombard, 1989). Tms is also found in the southeast part of our map area, where it is along strike with, and stratigraphically just above, Tmt. This area was, presumably, also along the base of the fault scarp, but to the east of the depositional area of the Tmt avalanche, which was centered slightly west of Ross Canyon (based on maximum clast size of rhyolite-tuff breccia with diameter as much as 3 m and with maximum thickness).

The Mud Hills provide unique three-dimensional exposures of such relations owing to deep dissection of structurally repeated proximal and distal parts of the sedimentary wedges. These complex stratigraphic relations only make sense when the complex structural relations are discerned (and vice versa).

Barstow Formation

The Barstow Formation has been described extensively, most recently by MacFadden et al. (1990) and Woodburne et al. (1990). We discuss only aspects of the Barstow Formation which are directly relevant to structural development and basin evolution associated with the Waterman Hills detachment.

The basal unconformity is described in a preceding section. It is important to emphasize that the Owl Conglomerate Member is areally restricted on the north limb of the Barstow syncline (south side of our map area, see following discussion). In Owl Canyon, this member consists of 60 m of alternating alluvial conglomerate, sandstone, and mudrock (Fig. 3). Owl Conglomerate is not exposed west of Owl Canyon; it extends slightly east of Ross Canyon, but is absent on the divide between Ross and Solomon Canyons. Sporadic lenses of conglomerate occur near the base of the Barstow Formation in Solomon Canyon (Fig. 3); these are herein designated as Owl Conglomerate (Tbo), although they are finer grained and thinner bedded than exposures in Owl Canyon. Thus, the Owl Conglomerate is found primarily in the middle third of our map area; this area also contains the most deeply dissected exposures of the coarsest and thickest parts of the Mud Hills Formation (Fig. 4). This is not a fortuitous coincidence; rather, the Owl Conglomerate is probably thickest and coarsest where its source rocks also are thickest and coarsest. This relation is observed only where modern erosion has exposed deeper levels of the buttress unconformity (e.g., Owl Canyon; Fig. 6).

Most cross sections (e.g., Dibblee, 1967, 1968; and many reproductions and modifications thereof) show a thicker section of Owl Conglomerate on the north limb of the Barstow syncline than is warranted by our mapping. A thickness of 60 m is the maximum; in many locations, no Owl Conglomerate is found at the base of the Barstow Formation. In contrast, the Owl Conglomerate in the south limb of the Barstow syncline consists of more than 200 m of conglomerate, sandstone, mudrock, and tuffaceous deposits (Woodburne et al., 1990). We have identified clasts of the Waterman Gneiss derived from the south in this area (see discussion in following section). None of these clasts can be found in the Owl Conglomerate on the north limb. Thus, the northern Owl Conglomerate was derived from the north and the southern from the south, following cessation of extensional faulting.

STRUCTURE

The complex stratigraphic relations illustrated in Figure 4 are discerned directly from the map and cross sections of Figures 10 and 11. Virtually without exception, every south-dipping surface in the study area is stratigraphic and every north-dipping surface is a normal fault, commonly with listric character (Fig. 12). Vertical right-slip faults cut both these strata and the normal faults. Previous workers (e.g., Dibblee, 1967, 1968, and derivatives) tended to map all NW to SE–striking faults as belonging to this class of right-slip faults. Our mapping, however, shows that several of these NW to SE–striking faults are listric normal faults that splay intricately, especially in the middle of our map area between Owl and Ross Canyons (Figs. 2 and 10). Left separation in map view results from normal slip on several of these faults, especially along the east side of Owl Canyon. Recognition of the kinematics of these faults is the key to deciphering the stratigraphy.

Figure 7. Photo mosaics showing decreasing dips upsection within Mud Hills Formation in Owl Canyon. (See Figures 2 and 10 for locations and relations). (A) Looking west, with Pickhandle Formation (Tp) in upper right, overlain by Mud Hills plutoniclastic breccia (Tmp), rhyolite-tuff breccia (white) (Tmt), sandstone (extreme left, at bottom) (Tms) and more plutoniclastic breccia (Tmp). K. A. Devaney for scale in bottom of canyon (2 circles). Just to south (left) of northern Devaney is a coherent basement block, which is probably a single clast in the megabreccia, although interpreting this block as a protruding piece of attached basement is permissible. B. Looking east at plutoniclastic breccia (Tmp), overlain by rhyolite-tuff breccia (white) (Tmt), sandstone (Tms) and more plutoniclastic breccia (Tmp). Undifferentiated Barstow Formation (Tbu) is poorly exposed in background along upper right edge of figure. Prominent white outcrop near center of photo consists of approximately 4 m thick rhyolite-tuff breccia.

Tmp
Tp
Tmp
Tms
Tmt
A
Tmp

Tbu
Tmp
Tbu
Tmp
Tmt
Tms
B

Figure 8. Depositional facies of Mud Hills (A–D) and Barstow (E–F) Formations. (A) Plutoniclastic crackle breccia. (B) Crackle breccia consisting of plutoniclastic detritus on left and predominantly rhyolite-tuff detritus on right. (C) Matrix-supported debris-flow deposit. (D) Alluvial sandstone and conglomerate. (E) Inversely graded, hyperconcentrated-flow deposit at Ross Canyon, looking east. (F) Imbricated plutoniclastic conglomerate. Same location as E, looking west, paleocurrent to south (left) (see Fig. 16).

Figure 9. Looking west at sandstone wedge (Tms) within Owl Canyon. Maximum thickness of sandstone wedge is approximately 20 m in bottom of canyon (to bottom left). Sandstone wedge pinches out in distance to upper right where it onlaps rhyolite-tuff breccia (white) (Tmt). Plutoniclastic breccia (Tmp) overlies the former and underlies the latter. (See Figs. 2 and 10 for locations and relations).

The key geometric observation consistent with this interpretation is based on the three-dimensional nature of exposures. Almost without exception, stratigraphic contacts "v" southward and fault contacts "v" northward in the canyons (Fig. 13). The apparent interfingering of stratigraphic units, as shown by Dibblee (1967, 1968), Dokka et al. (1991), and Travis (1992), results from the structural repetition of stratigraphic units by these normal faults. The tabular nature of far-traveled pyroclastic-flow deposits (e.g., Tp) is incompatible with the type of stratigraphic architecture suggested by these other workers, whereas imbricate normal-fault repetition of units is consistent with the map patterns and depositional mechanisms for the stratigraphic units.

Three NW to SE–trending faults are likely right-slip faults that postdate extensional faulting and deposition of the Mud Hills Formation. Fault 1 (the Mud Hills fault of Travis, 1992) (see Figs. 10 and 11) exhibits right separation of Tj of approximately 1,200 m and offsets all Mesozoic and Miocene units. Fault 8 has right separation of the Tmp-Tbo contact of a few meters in Owl Canyon. The main fault of Rainbow Canyon (Figs. 2 and 10) cuts the Tmp-Tb contact with right separation, but exhibits left separation of fault 7 and the Tp-Tms-Tmp contacts. We suspect that this fault began as a northeast-dipping normal fault or faults that merged with and/or interacted with fault 7. Right slip on the Rainbow Canyon fault occurred, in part, by reactivation of these faults, after deposition of the Barstow Formation. The NW-trending faults west of the Rainbow Canyon fault probably have similar multistage histories, involving early normal slip, followed by right slip.

The north- and northeast-dipping normal faults, which are mappable because of repetition of the Tp-Tm contact and Tm members, extend into the Tp, especially in the upper Owl

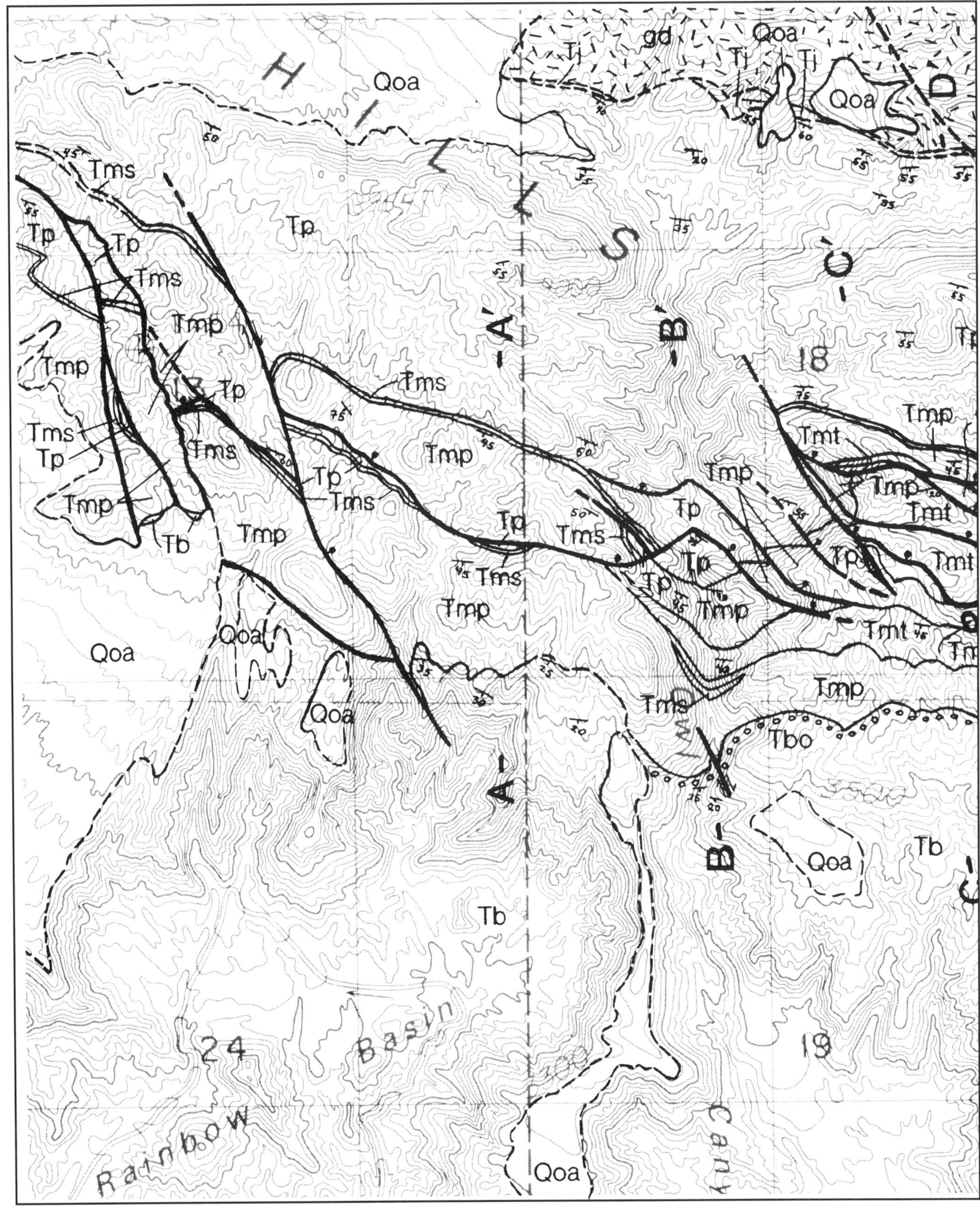

Figure 10. Geologic map of study area in Mud Hills. Topographic base from U.S. Geological Survey Mud Hills 1:24,000 Provisional Map (1988). Bedrock symbols explained in Figure 4. Qoa = Qua-

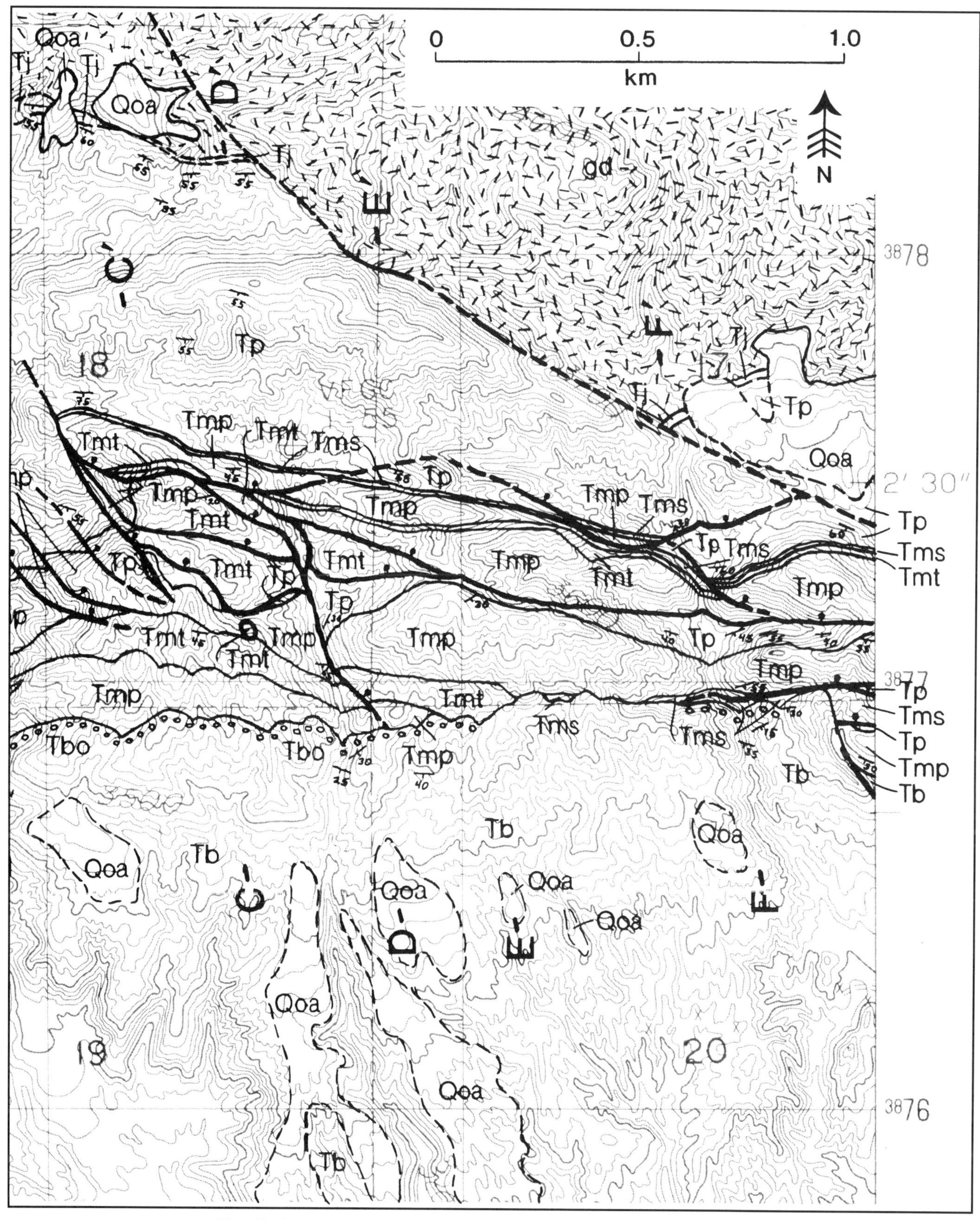

ternary older alluvium (pediment deposits); younger alluvium not shown. Locations of cross sections of Figure 11 are also indicated; see these cross sections for fault numbers referred to in text.

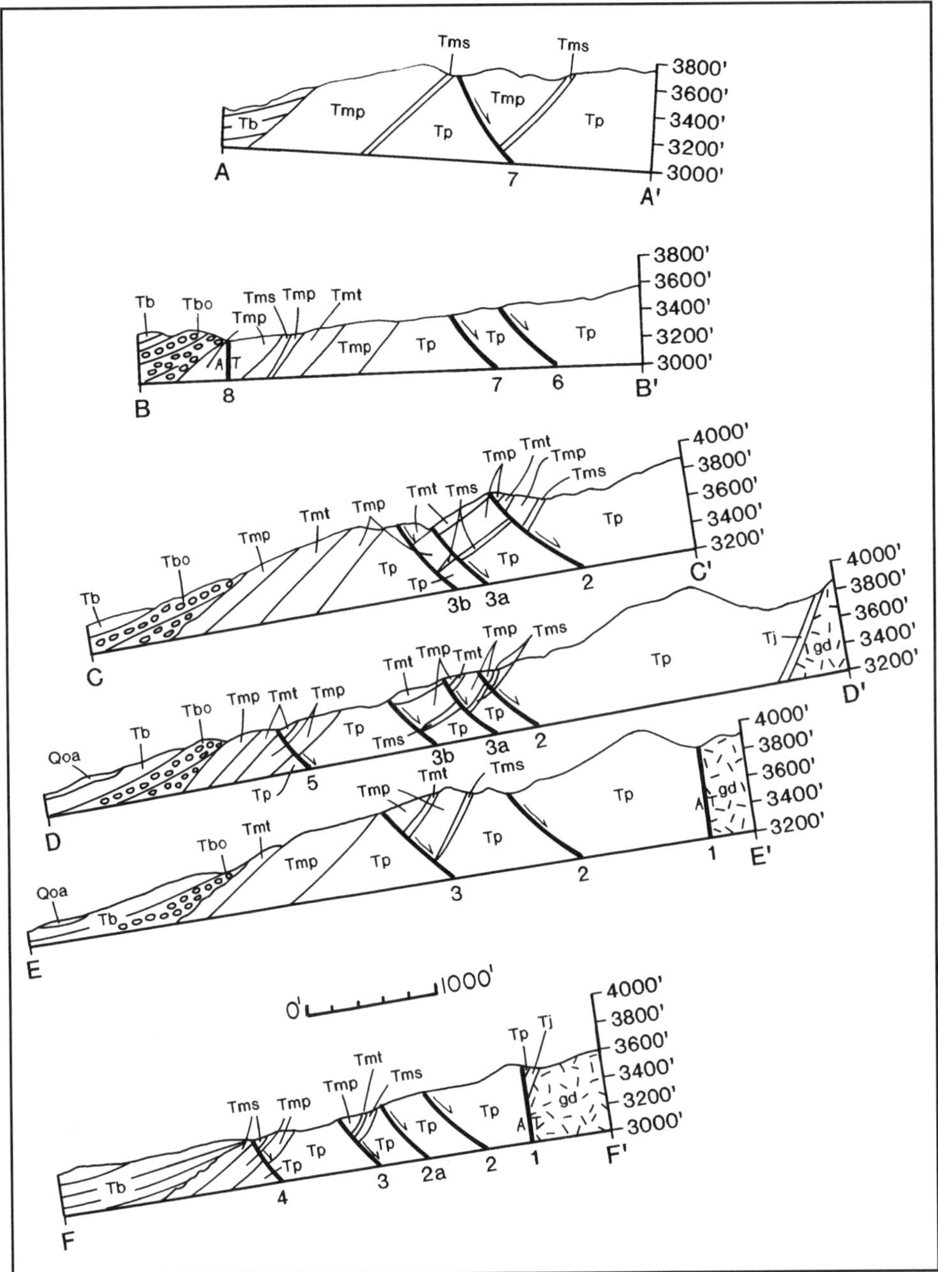

Figure 11. True-scale cross sections from Figure 10. Thin units (e.g., Tms) are slightly exaggerated for clarity. Basal Tms thins southward. Cross sections are oriented according to their positions on Figure 10. Fault numbers referred to in text are shown at bottom of each cross section.

Canyon area. In fact, this is the area where imbricate normal faulting is best expressed (Fig. 12); the ninefold repetition of a single tuffaceous unit is found here. Faults 2 and 3 (Figs. 10 and 11) merge toward the upper Owl Canyon area and diverge to the southeast of this area; we suspect that these faults splay in a complex way to the northwest of the end of cross section B–B′. The intricacy of fault repetition of Tp units in steep topography is difficult to convey even at the relatively large scale of our map (Fig. 10). Therefore, we do not extend fault 2-3 far into Tp.

South of our map area, on the south side of the Barstow syncline (east of the continuation of Ross Canyon) is a small exposure of andesite and related sediments in a z-fold (Dibblee, 1967, fig. 55). We agree with Dibblee's mapping of this as part of the Pickhandle Formation (excluding the Mud Hills Formation). Its presence beneath the Barstow Formation on the south limb of the Barstow syncline is consistent with the model for structural evolution described in a following section.

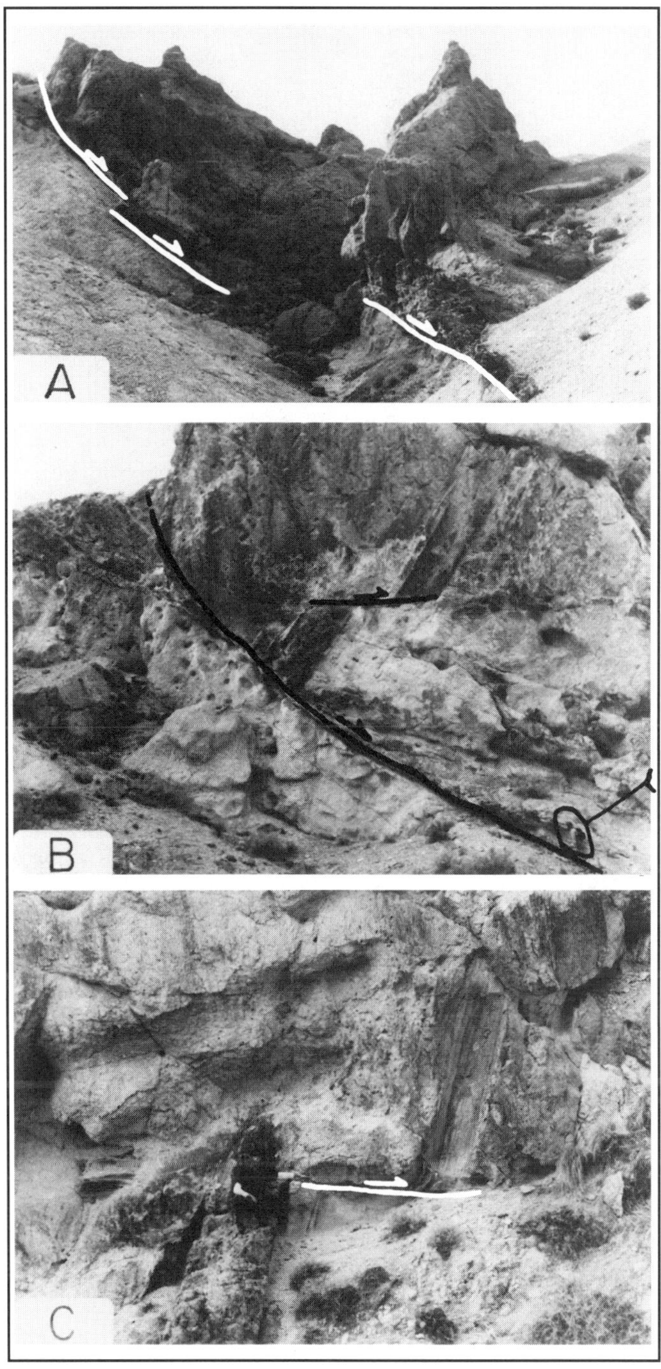

Figure 12. Normal faults in Pickhandle Formation, upper Owl Canyon. All views are to west, with normal faults dipping to north (right). (See Figs. 2 and 10 for locations and relations). (A) Wedge of dark tuff surrounded by light tuff. Left boundary (dipping to north) is fault; right boundary (dipping to south) is stratigraphic. (B) Listric normal fault dipping approximately 30° near J. K. Geslin [circled] at lower right; dip of fault is nearly 60° at upper left. (C) Nearly horizontal listric normal fault showing normal separation of stratigraphic unit. Location to the lower right of B.

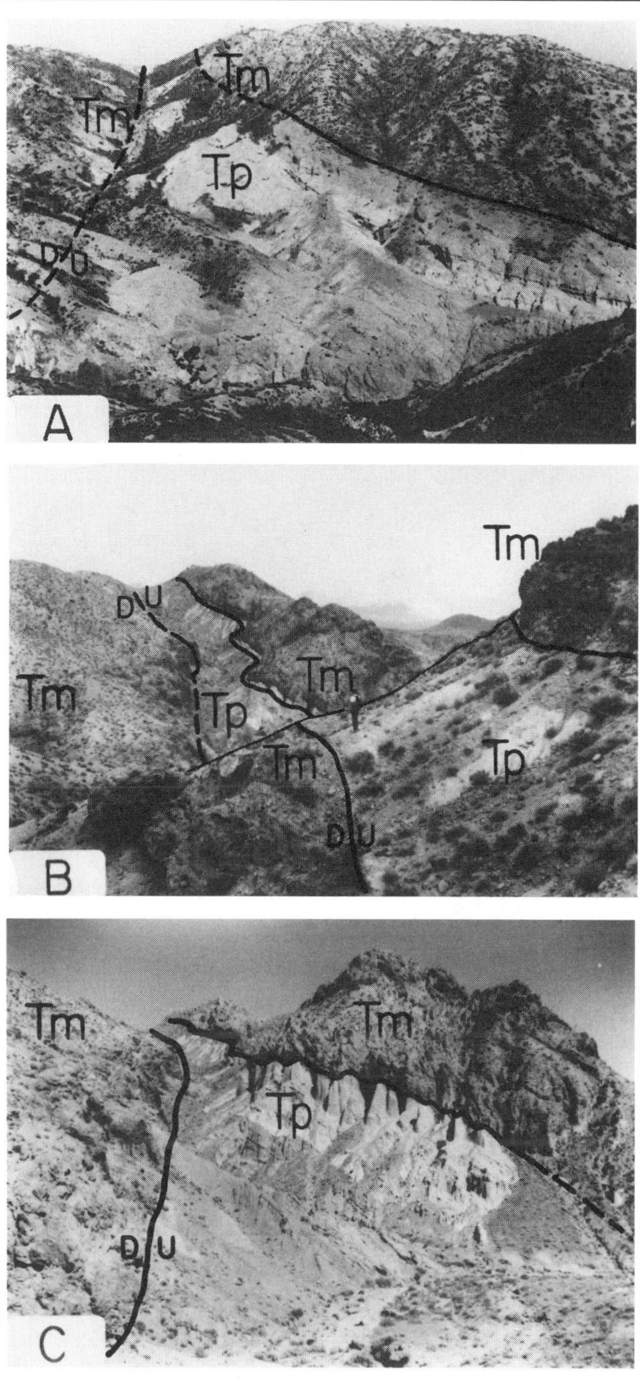

Figure 13. Views to east of Pickhandle (Tp)–Mud Hills (Tm) contact (dipping south), offset by north-dipping normal faults. (See Figs. 2 and 10 for locations and relations). (A) Middle of Ross Canyon. Light Tp overlain by darker and more vegetated Tm. Contact "vees" to south (right). Fault contact "vees" slightly to north on left side. Outcrop exposure of Tp is approximately 10 times wider in bottom of canyon than at distant saddle. (B) Looking into Solomon Canyon in distance. Light Tp overlain by darker Tm to right. J. K. Geslin is standing in foreground on normal fault, dipping to north, which drops Tm against Tp. Same fault can be traced into distance. Width of exposure of Tp increases approximately fivefold in bottom of Solomon Canyon. (C) Close-up of east side of Solomon Canyon. Photo taken from just beyond where J. K. Geslin is standing in B.

SEDIMENTOLOGY AND PETROLOGY

Depositional facies

All of the Tertiary units described herein are nonmarine. The Jackhammer Formation consists primarily of braided-stream deposits. The Pickhandle Formation consists primarily of lahars, pyroclastic-flow deposits, and braided-stream deposits (also, see Fillmore and Walker, this volume). A few fine-grained floodplain and lacustrine deposits also exist locally in both the Jackhammer and the Pickhandle Formations.

In contrast, the Mud Hills Formation consists primarily of rock-avalanche deposits (Tmp and Tmt), with local braided-stream deposits (Tms). The latter occur primarily either near the distal edges of Tmp and Tmt (i.e., northern and western exposures) or closest to the presumed fault scarp (i.e., in the deepest part of Owl Canyon) (Fig. 4). The rock-avalanche deposits exhibit all the characteristics described by Yarnold and Lombard (1989) and Yarnold (1993) (also, see Longwell, 1951; Burchfiel, 1966; Shreve, 1968; Travis, 1992), with the exception of substrate. The examples studied by Yarnold and Lombard (1989) involved emplacement on alluvial fans adjoining playas; as a result, their model includes significant deformation and mixing of poorly consolidated substrate, especially in distal settings. In contrast, the basal Mud Hills breccia was deposited primarily on well indurated Pickhandle volcaniclastic rocks; as a result, deformation of substrate was rare, although some mixing of underlying sediment likely occurred. The most proximal parts of this basal breccia (bottom of Owl Canyon and southernmost exposures in Ross and Solomon Canyons) rest directly on the Pickhandle Formation, whereas distal exposures (western and northern exposures) overly thin Tms. The best exposure of the proximal basal contact is in Owl Canyon (Fig. 5); here, north-south grooves in the base of the breccia indicate the direction of transport. The base of the breccia locally contains significant nonplutoniclastic detritus, probably reflecting incorporation and mixing of substrate. The absence of Tms beneath proximal parts of the basal breccia may reflect erosion and mixing of existing sand deposits into the rock avalanche. At some locations, especially west of Owl Canyon, flame-like structures up to several meters in height (Fig. 14) suggest loading and partial mixing of substrate into the basal avalanche.

The distinctive rhyolite-tuff breccia (Tmt) (Figs. 8b and 9) requires special explanation. The internal arrangement and textural character of clasts are identical to those of the plutoniclastic breccias; in fact, the only difference between Tmt and Tmp is the presence of white rhyolite-tuff clasts in the former and their absence in the latter. Thus, emplacement processes (i.e., rock avalanche) must have been identical. The enigmatic aspect of Tmt is the restriction of rhyolite-tuff clasts to one stratigraphic horizon; almost all other breccias in the Mud Hills Formation are monolithologic (granodiorite clasts). Our paleogeographic and palinspastic reconstruction includes synextensional eruption of rhyolite tuff to the south of the Mud Hills basin, which resulted in deposition on granodiorite at the top of the fault scarp (Fig. 15; 20.0 Ma). Collapse of this fault scarp then resulted in mixing of granodioritic and rhyolite-tuff breccia in a rock avalanche. Our attempt to date this rhyolite radiometrically has been futile, because it consists of quartz phenocrysts in a vitric matrix. We tentatively correlate it with scattered exposures of rhyolite in the Waterman Hills area (see Dibblee, 1968; Glazner et al., 1988, 1989a; Walker et al., 1990), which Glazner et al. (1994) have dated at 20 Ma.

Figure 14. Flame structure at base of plutoniclastic breccia (Tmp) at base of Mud Hills Formation. Flame consists of mudstone and sandstone. Underlying Pickhandle Formation (Tp) at lower left. Location in western part of map area. Hammer and backpack for scale.

Depositional facies of the Barstow Formation have been described by Link (1980) and Woodburne et al. (1990). The Barstow was deposited in lacustrine, fluvial, and alluvial-fan environments, following cessation of extension in the Mud Hills area. Our study concerns only the basal Barstow, primarily consisting of the Owl Conglomerate. Locally, sandstone and lacustrine limestone rest directly on Mud Hills and Pickhandle Formations, without intervening conglomerate. In fact, Owl Conglomerate is exposed in our study area only from the west side of Owl Canyon to the east side of Ross Canyon, and in Solomon Canyon (Fig. 10). These areas are currently the most deeply dissected; they also contain the most proximal parts of the Mud Hills breccia. This suggests that the Owl Conglomerate is an alluvial-fan deposit derived primarily from the coarsest parts of the older deposits. The Owl Conglomerate consists predominantly of well imbricated alluvial deposits derived from the north, and consisting primarily of recycled Mud Hills and Pickhandle detritus.

Paleocurrents

Paleocurrent measurements are difficult to obtain in any of the sub-Barstow stratigraphic units. Figure 16 summarizes paleocurrent measurements for the Mud Hills and Barstow Formations, including two locations from Travis (1992). The data set is robust for the Owl Conglomerate. It is moderately robust for the basal Mud Hills Formation (both Tmp and Tms), but no paleocurrent indicators could be found higher in the megabreccia and breccia units, in spite of much effort. Combining the paleocurrent data of Figure 16 with other sedimentologic, stratigraphic, and structural characteristics described herein, the following picture emerges. The Pickhandle Formation was deposited as widespread volcaniclastic aprons. The Mud Hills Formation was derived primarily from an uplifted fault block to the south, with local longitudinal east-west transport along the subsiding trough. Following cessation of extension, the uplifted Pickhandle and Mud Hills Formations were eroded and recycled into the Owl Conglomerate, with southerly transport on the north side of the Barstow basin. Northerly transport of sediment derived from the Waterman Hills occurred on the south side of the Barstow basin.

Petrology and provenance

Dibblee (1967, p. 42) described extensive quartz monzonite in the western Mojave Desert, but pointed out that "a slightly mafic facies, having composition of granodiorite" forms a transitional zone in several places, including the hills north of Barstow. Dibblee (1967, plate 1) included this granodiorite in his general map unit of "quartz monzonite," including the northern part of our map area (Fig. 10). We determined modal compositions of five nonmylonitic samples, both directly from nonmylonitic granitoid basement and from clasts in the Mud Hills Formation, as well as one sample of mylonitic granitoid from the Waterman Hills (lower plate) (Fig. 17). All six samples are within Streckeisen's (1976) granodiorite field. We conclude, therefore, that upper and lower plates of the Waterman Hills detachment fault both consist dominantly of "granodiorite" in the Mud Hills area, and that the primary (but not only) difference between the exposures is the presence of mylonitic fabric in the lower plate (e.g., Glazner et al., 1988; Bartley et al., 1990). In any case, Tmp consists of monolithologic breccia containing Mesozoic nonmylonitic granodiorite derived from the upper plate, whereas mylonitic granodiorite of indeterminate age is found only as clasts along the south limb of the Barstow syncline in the Barstow Formation.

Dibblee (1968) pointed out that the granitoid basement north of our study area contains quartz diorite and hornblende diorite, neither of which is found in the Mud Hills breccias. This observation and especially the lack of erosion of Tj and Tp overlying the basement eliminate the possibility of the Mud Hills Formation having been derived from the north.

Two types of sedimentary-provenance data were gathered: conglomerate clast types and sandstone petrography. Conglomerate data were gathered by placing a square grid on each measured outcrop; the lithology of 100 conglomerate clasts intersected by the nodal points of this grid was then determined. Identification of several clasts was refined through petrographic study. Figure 18 summarizes clast counts, both individually and grouped by formation and member. Table 1 summarizes the provenance of each unit.

Basement clasts and sand(stone) were point-counted, using the Gazzi-Dickinson method (Ingersoll et al., 1984), as adapted by Devaney and Ingersoll (1993) and Heins (1993). Thin sections were etched with hydrofluoric acid and stained for potassium feldspar (see Ingersoll and Cavazza, 1991); 500 points were counted on each section. The results are summarized in Figure 19. These petrographic results are consistent with the provenance conclusions based on conglomerate clasts. QFL compositions are bimodal. Basement (both Mesozoic nonmylonitic granodiorite from the Mud Hills and mylonitic granodiorite from the Waterman Hills) is quartzofeldspathic, whereas volcaniclastic sand(stone) (Tj, Tp, and Tmt) is lithic. Mud Hills sandstone (Tms) and Barstow sand(stone) (Tb) are bimodal, reflecting derivation from both of these older petrosomes. Subsidiary modes (e.g., QmKP, LmLvLs) show similar bimodality, with plutoniclastic sand(stone) higher in K and Lm, and volcaniclastic sand(stone) higher in P and Lv. Tmt has the highest QmKP%Qm of any unit.

An interesting sidelight is the observation that the Barstow Formation probably owes much of its varicolored character (e.g., Rainbow basin area) to inheritance of color from its source rocks. The Pickhandle Formation is predominantly green and the Mud Hills Formation is predominantly red, whereas the Barstow Formation is green, red, brown, and other intermediate colors. A test of this hypothesis would be interesting.

EVOLUTION OF THE MUD HILLS AREA

All data and observations presented herein are consistent with the sequential palinspastic reconstruction shown in Figure 15. Key timing constraints include the following: (1) eruption of andesite and rhyodacite found in the Jackhammer and Pickhandle Formations occurred at 24–21 Ma (Burke et al., 1982; Woodburne et al., 1982; Fillmore and Walker, this volume); (2) rhyolite in the upper plate of Waterman Hills erupted at 20 Ma (Glazner et al., 1994); (3) basalt near the base of the Barstow Formation erupted at 16.5 Ma (Burke et al., 1982). All other times shown in Figure 15 are interpolated at arbitrary equal intervals. The sequence of events is well constrained by the structure and stratigraphy; the quantitative ages are poorly constrained in detail.

We illustrate the initiation of extensional faulting in the Mud Hills area as occurring between 23 Ma and 22 Ma (Fig. 15). In detail, this is poorly constrained, and it is impossible to say, based on our study, whether extension had already begun in surrounding areas, and/or whether the Waterman Hills detachment was only one strand of a complex detachment system (e.g., Davis and Lister, 1988). Our palinspastic reconstruction utilizes the rolling-hinge model for detachment evolution (e.g., Buck, 1988; Hamilton, 1988; Wernicke and Axen, 1988), although our data are insufficient to rule out other modes of detachment evolution.

Early in the evolution of the Waterman Hills detachment, a supradetachment basin formed and was quickly dismembered

Figure 15 (on this and facing page). Sequential development of Waterman Hills detachment and Mud Hills area (see Fig. 1 for location). Palinspastic reconstruction is at true scale, with right side (north) held fixed. Some thin units (e.g., distal Tms) are omitted because of scale. Topography and geologic cross section of study area in Mud Hills in modern cross section (bottom right) are well constrained (e.g., cross sections of Fig. 11); other aspects of reconstruction are speculative, but consistent with all observations and data. Arched detachment fault after Walker et al. (1990). Two right-slip vertical faults are shown for modern cross section, after Dibblee (1967, fig. 55). Ages are interpolated between well-constrained times. Extension may have begun earlier than shown in upper left (e.g., Fillmore and Walker, this volume); 21 Ma is latest possible time in Mud Hills area. Symbols: gd = Mesozoic granodiorite; Tj = Jackhammer Fm.; Tp = Pickhandle Formation; Tmp = plutoniclastic breccia of Mud Hills Formation; Tmt = rhyolite-tuff breccia of Mud Hills Formation; Tms = sandstone of Mud Hills Formation; Tbo = Owl Canyon Conglomerate Member of Barstow Formation; Tb = other members of Barstow Formation; Tr = rhyolite of Waterman Hills; Tw = sedimentary strata of Waterman Hills; Q = Quaternary units. Long-dashed lines indicate mylonitic gneiss of lower plate of Waterman Hills detachment. Short-dashed lines indicate inferred eroded detachment fault.

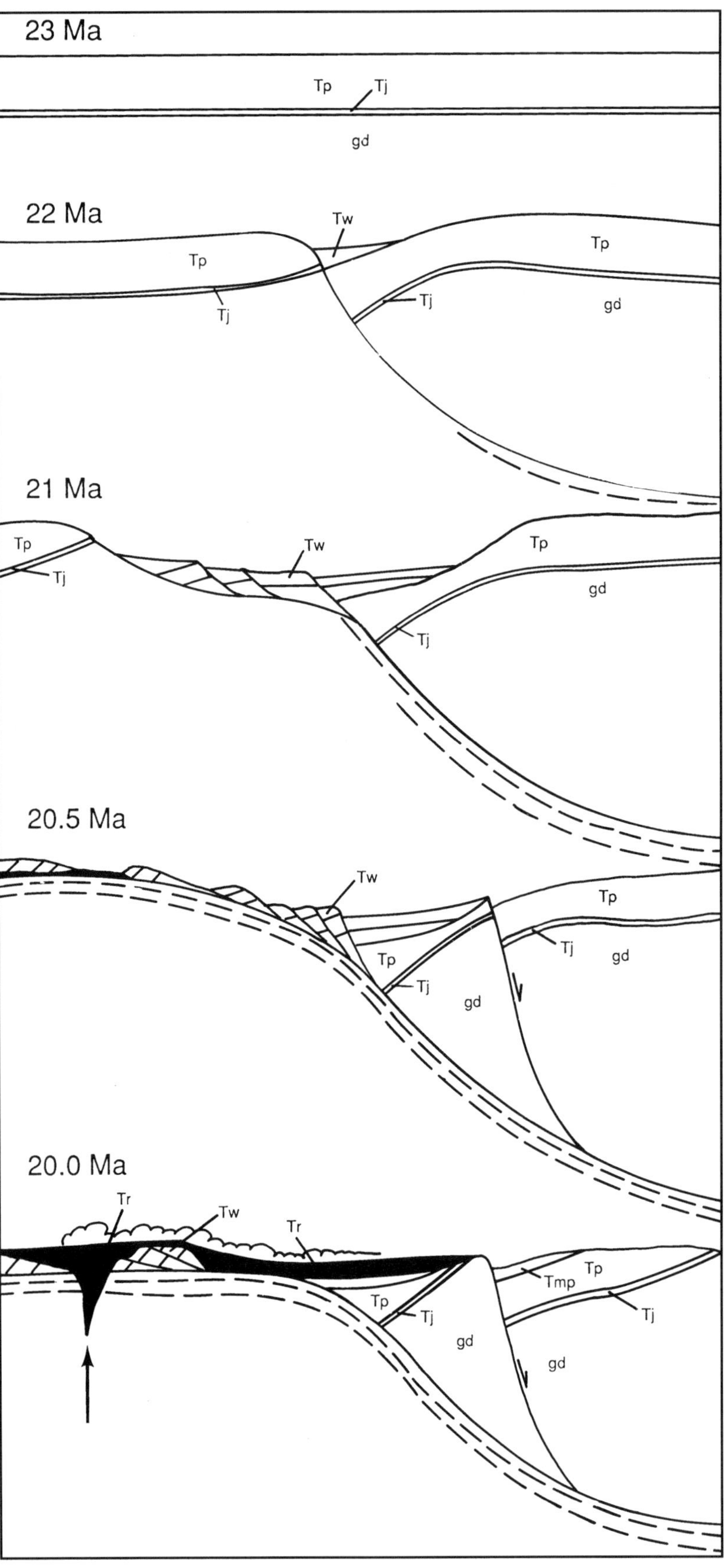

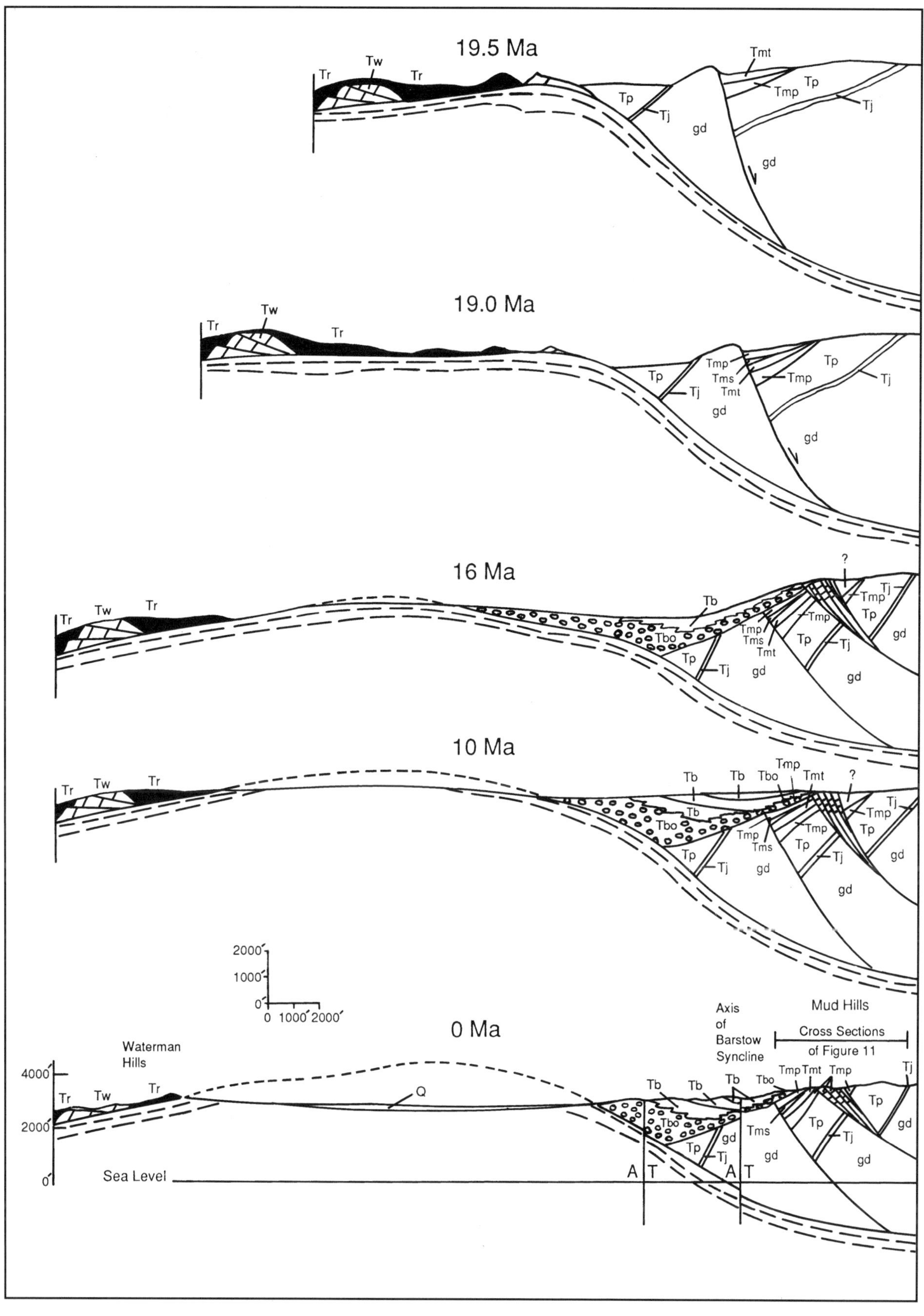
19.5 Ma
Tr
Tw
Tr
Tmt
Tp
Tmp
Tp
Tj
Tj
gd
gd
19.0 Ma
Tr
Tw
Tr
Tp
Tj
Tmp
Tms
Tmt
gd
Tp
Tmp
Tj
gd
16 Ma
Tr
Tw
Tr
Tb
Tbo
Tmp
Tms
Tmt
Tp
Tj
gd
Tmp
Tp
Tj
gd
10 Ma
Tr
Tw
Tr
Tb
Tb
Tbo
Tmp
Tmt
Tbo
Tms
Tp
Tj
gd
2000′
1000′
0′
0 1000′ 2000′
0 Ma
Waterman Hills
4000′
2000′
0′
Tr
Tw
Tr
Q
Sea Level
Axis of Barstow Syncline
Mud Hills
Cross Sections of Figure 11
Tb
Tb
Tb
Tbo
Tmp
Tmt
Tmp
Tj
Tbo
Tp
Tj
gd
Tms
gd
Tp
Tp
Tj
gd
gd
A
T
A
T

(21–20 Ma). Resulting strata are preserved on the south side of the Waterman Hills in the form of a highly faulted syncline, plunging steeply to the south (Walker et al., 1990). This section is intruded by rhyolite, which is intensely potassium-metasomatized (Walker et al., 1990). Based on our reconnaissance, these strata are distinct from either the volcaniclastic Pickhandle or the plutoniclastic Mud Hills Formations; our model shown in Figure 15 predicts that these strata were deposited after the Pickhandle Formation and before the Mud Hills Formation. They were deformed immediately prior to, during, and following eruption and intrusion of the rhyolite. (See Fillmore and Walker, this volume, for an alternative interpretation.)

Just prior to eruption of the rhyolite (20.0 Ma; Fig. 15), an upper-plate listric normal fault initiated down-to-the-north dropping of a half graben, in which the basal Mud Hills breccia was deposited. This breccia was derived from Mesozoic granodiorite uplifted in the footwall of this fault; most of the overlying Pickhandle Formation had previously been eroded from the footwall block. Following eruption of rhyolite at 20 Ma, Tmt was deposited as a mixture of granodiorite and neovolcanic (i.e., Zuffa, 1985) rhyolite-tuff breccia as motion continued along the listric normal fault. Tms in Owl Canyon and the overlying Tmp were deposited subsequently (19.0 Ma; Fig. 15), as motion continued on both the Waterman Hills detachment and the upper-plate fault. Late-stage listric normal faulting repeated the section in the study area and rotated the strata further. Deposition of basal Barstow probably was coincident with movement along these faults.

By 16 Ma, extension had ceased, and the residual basin south of the Mud Hills began filling with the Barstow Formation. Isostatic adjustments elevated both the exposed lower plate and the extended upper plate, resulting in detritus derived from both north and south. Strike-slip motion and related folding after 10 Ma further disrupted the area and formed the Barstow syncline.

CONCLUSIONS

This study illustrates how integration of structural, stratigraphic, and sedimentologic information is essential for correct reconstruction of highly extended terranes. Previous workers had misinterpreted faults as stratigraphic contacts; their resulting pseudostratigraphy resulted in errors in paleogeographic and

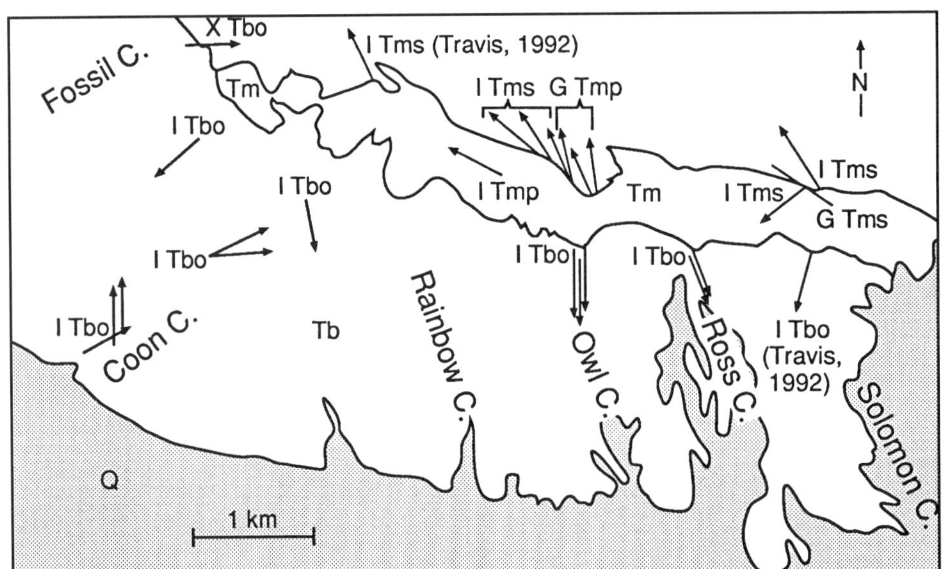

Figure 16. Paleocurrent measurements. Base map (after Dibblee, 1967, fig. 55) shows generalized outcrop area of Mud Hills Formation (Tm), Barstow Formation (Tb), and Quaternary deposits (Q). Area northeast of Tm is underlain by Mesozoic granodiorite and Miocene Jackhammer and Pickhandle Formations (see Fig. 10 for details). Measurements are of three types: gravel imbrications (I), basal grooves (G), and cross beds (X). Shown at each location is paleocurrent orientation, restored to horizontal by rotation around strike of bedding; tail of each arrow is at measured location. All imbrication arrows show mean values based on measurements of from 5 to 12 clasts. Grooves and crossbed orientations represent individual measurements. Most grooves show some sense of asymmetry or are intimately related to imbrication measurements, thus allowing azimuth to be determined; one exception is the groove measured in basal Tms near the head of Solomon Canyon (no arrow shown here). Two imbrication measurements are included from Travis (1992, fig. 3.15) for locations not included in our study. Some of his other data duplicate our results.

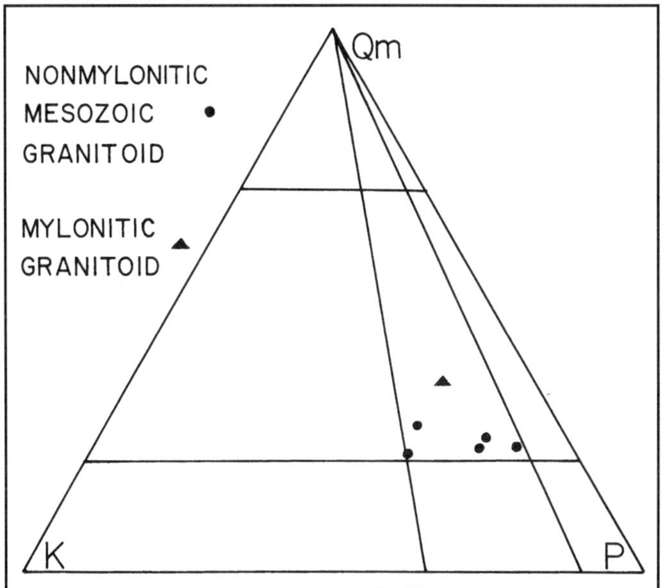

Figure 17. Modal composition of basement from Mud Hills and Waterman Hills, as plotted on modified Streckeisen (1976) triangle. Symbols: Qm = monocrystalline quartz; K = potassium feldspar; P = plagioclase feldspar. Each sample represents 500 points, determined by method of Heins (1993). All samples lie within granodiorite field.

palinspastic reconstruction. Even though the overall stratigraphy is simpler than these workers have proposed (e.g., we conclude that Tm is everywhere younger than Tp), facies relations within the study area are exceedingly complex within Tm as a result of synextensional tilting and sedimentation (Fig. 4). Deciphering this complexity was possible only through several iterations of stratigraphic and sedimentologic analysis conducted at the same time as structural mapping.

Our experience with this small part of the Mojave Desert persuades us to be skeptical about many facile interpretations derived from cursory examination of other highly extended terranes. Even though our study area is easily accessible and excellently exposed, previous models failed to appreciate many aspects of the area. We hope that our work will inspire others to undertake similar integrated studies of complex extensional terranes.

ACKNOWLEDGMENTS

Financial assistance was provided by the Committee on Research of the Academic Senate of the Los Angeles Division of the University of California.

We thank P. A. Craig, R. J. Hill, E. A. Large, P. E. Rumelhart, and An Yin for assistance in the field.

We thank K. K. Beratan, A. F. Glazner, J. D. Walker, and An Yin for reviewing the manuscript.

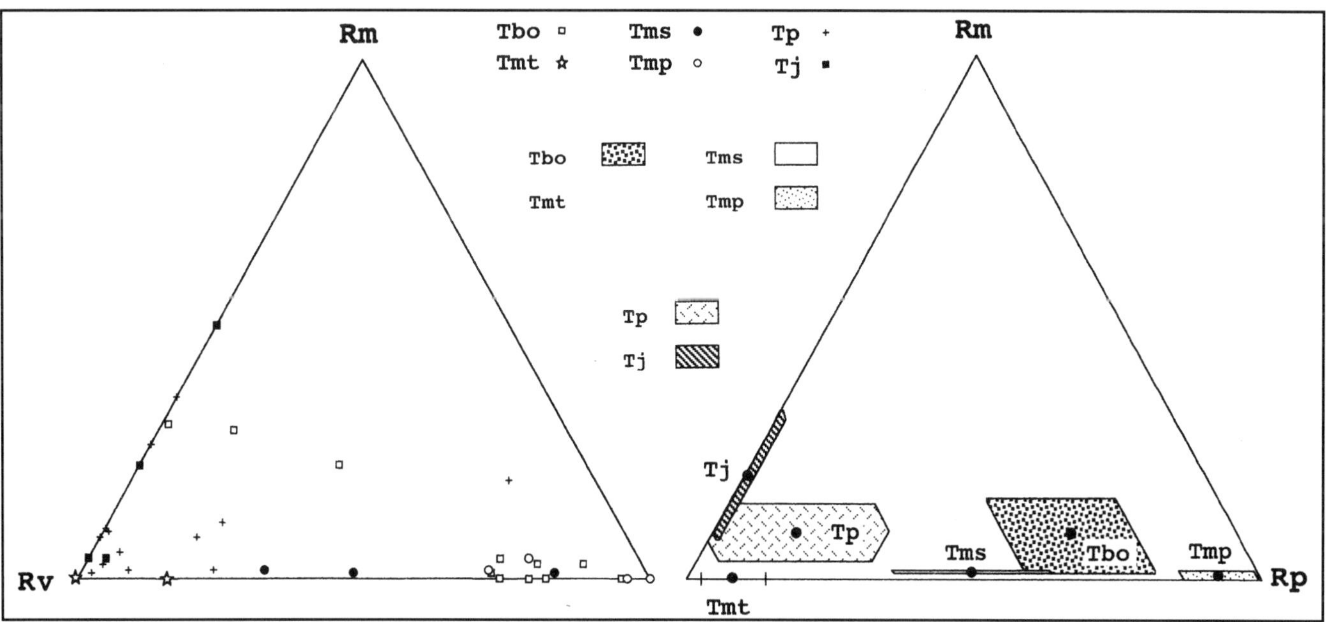

Figure 18. Conglomerate petrology. Each data point represents approximately 100 counts on a square grid at individual sites. Most clasts were identified in field; ambiguous clasts were thin-sectioned and identified petrographically. Symbols: Rm = metamorphic rocks; Rv = volcanic rocks; Rp = plutonic rocks. Individual counts on left; means and fields of variation on right. See text for discussion.

TABLE 1. PROVENANCE OF TERTIARY UNITS

Unit	Types of Source Rocks	Interpretation
Tj	Volcanic and metamorphic rocks, not locally exposed.	Many of the volcanic clasts may be paleovolcanic and metavolcanic; granodiorite basement not exposed; distant source(s).
Tp	Primarily volcanic rocks; some of the same metamorphic rocks as Jackhammer; locally some granodiorite	Most of the volcanic rocks were newly erupted (interbedded neovolcanic); one granodiorite-rich conglomerate skews field away from the Rv apex; distant source(s).
Tmp	Almost 100% Mesozoic granodiorite; locally some volcanic rock.	Primarily monolithologic rock avalanches; nearby source.
Tmt	Almost 100% rhyolite tuff, locally with variable amounts of granodiorite.	Monolithologic nature and large clast size (as much as several meters) argue for neovolcanic origin; Tmt is intimately intermixed with Tmp; nearby source.
Tms	Mixture of Tp and Tmp clasts, primarily volcanic and granodiorite.	Local recycling possible; nearby source(s).
Tbo	Mixture of all older units, primarily Tm, Tp, Tj, in descending order of abundance.	Recycling of all older units, with greatest contribution from youngest of these; nearby source(s).

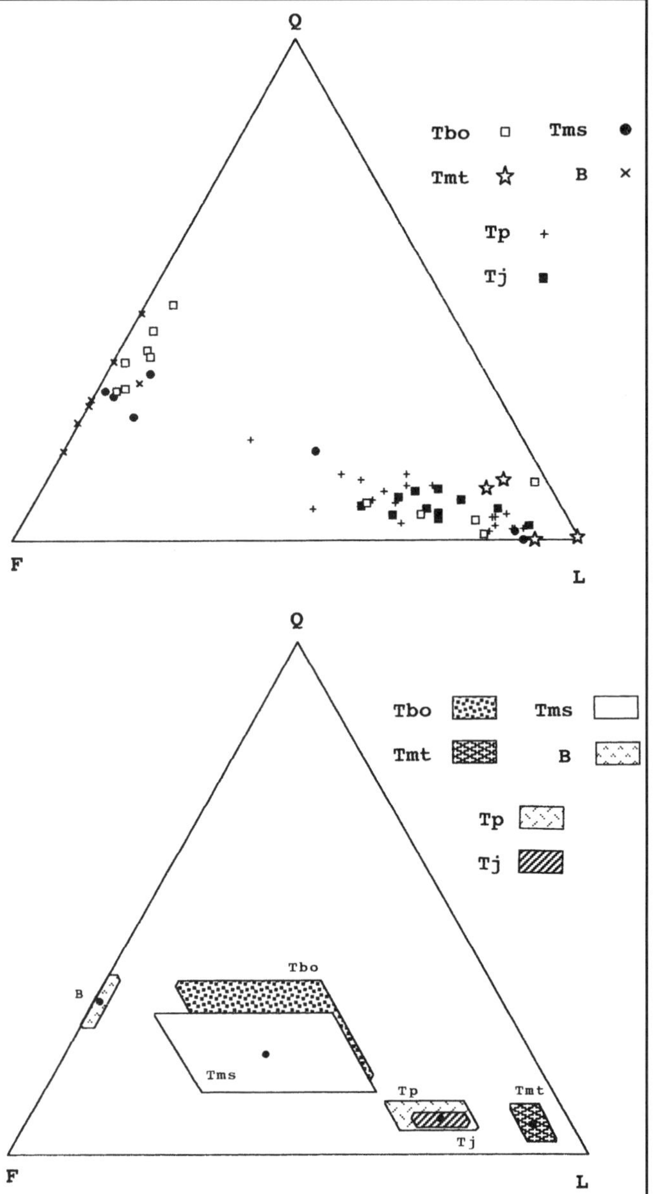

Figure 19. Sandstone petrography. Each data point represents approximately 500 counts in thin section, using the Gazzi-Dickinson method (i.e., Ingersoll et al., 1984). Q = monocrystalline and polycrystalline quartz; F = potassium and plagioclase feldspar; L = nonquartzose aphanitic lithic fragments. Individual counts on top; means and fields of variation on bottom. See text for discussion.

REFERENCES CITED

Bartley, J. M., Fletcher, J. M., and Glazner, A. F., 1990, Tertiary extension and contraction of lower-plate rocks in the central Mojave metamorphic core complex, southern California: Tectonics, v. 9, p. 521–534.

Bowen, O. E., 1954, Geology and mineral deposits of Barstow quadrangle, San Bernardino County, California: California Division of Mines and Geology Bulletin 165, 208 p.

Buck, W. R., 1988, Flexural rotation of normal faults: Tectonics, v. 7, p. 959–973.

Burchfiel, B. C., 1966, Tin Mountain landslide, southeastern California, and the origin of megabreccia: Geological Society of America Bulletin, v. 77, p. 95–99.

Burke, D. B., Hillhouse, J. W., McKee, E. H., Miller, S. T., and Morton, J. L., 1982, Cenozoic rocks in the Barstow basin area of southern California—Stratigraphic relations, radiometric ages, and paleomagnetism: U.S. Geological Survey Bulletin 1529-E, p. E1–E16.

Davis, G. A., and Lister, G. S., 1988, Detachment faulting in continental extension: perspectives from the southwestern U. S. Cordillera, *in* Clark, S. F., Jr., ed., Processes in continental lithospheric deformation: Boulder, Colorado, Geological Society of America Special Paper 218, p. 133–159.

Devaney, K. A., and Ingersoll, R. V., 1993, Provenance evolution of upper Paleozoic sandstones of north-central New Mexico, *in* Johnsson, M. J., and Basu, A., eds., Processes controlling the composition of clastic sediments: Boulder, Colorado, Geological Society of America Special Paper 284, p. 91–108.

Dibblee, T. W., Jr., 1967, Areal geology of the western Mojave Desert, California: U.S. Geological Survey Professional Paper 522, 153 p.

Dibblee, T. W., Jr., 1968, Geology of the Fremont Peak and Opal Mountain quadrangles, California: California Division of Mines and Geology Bulletin 188, 64 p.

Dibblee, T. W., Jr., 1980a, Cenozoic rock units of the Mojave Desert, *in* Fife, D. L., and Brown, A. R., eds., Geology and mineral wealth of the California Desert [Dibblee Volume]: Santa Ana, California, South Coast Geological Society, p. 41–68.

Dibblee, T. W., Jr., 1980b, Geologic structure of the Mojave Desert, *in* Fife, D. L., and Brown, A. R., eds., Geology and mineral wealth of the California Desert [Dibblee Volume]: Santa Ana, California, South Coast Geological Society, p. 69–100.

Dibblee, T. W., Jr., 1980c, Pre-Cenozoic rock units of the Mojave Desert, *in* Fife, D. L., and Brown, A. R., eds., Geology and mineral wealth of the California Desert [Dibblee Volume]: Santa Ana, California, South Coast Geological Society, p. 13–40.

Dokka, R. K., 1986, Patterns and modes of early Miocene crustal extension, central Mojave Desert, California: Boulder, Colorado, Geological Society of America Special Paper 208, p. 75–95.

Dokka, R. K., 1989a, Magnitude and significance of Miocene crustal extension in the central Mojave Desert, California: Comment: Geology, v. 17, p. 1061.

Dokka, R. K., 1989b, The Mojave extensional belt of southern California: Tectonics, v. 8, p. 363-390. (Also, see Correction, v. 8, p. 937.)

Dokka, R. K., and Woodburne, M. O., 1986, Mid-Tertiary extensional tectonics and sedimentation, central Mojave Desert, California: Louisiana State University Publications in Geology and Geophysics 1, 55 p.

Dokka, R. K., McCurry, M., Woodburne, M. O., Frost, E. G., and Okaya, D. A., 1988, A field guide to the Cenozoic crustal structure of the Mojave Desert, *in* Weide, D. L., and Faber, M. L., eds., This extended land: geological journeys in the southern Basin and Range: Las Vegas, University of Nevada, Department of Geoscience Special Publication 2, p. 21–44.

Dokka, R. K., and 10 others, 1991, Aspects of the Mesozoic and Cenozoic geologic evolution of the Mojave Desert, *in* Walawender, M. J., and Hanan, B. B., eds., Geological excursions in southern California and Mexico, Geological Society of America, 1991 Annual Meeting, Guidebook: San Diego, California, San Diego State University Guidebook, p. 1–43.

Durrell, C., 1953, Geological investigations of strontium deposits in southern California: California Division of Mines and Geology Special Report 32, 48 p.

Fillmore, R. P., and Walker, J. D., 1993, Three extensional basin types associated with detachment-style faulting, early Miocene of the central Mojave Desert: Geological Society of America Abstracts with Programs, v. 25, no. 5, p. 37.

Glazner, A. F., Bartley, J. M., and Walker, J. D., 1988, Geology of the Waterman Hills detachment fault, central Mojave Desert, California, *in* Weide, D. L., and Faber, M. L., eds., This extended land: Geological journeys in the southern Basin and Range: Las Vegas, University of Nevada, Department of Geoscience Special Publication 2, p. 225–237.

Glazner, A. F., Bartley, J. M., and Walker, J. D., 1989a, Magnitude and significance of Miocene crustal extension in the central Mojave Desert, California: Geology, v. 17, p. 50–53.

Glazner, A. F., Bartley, J. M., and Walker, J. D., 1989b, Magnitude and significance of Miocene crustal extension in the central Mojave Desert, California: Reply: Geology, v. 17, p. 1061–1062.

Glazner, A. F., and 9 others, 1994, Reconstruction of the Mojave block, *in* McGill, S. F., and Ross, T. M., eds., Geological investigations of an active margin: Redlands, California, San Bernardino County Museum Association, p. 3–30.

Hamilton, W. B., 1988, Detachment faulting in the Death Valley region, California and Nevada: U.S. Geological Survey Bulletin 1790, p. 51–85.

Heins, W. A., 1993, Source rock texture versus climate and topography as controls on the composition of modern, plutoniclastic sand, *in* Johnsson, M. J., and Basu, A. eds., Processes controlling the composition of clastic sediments: Boulder, Colorado, Geological Society of America Special Paper 284, p. 135–146.

Hewett, D. F., 1954, General geology of the Mojave Desert region, California: California Division of Mines and Geology Bulletin 170, p. 5–20.

Ingersoll, R. V., and Cavazza, W., 1991, Reconstruction of Oligo-Miocene volcaniclastic dispersal patterns in north-central New Mexico using sandstone petrofacies: Tulsa, Oklahoma, SEPM Special Publication 45, p. 227–236.

Ingersoll, R. V., and Diamond, D. S., 1990, The dangers and uses of pseudostratigraphy in the reconstruction of highly extended terranes: Geological Society of America Abstracts with Programs, v. 22, no. 7, p. A227.

Ingersoll, R. V., Bullard, T. F., Ford, R. L., Grimm, J. P., Pickle, J. D., and Sares, S. W., 1984, The effect of grain size on detrital modes: A test of the Gazzi-Dickinson point-counting method: Journal of Sedimentary Petrology, v. 54, p. 103–116.

Ingersoll, R. V., and 8 others, 1993, The Mud Hills, Mojave Desert, California: Structure, stratigraphy and sedimentology of a rapidly extended terrane: Geological Society of America Abstracts with Programs, v. 25, no. 5, p. 56.

Leeder, M. R., and Gawthorpe, R. L., 1987, Sedimentary models for extensional tilt-block/half-graben basins: London, Geological Society of London Special Publication 28, p. 139–152.

Link, M. H., 1980, Sedimentary facies and mineral deposits of the Miocene Barstow Formation, California, *in* Fife, D. L., and Brown, A. R., eds., Geology and mineral wealth of the California Desert [Dibblee Volume]: Santa Ana, California, South Coast Geological Society, p. 191–203.

Longwell, C. R., 1951, Megabreccia developed downslope from large faults: American Journal of Science, v. 249, p. 343–355.

MacFadden, B. J., Swisher, C. C., III, Opdyke, N. D., and Woodburne, M. O., 1990, Paleomagnetism, geochronology, and possible tectonic rotation of the middle Miocene Barstow Formation, Mojave Desert, southern California: Geological Society of America Bulletin, v. 102, p. 478–493.

Martin, M. W., and Walker, J. D., 1991, Upper Precambrian to Paleozoic paleogeographic reconstruction of the Mojave Desert, California, *in* Cooper, J. D., and Stevens, C. H., eds., Paleozoic paleogeography of the western United States—II: Los Angeles, California, SEPM, Pacific Section, p. 167–191.

Martin, M. W., and Walker, J. D., 1992, Extending the western North American Proterozoic and Paleozoic continental crust through the Mojave Desert: Geology, v. 20, p. 753–756.

McCulloh, T. H., 1952, Geology of the southern half of the Lane Mountain quadrangle, California [Ph.D. thesis]: Los Angeles, California, University of California, 182 p.

Merriam, J. C., 1919, Tertiary mammalian faunas of the Mohave Desert: University of California Bulletin of the Department of Geology, v. 11, p. 437–585.

Miller, E. L., 1981, Geology of the Victorville region, California: Geological Society of America Bulletin, v. 92, Part 1, p. 160-163, Part 2, p. 554–608.

Schlische, R. W., 1991, Half-graben basin filling models: New constraints on continental extensional basin development: Basin Research, v. 3, p. 123–141.

Shreve, R. L., 1968, The Blackhawk landslide: Boulder, Colorado, Geological Society of America Special Paper 108, 47 p.

Streckeisen, A. L., 1976, To each plutonic rock its proper name: Earth-Science Reviews, v. 12, p. 1–33.

Travis, C. J., 1992, Stratigraphic architecture of an extensional orogen: the Mojave extensional belt, southern California [Ph.D. thesis]: Baton Rouge, Louisiana, Louisiana State University, 353 p.

Travis, C. J., Dokka, R. K., and Ross, T. M., 1990, Stratigraphy and tectonic significance of early Miocene coarse-grained sediments in the Mojave extensional belt, Mud Hills, California: Geological Society of America Abstracts with Programs, v. 22, no. 3, p. 89.

U.S. Geological Survey, 1988, Mud Hills Quadrangle California—San Bernardino County 7.5 minute series (topographic) [provisional edition 1988]: Denver, Colorado, U.S. Geological Survey, scale 1:24,000.

Walker, J. D., 1988, Permian and Triassic rocks of the Mojave Desert and their implications for timing and mechanisms of continental truncation: Tectonics, v. 7, p. 685–709.

Walker, J. D., Bartley, J. M., and Glazner, A. F., 1990, Large-magnitude Miocene extension in the central Mojave Desert: Implications for Paleozoic to Tertiary paleogeography and tectonics: Journal of Geophysical Research, v. 95, p. 557–569.

Wernicke, B., 1985, Uniform-sense normal simple shear of the continental lithosphere: Canadian Journal of Earth Sciences, v. 22, p. 108–125.

Wernicke, B., and Axen, G. J., 1988, On the role of isostasy in the evolution of normal fault systems: Geology, v. 16, p. 848–851. (Also, see Comment and Reply, v. 17, p. 774–776.)

Woodburne, M. O., Miller, S. T., and Tedford, R. H., 1982, Stratigraphy and geochronology of Miocene strata in the central Mojave Desert, California, *in* Geologic excursions in the California desert: Geological Society of America, Cordilleran Section, Annual Meeting, 78th, Anaheim, Volume and Guidebook, p. 47–64.

Woodburne, M. O., Tedford, R. H., and Swisher, C. C., III, 1990, Lithostratigraphy, biostratigraphy, and geochronology of the Barstow Formation, Mojave Desert, southern California: Geological Society of America Bulletin, v. 102, p. 459–477.

Yarnold, J. C., 1993, Rock-avalanche characteristics in dry climates and the effect of flow into lakes: Insights from mid-Tertiary sedimentary breccias near Artillery Peak, Arizona: Geological Society of America Bulletin, v. 105, p. 345–360.

Yarnold, J. C., and Lombard, J. P., 1989, A facies model for large rock-avalanche deposits formed in dry climates, *in* Colburn, I. P., Abbott, P. L., and Minch, J., eds., Conglomerates in basin analysis: a symposium dedicated to A. O. Woodford: Los Angeles, California, Society of Economic Paleontologists and Mineralogists, Pacific Section, p. 9–31.

Zuffa, G. G., 1985, Optical analyses of arenites: Influence of methodology on compositional results, *in* Zuffa, G. G., ed., Provenance of arenites: Dordrecht, D. Reidel Publishing Co., p. 165–189.

Manuscript Accepted by the Society April 21, 1995

Printed in U.S.A.

Geological Society of America
Special Paper 303
1996

Geometry, paleodrainage, and geologic rates from the Miocene Shadow Valley supradetachment basin, eastern Mojave Desert, California

S. Julio Friedmann,* Gregory A. Davis, and T. Kenneth Fowler
Department of Geological Sciences, University of Southern California, Los Angeles, California 90089-0740

ABSTRACT

The Shadow Valley basin of the eastern Mojave Desert, California, formed above the Kingston Range/Halloran Hills detachment fault during late Miocene continental extension. Structural, stratigraphic, and geochronologic data from the basin allow for detailed reconstruction of its evolution. Four unconformity-bounded members define the stratigraphy, each member consisting chiefly of alluvial fan and lacustrine facies. Rock-avalanche breccias, glide-blocks, volcanic, and epiclastic facies are also common. The unconformities are linked to short-lived tectonic episodes, only one of which (the last) can be attributed to tilt-block, domino-style normal faulting within the upper plate.

Basin sedimentation and detachment faulting began approximately between 13.5 Ma and 13.0 Ma. Following shallow-level intrusion of the Kingston Peak pluton across the Kingston Range segment of the detachment, fault sedimentation ceased in the Kingston Range but continued to the south in the Halloran Hills area. Translation above the detachment dominated regional extensional deformation during the first 2–3 m.y., and 2–3 km of sediment, predominantly alluvial fan, mass wasting, and lacustrine deposits, accumulated during this phase. Extension within the upper plate did not begin until approximately 3 m.y. after initial deposition and detachment faulting.

The primary geometry of the detachment fault was curviplanar and corrugated, with corrugation wavelengths of 10–15 km and amplitudes as much as 1.5 km. Sediment entered the basin chiefly through negative corrugations along the eastern detachment breakaway. These corrugations also served as likely source regions for the abundant rock avalanche breccias found throughout the basin.

Paleocurrent, provenance, and facies distribution data indicate that most basin fill was derived from the detachment footwall. Sediments traversed across the entire basin (currently more than 50 km from the detachment breakaway), and the depocenter remained far (>15 km) from the range front during active faulting and deposition. Long, transverse drainage and a distal depocenter are not compatible with popular half-graben models for extensional basin filling, and suggest a different geometry and fill history for the Shadow Valley basin. The differences are most likely the result of rapid extension common to detachment systems and the associated uplift of the detachment footwall.

*Present address: Exxon Production Research, P.O. Box 2189, Houston, Texas 77252-2189.

Friedmann, S. J., Davis, G. A., and Fowler, T. K., 1996, Geometry, paleodrainage, and geologic rates from the Miocene Shadow Valley supradetachment basin, eastern Mojave Desert, California, *in* Beratan, K. K., ed., Reconstructing the History of Basin and Range Extension Using Sedimentology and Stratigraphy: Boulder, Colorado, Geological Society of America Special Paper 303.

INTRODUCTION

The recognition of low-angle normal faults, or detachment faults, and their importance in the tectonics of highly extended terrains has led to numerous studies of continental extension. Over the past six or seven years, this work has included detailed sedimentology and stratigraphy of basins which formed above active detachments (e.g., Miller and John, 1988; Fedo and Miller, 1992, Topping, 1993; Yarnold, 1994; Holm et al., 1994). Studies of supradetachment basins (Friedmann and Burbank, 1992, 1995) are significant in that reconstructions of basin geometry, paleodrainage, and fill history place constraints on the surficial processes and paleotopography of detachment basins and the timing of structural events.

The upper Miocene Shadow Valley basin (Hewett, 1956; Davis et al., 1993) provides an extraordinary field laboratory to reconstruct the Cenozoic history and process in the eastern Mojave Desert. This large basin (minimum area 1,500 km^2) formed above the Kingston Range/ Halloran Hills detachment fault and records rapid continental extension beginning at ca. 13.5 Ma. Structural and stratigraphic data from the basin provide new insights into detachment faults and their associated basins, in particular the geometry of the master fault, the evolution of paleotopography and drainage, rates of sediment accumulation, and the timing and sequence of deformational episodes.

Conclusions drawn from the Shadow Valley basin can be compared with those of other areas to understand which parameters are most important to large-scale and meso-scale basin processes. Such a comparison has led to the conclusion that the history of the Shadow Valley basin differs substantially from that predicted from current models of basin geometry and drainage in extended terranes. Most of the basin sediments were derived from the footwall of the basin-bounding fault, and this sediment was transported across the basin to a depocenter located, not adjacent to main fault scarp, but at least 15 km from it. As such, many predictions of half graben depositional models (Leeder and Gawthorpe, 1987; Rosendahl, 1987) are incompatible with observations from Shadow Valley and other supradetachment basins (Friedmann and Burbank, 1995).

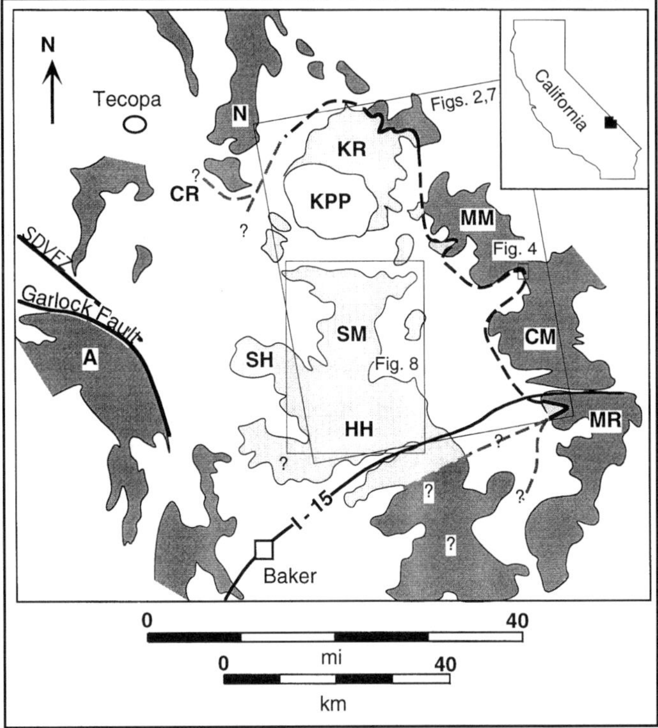

Figure 1. Location map showing local ranges. Kingston/Halloran detachment fault indicated by thick dashed and solid line; light shading indicates inferred extent of the Shadow Valley basin. A = Avawatz Mountains; CM = Clark Mountain; CR = China Ranch basin; HH = Halloran Hills; I-15 = Interstate highway 15; KPP = Kingston Peak Pluton; KR = Kingston Range; MM = Mesquite Mountains; MR = Mescal Range; N =Nopah Range; SDVFZ = Southern Death Valley Fault Zone; SH = Silurian Hills, SM = Shadow Mountains.

GEOLOGIC SETTING

The Shadow Valley basin, located in the easternmost Mojave Desert (Fig. 1), lies at the north-east corner of the Mojave structural province (Dokka, 1989) and the southeast corner of the Death Valley extensional corridor (Wernicke et al., 1988), two regions characterized by Neogene detachment faulting and coincident magmatism and sedimentation. The basin includes parts of the Kingston Range, the Shadow Mountains, the Halloran and Silurian Hills, and possibly the China Ranch basin (Fig. 1). The basin lies just east of the eastern terminus to the modern Garlock Fault and may have once continued south of Interstate 15 (Skirvin, 1989). The basin formed above an active detachment system, the Kingston Range/Halloran Hills detachment fault (Davis et al., 1993).

Along the eastern edge of the basin, the Clark Mountain thrust complex is a major, long-lived geologic boundary which separates two distinctly different geologic terranes. East of the thrust complex, Proterozoic gneiss and garnetiferous amphibolites underlie a cratonal sedimentary succession capped by Cretaceous volcanic rocks. In contrast, thrust sheets of the Clark Mountain complex contain Precambrian quartzofeldspathic gneiss overlain by Neoproterozoic to Paleozoic miogeoclinal and craton-transitional strata (Burchfiel and Davis, 1975). In addition, members of the Neoproterozoic Pahrump Group (Wright and Troxel, 1967) only occur west of the thrust belt (Burchfiel and Davis, 1988). Thus, the eastern edge of the Shadow Valley basin, the Kingston/Halloran detachment, the Clark Mountain thrust belt, the miogeoclinal hinge, and the edge of the Pahrump basin occur within a few kilometers of each other, separating two geologic regions with very different histories and lithologies. Despite the complicated regional history, Shadow Valley strata and subjacent Eocambrian, Cambrian, and Miocene strata are generally subparallel (Hewett, 1956; Fowler,

1992), indicating no significant pre-basin tilting. The distinctive nature of geologic units across the Clark Mountain thrust belt is important in understanding the basin provenance (see discussion in the following section).

MIOCENE STRUCTURAL GEOLOGY

Important aspects of the structural geology are summarized here. More detailed descriptions are presented in Fowler (1992), Davis et al. (1993), Bishop (1994), and Fowler et al. (1995). The major, basin-bounding fault of the Shadow Valley basin is the Kingston Range/Halloran Hills detachment fault (Davis et al., 1993; Burchfiel et al., 1983), which extends from the northern Kingston Range for more than 60 km south to the Mescal Range (Figs. 1 and 2). The basin-bounding detachment fault occurs within and locally cuts across the Mesozoic Clark Mountain thrust complex of the Sevier fold-thrust belt (Figs. 3, 4). The detachment fault reactivated portions of individual thrusts, but generally is discordant to Mesozoic structures.

Detachment faulting probably began about 13.5 Ma (see following discussion; Table 1). The Kingston Peak pluton (Fig. 1) intruded the detachment fault at a shallow level between 12.8 Ma and 12.1 Ma (Calzia, 1991) and effectively terminated motion along northern portions of the detachment fault in the Kingston Range. South of the Kingston Range, an accommodation zone developed in or near Kingston Wash, allowing for continued extension above the southern, Halloran Hills segment of the detachment fault (Fig. 2). This accommodation zone is locally exposed southeast of Kingston Peak, where it is called the Blacksmith Hills fault (Fig. 2; McMackin, 1992; Davis et al., 1993). Along the western end of Kingston Wash (Fig. 2), portions of the fault zone were reactivated with a sinestral sense (Davis and Burchfiel, 1993).

After plutonic pinning of the detachment in the Kingston Range, extension and deposition continued south of Kingston Peak in the Shadow Valley basin (Davis et al., 1993). Shadow Valley structures record a two-phase extensional history (Fowler et al., 1995). First, the detachment fault's upper plate slipped between 5 km and 9 km west-southwestward above the Kingston Range/Halloran Hills detachment fault until approximately 10.5 Ma (Friedmann et al., 1994a), when it began to be internally extended by domino-style tilt-block faulting (Reynolds, 1990; Bishop, 1994). Termination of Shadow Valley deformation is constrained by basalt flows that cap tilted strata. These flows, dated from 5.1 to 4.48 Ma, are part of the Cima Volcanic field which produced basalts as early as 7 Ma (Table 1; Turrin et al., 1985).

Structural studies (Fowler, 1992; Bishop, 1994; Friedmann et al., 1994a) constrain the amount of extension associated with detachment faulting and the initial dip of the fault system. *Translation* of the upper plate above the detachment is constrained between 5 km and 9 km based on regional geologic relationships. Extension *within* the upper plate is limited to 11%–42%, or 3–8 km, based on bed dips and preserved cut-off angles. The initial dip of the detachment at the latitude of the Kingston Range was 38° ± 3°, as constrained by structural relations there and paleomagnetic data from the Kingston Peak pluton (Jones, 1983).

All members of the Shadow Valley basin are cut by upper plate normal faults and transverse, east-west–trending to southeast-northwest–trending folds. Folding appears to have occurred to greater or lesser amounts through much of the basin history (see subsequent discussion), although the latest folding cuts late faults and probably represents the last major pulse of deformation. Upper plate faulting is linked to late-stage uplift of the detachment footwall (Fowler et al., 1995); the folding is less well understood, although folds within several kilometers of Kingston Wash probably formed from late-stage sinestral faulting (Davis and Burchfiel, 1993; G. Davis, 1994 and 1995, unpublished mapping).

MIOCENE STRATIGRAPHY

The Shadow Valley basin contains more than 3 km of middle to late Miocene strata that were deposited during and as a consequence of regional extension. More than 11 measured sections serve as the primary database (Friedmann and Burbank, 1994; Friedmann, 1995). These strata unconformably or nonconformably overlie various units, including older Miocene volcanic and sedimentary rocks. At least four unconformity-bounded members comprise the basin (Fig. 5), designated from oldest to youngest members I through IV, and they are defined by their bounding unconformities and their stratal character (unpublished mappings: G. A. Davis, 1992–1995; K. Bishop, 1994; T. K. Fowler, 1992; S.J. Friedmann, 1995).

The unconformity at the base of member I juxtaposes member I deposits against Precambrian to Middle Miocene rocks. Although the unconformity is generally recognized easily, at several locations member I overlies slightly older Miocene strata. In the Kingston Range, for example, member I volcanic and coarse clastic rocks overlie several hundred meters of prebasinal lacustrine dolomite (16.0 Ma; Table 1) with gentle discordance. In the Halloran Hills, a 19.7 Ma welded ignibrite and several thin lacustrine dolomite beds are disconformably overlain by thick red-bed conglomerates and breccias. A complete description of the older Miocene strata can be found in Friedmann (1995).

The lower part of member I contains 600–900 m of carbonate-rich fanglomerate, carbonate and quartzite megabreccia, lacustrine limestone and siltstone, volcanic flows, and volcanic breccias. Volcanic deposits include lahars, mafic to andesitic flows, and abundant volcanic breccia. Lacustrine mudstones and carbonates of limited thickness are intercalated with the volcanics. Several volcanic units serve as marker beds, including an andesite flow containing flow-aligned plagioclase plates, a violet andesite breccia, and a felsic ash-flow tuff. These

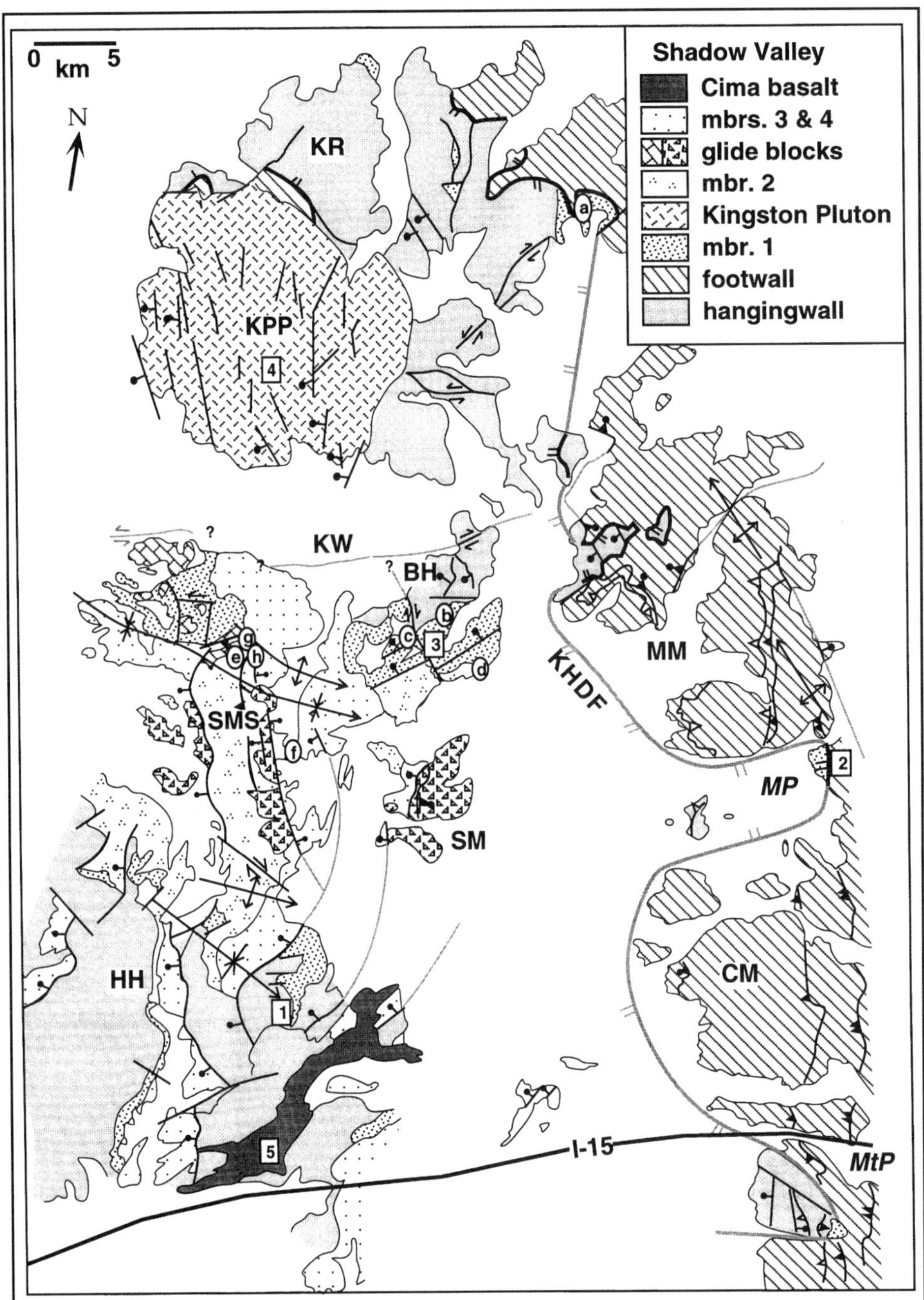

Figure 2. Simplified geologic map of the Shadow Valley basin and surrounding region. Kingston/Halloran detachment fault indicated as thick solid or gray line (gray where inferred in subsurface). Numbers and letters correspond to locations of rocks radiometrically dated by workers (listed in Table 1; boxed numbers are from previous workers, circled letters are samples presented in this study). BH =Blacksmith Hills; CM = Clark Mountain; HH = Halloran Hills; I-15 = Interstate highway 15; KHDF = Kingston Halloran detachment Fault; KPP = Kingston Peak Pluton; KR = Kingston Range; KW = Kingston Wash; MM = Mesquite Mountains; MP = Mesquite Pass; MtP = Mountain Pass; SM = Shadow Mountain; SMS = Shadow Mountains.

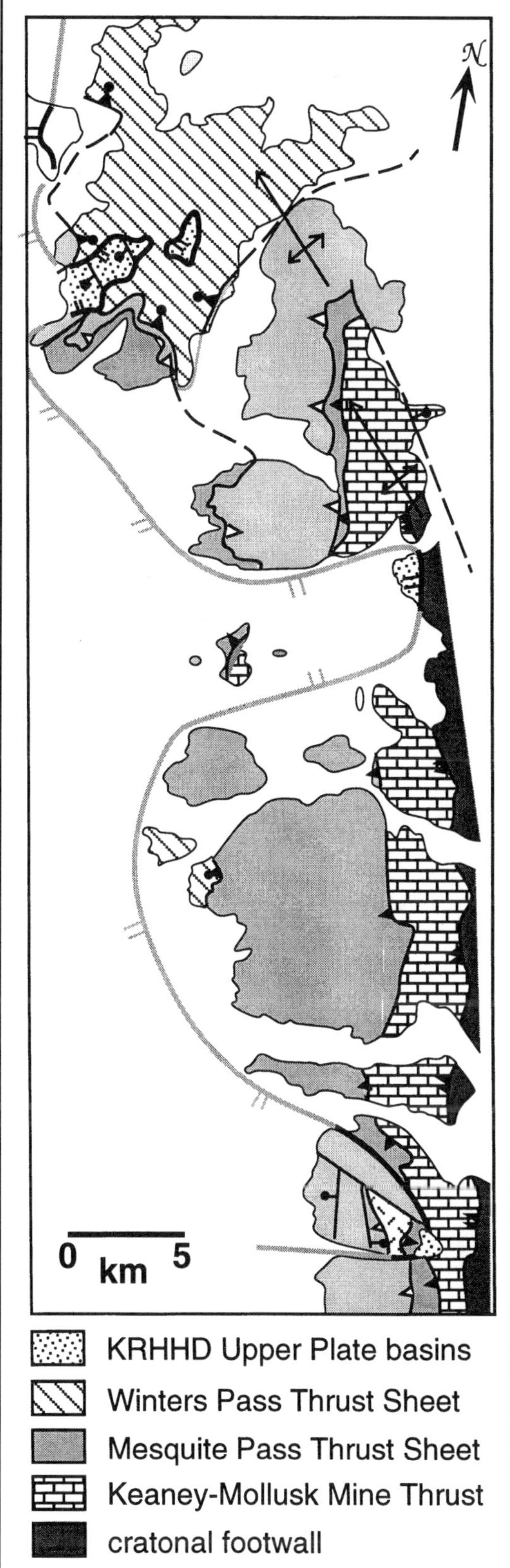

Figure 3. Simplified map of Clark Mountain thrust complex and Halloran Hills segment of the Kingston/Halloran detachment. After B. C. Burchfiel and G. A. Davis (1988; 1970, unpublished mapping) and S. J. Friedmann (1993, unpublished mapping).

markers allow correlation between the northern Kingston Range and the rest of the Shadow Valley basin. In the Kingston Range, the only Shadow Valley strata preserved belong to member I.

The unconformity separating members I and II represents approximately one million years of time between the end of member I deposition and the beginning of member II (Fig. 5). Although strikes can differ locally as much as 30°, chiefly in the Blacksmith Hills region, member I and member II beds are generally subparallel. Locally, small, subvertical faults that cut member I are truncated by the member II unconformity. Volcaniclastic conglomerates and sandstones derived from member I occur locally just above the unconformity and include clasts of the platy andesite.

Member II contains 600–1,000 m of interfingered fanglomerates and fine-grained lacustrine deposits, as well as minor ash beds. Subaqueous deposits include matrix-rich to matrix-poor debris-flow conglomerates, turbiditic sandstones, shales, soft-sediment deformation, and at least one Gilbert-type fan delta (Friedmann et al., 1994a). Rock avalanche breccias (c.f. Keefer, 1984; Yarnold and Lombard, 1989) and glide blocks (c.f. Varnes, 1978; Boyer and Hossack, 1992; Friedmann and Davis, 1994) are abundant in outcrop although laterally discontinuous. Member II generally coarsens upward into a thick (300 m) alluvial fan complex.

The unconformity between members II and III is reflected in a subtle (<10°) discordance and a change in depositional style. Where both members II and III are folded in the northern Shadow Mountains, fold limbs can be 20° steeper in member II than in member III, indicating a growth folding relationship (c.f. Suppe et al., 1992).

Member III consists predominantly of fluvial sandstones and conglomerates, lacustrine mudstones, and volcanic-clast dominated, coarse sandstones and conglomerates. Member III strata both thin across and lap onto the crest of growth anticlines. A very large gravity-driven glide block consisting of Precambrian gneiss, diabase, quartzite, and carbonate occurs near the top of member III (Hewett, 1956; Parke and Davis, 1990; Friedmann, 1995). Conglomerates contain clasts derived from the Kingston Peak pluton, providing a minimum age constraint on the timing of its unroofing. Member III is at least 500 m thick.

The unconformity that separates members III and IV is marked by both gentle and dramatic angular discordance. The lithologic similarity of upper member III and lower member IV can sometimes make recognition of the unconformity difficult. Member IV is almost exclusively cobble-boulder fanglomerate. Regionally restricted lacustrine environments were preserved in small sub-basin strata during upper plate faulting. Deposits containing blocks shed from north of Kingston Wash and from south of the Halloran Hills fill troughs between fault blocks and are locally derived from adjacent tilt-blocks (Bishop, 1994). Dips of beds within member IV fan decrease upsection (Bishop, 1994; Fig. 6) and show onlap and toplap with respect to older beds. Bedding dips are less than 20°, substantially gentler than those of underlying strata.

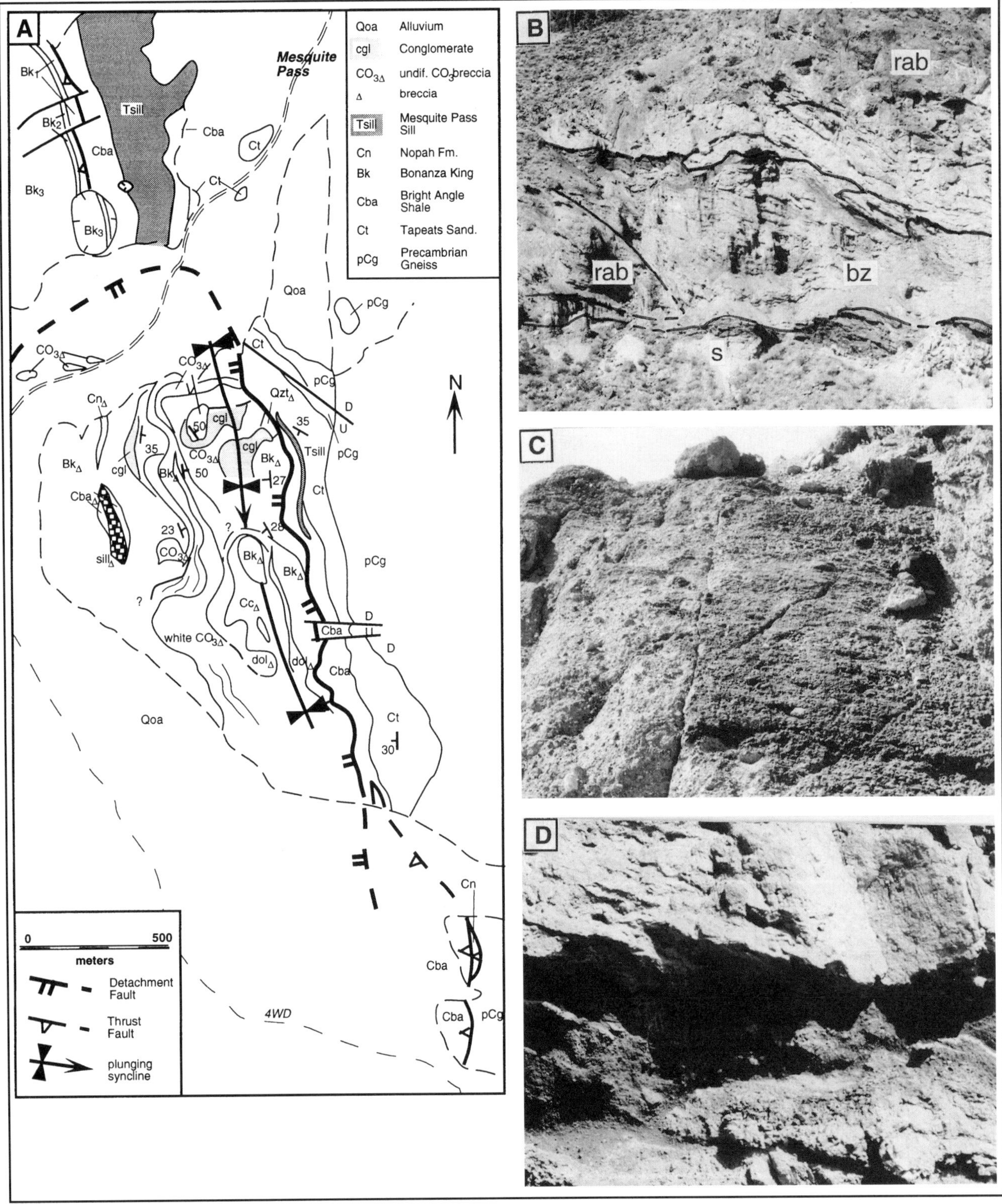

Figure 4. Details of the Mesquite Pass breakaway and synformal corrugation. (A) Detailed geologic map (B. C. Burchfiel and G. A. Davis, 1970, unpublished mapping; S. J. Friedmann, 1993, unpublished mapping). Thick dashed lines are inferred fault traces. (B) Down-plunge view of the Halloran Hills detachment fault at the breakaway. Solid lines represent boundaries of distributed sheer zone. Note wedge-shaped lateral ramp at left side of photo. Abbreviations: s = hypabyssal 13.4 Ma felsic sill; rab = rock avalanche breccias in hanging wall; bz = breakaway fault zone. (C) Striae on lateral ramp within shear zone; trend is 250°, plunge 8°. (D) Flame-like structures along upper boundary of sheer zone, indicating asymmetric extension top to the left (WSW).

TABLE 1. RADIOMETRIC AGES FROM THE SHADOW VALLEY BASIN*

	Age (Ma)	Method and Lithology	Member
A. From earlier studies			
1. R. Fleck, unpublished	19.7	K/Ar hornblende, welded tuff	Pre
2. R. Fleck, unpublished	13.4	K/Ar hornblende, felsic sill	Pre
3. Bell, 1971	13.1	K/Ar biotite, mafic breccia	I
Calzia, 1991	11.6 ± 0.3	K/Ar plagioclase, rhyodacite breccia	I
	11.1 ± 0.3	K/Ar whole rock, basalt	I
4. Armstrong, 1970	12.8	K/Ar, Kingston Peak pluton	
Calzia, 1991	12.4 ± 0.03	K/Ar biotite, Kingston Peak pluton	
	12.1 ± 0.1	K/Ar whole rock, Kingston Peak pluton	
5. Turrin et al., 1985	5.1–4.5	K/Ar, Cima basalt	Post
B. Present study			
a.	16.0 ± 0.2	Ar/Ar biotite†	Pre
b.§	13.12 ± 0.26	Ar/Ar biotite, intermediate flow	I
c.§	13.07 ± 0.27	Ar/Ar sanidine†	I
d.§	12.05 ± 0.25	Ar/Ar sanidine†	II
e.§	11.78 ± 0.24	Ar/Ar sanidine†	II
f.§	11.4 ± 0.3	Ar/Ar sanidine†	II
g.	11.2 ± 0.3	Ar/Ar K-feldspar†	II(?)
h.	10.8 ± 0.1	Ar/Ar biotite†	III

*Localities shown on Figure 2.
†From air fall tuff.
§Presented in Friedmann, 1993.

GEOCHRONOLOGY

Eight new radiometric ages from the Shadow Valley members I through III (Table 1, Fig. 5) were obtained in this study. All dates are from single-crystal fusion analyses performed at the Massachusetts Institute of Technology CLAIR facility. The argon release ratios from individually fused crystals were plotted on isotope ratio correlation diagrams (Fig. 7A). This technique yields an isochron for a given suite of crystals, with analytical errors of ~0.25 m.y.

In Mesquite Pass, the detachment cuts a hypabyssal felsic sill (Fig. 4, A and B) dated at ~13.4 Ma (R. Fleck, K-Ar on hornblende, 1992, personal communication), providing a lower age on initial detachment motion at this locality. Blocks of the sill are present in the Miocene sediments deposited above the detachment (Fig. 4A). The base of member I lies between the age of the Mesquite Pass sill and a volcanic flow near the base of member I (13.1 Ma). The top of member I is dated at 13.0 Ma. Our ages for member I are compatible with those of Bell (1971), but not those of Calzia (1991; see Table 1). The disparity between our data and Calzia's may reflect the sample site location. Unpublished mapping (Friedmann, 1995) reveals a dextral shear zone along eastern Kingston Wash, where Calzia collected his samples (J. Calzia, personal communication, 1991). His samples may have lost argon during post-depositional faulting.

An ash bed dated at 12.0 Ma occurs at the base of member II, less than 2 m above the unconfomity (Fig. 5). These dates are consistent with the observation that the Kingston Peak pluton (12.8–12.1 Ma) cuts and deforms member I and its associated faults yet does not cut member II. Member II contains several ash beds which yield datable material (Table 1). The youngest, 11.4 ± 0.3 Ma, lies in the middle of member II, 200–300 m from the overlying unconformity. Along strike to the north, an ash dated at 11.2 ± 0.3 Ma lies in a package of conglomerates with unknown member affinity, probably also member II.

The only dated ash from member III, 10. 8 ± 0.2 Ma, lies at the base of member III. The age is consistent with the member II age data and suggests that the unconformity between members II and III was short lived (<500 k.y.).

In addition to radiometric age dating, several sections were sampled to construct magnetic polarity stratigraphies. These samples were processed in the cryogenic magnetometer lab at the University of Southern California. All samples underwent thermal degmagnetization, the specific treatments determined by pilot studies for each section. Pilot studies consisted of 50°–25 °C temperature steps from 0 °C to 700 °C, with magnetic susceptibility measured every 200 °C. AF demagnetization was not effective because of the rock magnetic properties of the samples. Several vector components were found in many samples, with characteristic remanence typically between 400 °C and 600 °C. Zidjervelts, all reversal stratigraphies, and complete methods are presented in Friedmann (1995).

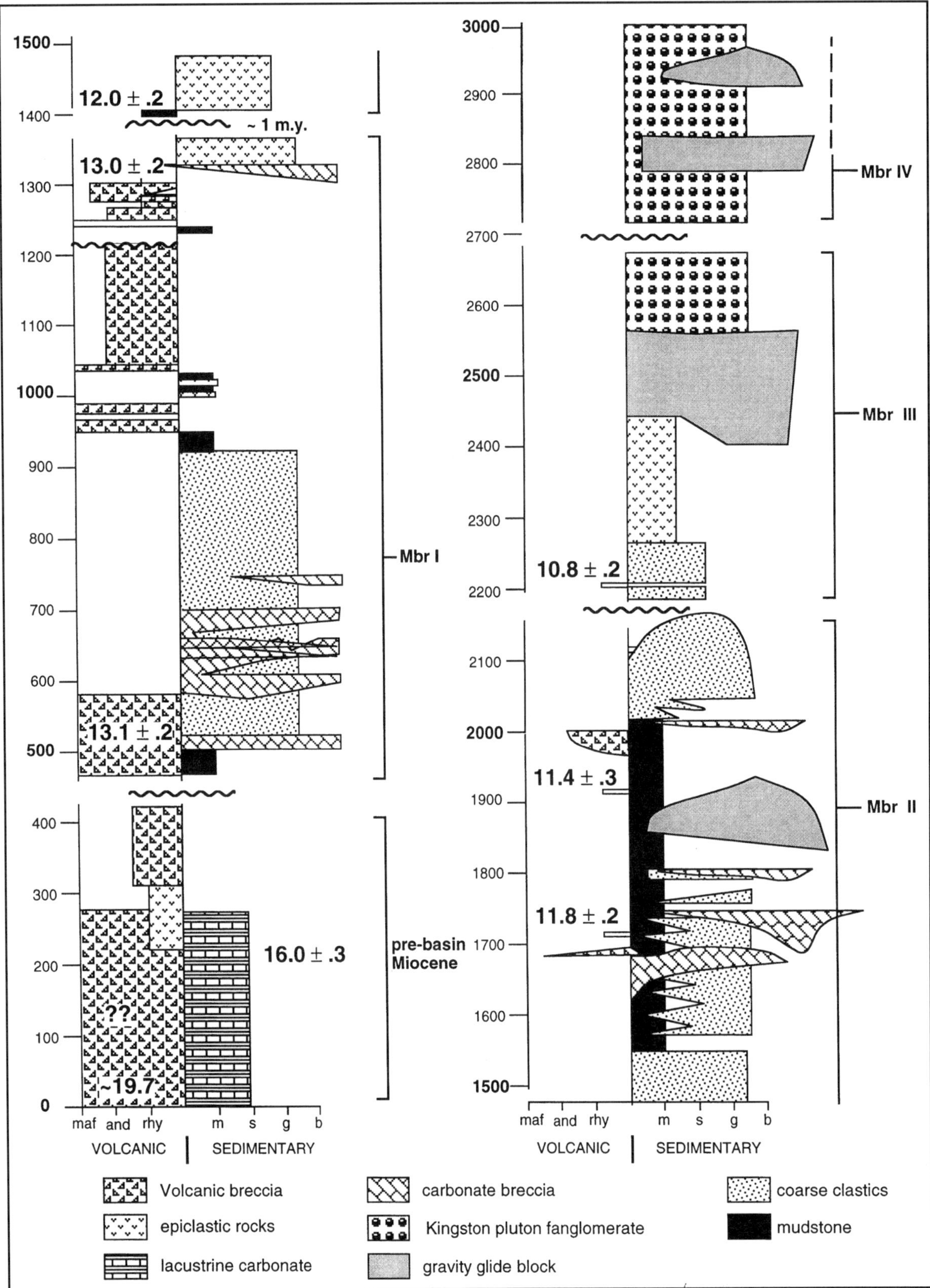

Figure 5. Simplified composite stratigraphic column; thickness in meters. All age dates are from this study (see Table 1). The right side of column represents sedimentary rocks, the left side volcanic rocks (maf = mafic, and = andesitic, rhy = rhyolitic, m = mud, s = sand, g = gravel, b = blocks). Note multiple unconformities, the presence of large glide blocks in members II, III, and IV, and abundant volcanic detritus in member III.

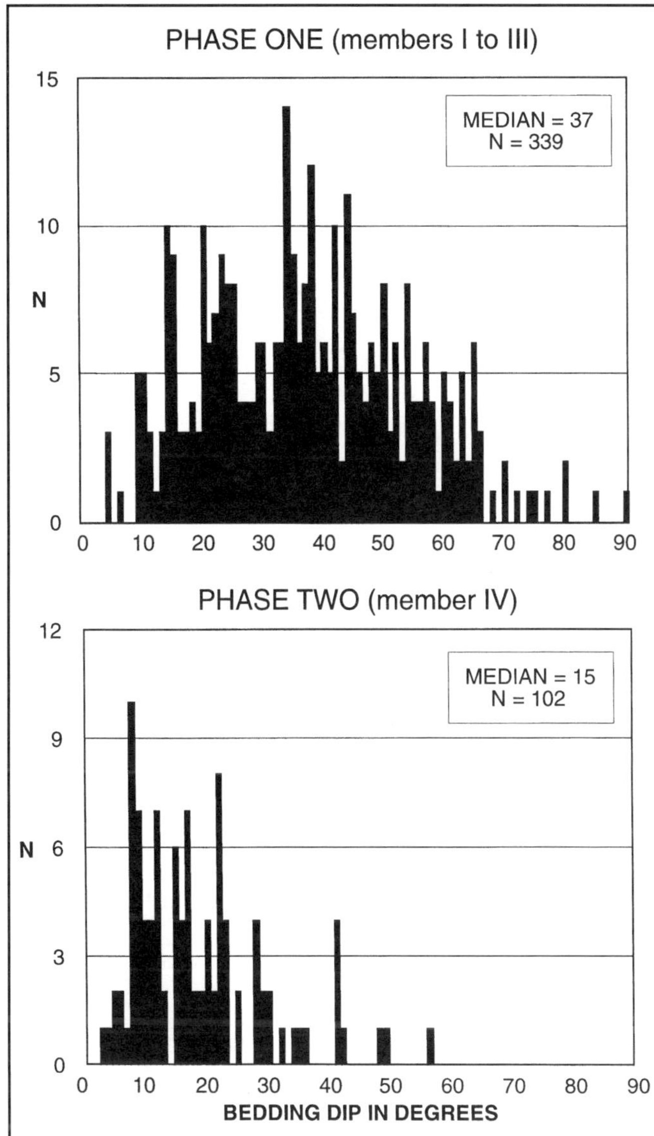

Figure 6. Bedding dip data from the Shadow Valley basin.

Current results are compatible with the radiometric ages (Fig. 7B). Chron 5n, a long normal polarity zone (Cande and Kent, 1995), appears to be absent from member II sections, which suggests deposition before Chron 5n. Similarly, virtually all samples taken from member III show normal polarity, compatible with Chron 5n deposition (Fig. 7B; Friedmann, 1995).

The paleomagnetic data from the basin also show that the Blacksmith Hills area is the only area which has undergone vertical axis rotation (Friedmann, 1995), with a rotation equal to ~60° clockwise. The rest of the basin yields virtual geomagnetic poles close to that predicted for the latitude of the basin; however, the data are incomplete, and their rock magnetism is complicated. Further tests of the magnetic reversal stratigraphies require additional sampling and analyses.

MASTER FAULT GEOMETRY

The geometry of the detachment fault is curviplanar, with corrugation amplitudes up to 1.5 km and wavelengths of 10–15 km (Fig. 3). The detachment fault cuts Mesozoic faults and fabrics of the Clark Mountain thrust belt, locally at a high angle. At Mesquite Pass (Fig. 3), corrugations cut downward across the entire Clark Mountain thrust belt into the structurally lower cratonal section (Burchfiel and Davis, 1988). The corrugations are primary features and cannot be attributed to later transverse folding, as evidenced by the unfolded, planar geometry of thrusts in the detachment footwall (Fig. 3; Davis et al., 1993). Similar curviplanar detachment geometries have been recognized in many other areas (e.g., John, 1987; Davis and Lister, 1988; Spencer and Reynolds, 1991; Dinter and Royden, 1993).

The detachment fault crops out in several locations. It is spectacularly exposed east of Mesquite Pass (Fig. 1) as an anastamosing shear zone 1–15 m thick (Fig. 4, A and B). Here, the fault juxtaposes a Cambrian cratonal stratigraphy including Tapeats Sandstone and Bright Angel Shale against locally derived Miocene megabreccias. Asymmetric folds, tear faults, striae on the fault surface and other kinematic indicators (Fig. 4, C and D) show top-to-the-west-southwest transport (~250°) with a shallow plunge (8°–20°). In the Mesquite Mountains and Kingston Range, the detachment fault crops out as a subhorizontal, microbrecciated, striated surface that separates various upper- and lower-plate lithologies (Buchfiel and Davis, 1988).

PALEODRAINAGE PATTERNS

Data from provenance, paleocurrent, and facies analyses (see details in following sections) strongly suggest that west-directed transverse drainage dominated during deposition of members I through III. During deposition of members II through IV, a secondary, axial drainage, centered around Kingston Peak, shed detritus across the northern accommodation zone (Friedmann and Burbank, 1992; Friedmann et al., 1994a). Most of the sediment that entered the basin, especially detritus within the east-west transverse system, was shed from the footwall of the detachment system. No evidence for a significant hanging wall source exists until upper plate faulting occurs during member IV deposition.

Provenance

Many clast types found within Shadow Valley strata have a restricted provenance. In particular, many rocks found in the footwall of the Kingston/Halloran detachment have no counterpart in the hanging wall and vice versa because of older structures (see previous discussion). Some lithotypes have source regions less than a few kilometers in area and thus are robust provenance indicators. Over 30 clasts counts, as well as more qualitative clast recognition during mapping and section measuring, are the provenance data base.

Six lithologies serve as particularly useful provenance indicators (Fig. 8). These include the following: (1) garnetiferous

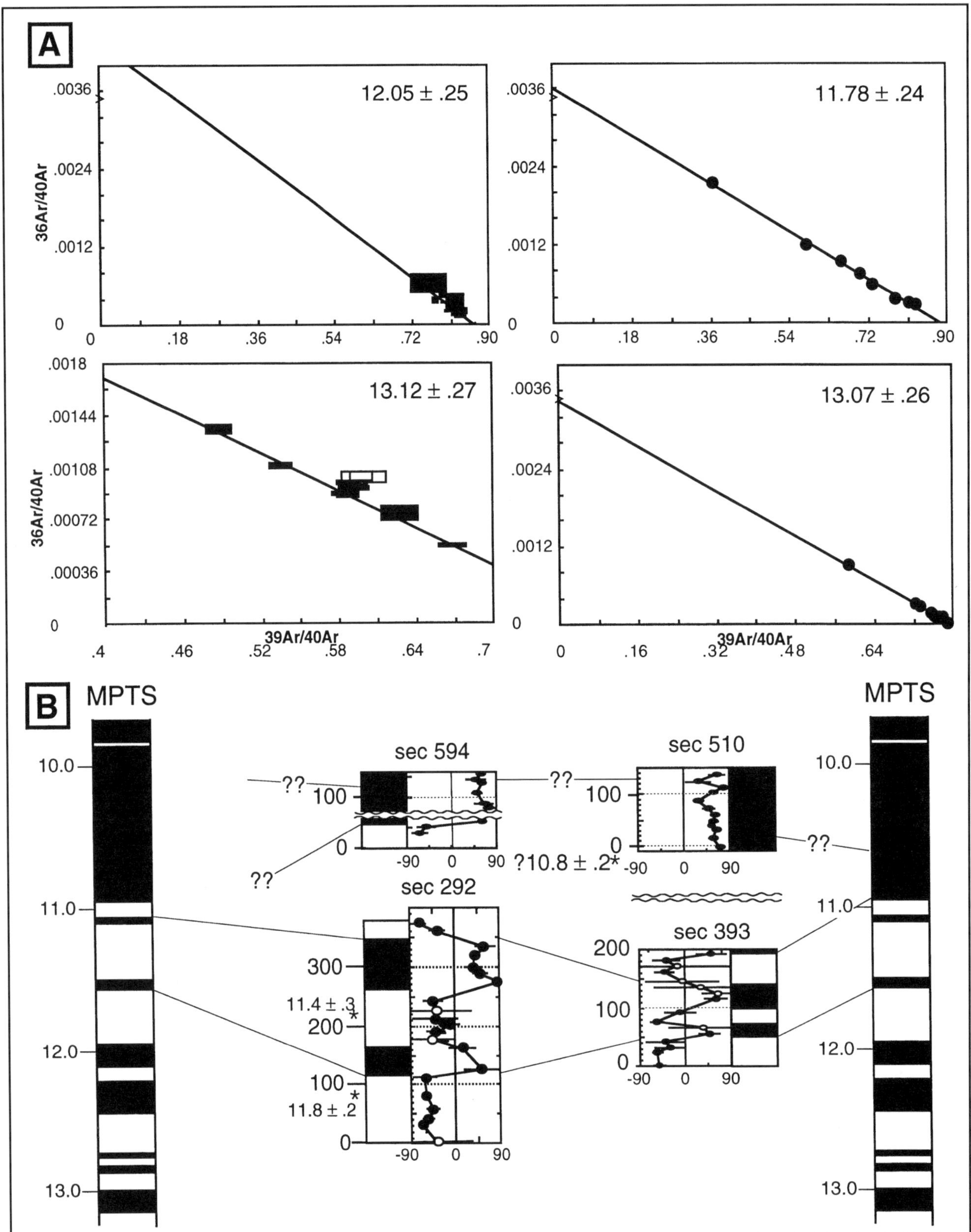

Figure 7. (A) Four isotope correlation diagrams from Shadow Valley samples. Chevrons on verticle axis represent the modern atmosphere. Black dots represent samples with invisibly small error bars; white boxes are samples which were excluded in the analysis. (B) Magnetic reversal stratigraphies for sections 292, 393, and 510; wavy lines represent unconformity between members II and III. Black circles represent class I sites, white circles class II. Magnetic polarity time scale (MPTS) from Cande and Kent (1995). Note long reversed intervals suggesting subchrons of Chron 5r in sections 292 and 393.

quartzofeldspathic and amphibolitic gneisses, found at Mountain Pass (Reynolds and McMackin, 1988); (2) Cretaceous Delfonte volcanics, also found only at Mountain Pass; (3) jasper- and chalcedony-bearing Cambrian Tapeats Sandstone. These three rock types occur only beneath the forwardmost (easternmost) Mesozoic thrust (Fig. 3; Reynolds and McMackin, 1988). Middle and Upper Paleozoic, unmetamorphosed carbonates (4), including a variety of distinct lithofacies, occur both within the thrust belt and farther east. These four lithologies occur only in the footwall of the Kingston Range/Halloran Hills detachment and do not occur in the detachment hanging wall.

In addition, two distinctive Tertiary rock types have a restricted outcrop distribution. They are (5) the Kingston Peak pluton granite (Calzia, 1991), and (6) some volcanic rocks from

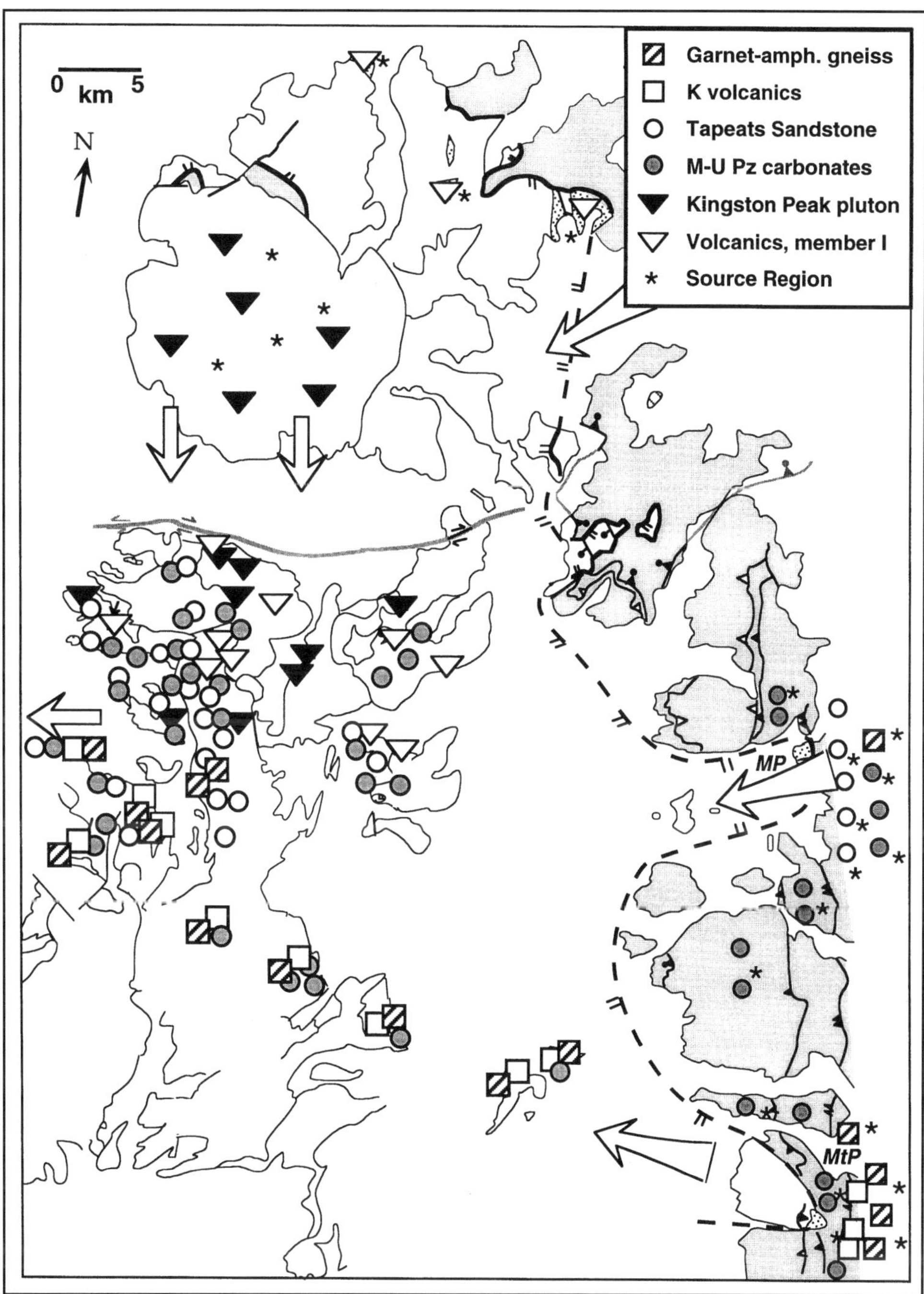

Figure 8. Map of source lithologies, probable paleodrainage pathways (large arrows), and basin clast occurrences. Note footwall source and discrete zones of provenance mixing. Arrow at left center indicates occurrence in the northernmost and westernmost Silurian Hills (Fig. 1). *MP* = Mesquite Pass; *MtP* = Mountain Mass. Shaded area represents detachment footwall.

Shadow Valley member I, in particular the flow-aligned, platy plagioclase andesite. These two rock types are exposed mostly north of the main part of the basin and fingerprint transport from north of the Kingston Wash accommodation zone.

Basement rocks of the detachment hanging wall block are equally distinctive. These include the Neoproterozoic Pahrump Group, including various facies of Beck Spring Dolomite and Kingston Peak Formation and marbles of probable middle to upper Paleozoic protolith. Teutonia Quartz Monzonite, although a common hanging wall basement rock, is not a diagnostic source rock because of its regional outcrop patterns.

Figure 8 shows the source areas for the six distinctive footwall lithologies, probable transport path, and deposits containing those lithologies as clasts. Several trends can be readily observed. First, several suites of clasts have a limited depositional range. For example, the Tertiary volcanic and plutonic clasts (triangles on Fig. 8) only occur in the northern part of the basin, whereas the gneisses and Cretaceous volcanic clasts (squares on Fig. 8) only occur in the southern part. Second, footwall-derived clasts occur across the full east-west extent of basin outcrop. Clasts of Delfonte volcanics, garnetiferous gneiss, and Tapeats Sandstone occur as far west as the westernmost Silurian Hills, more than 50 km from their modern source terrane (see subsequent discussion).

Clasts indicative of hanging wall sources are extremely uncommon in members I through III. They generally occur as single small clasts comprising less than 0.5% of the clast population. The most common rock types in this suite are black sandstones from the Kingston Peak Formation and some white marbles. Although their occurrence is not well understood, these clasts are found in beds with west-directed paleocurrent directions and abundant footwall-derived clasts. Member IV contains large volumes and large blocks (one >200 m thick) of these lithologies, but these clasts were derived from near Kingston Peak and transported southwestward.

Paleocurrents

Approximately 600 paleocurrent indicators have been measured in the main outcrop belt of the Shadow Valley basin (Fig. 9). Measurements were collected from each stratigraphic unit and from 17 locations spread throughout the basin. They include measurements of both uni-directional and bi-directional indicators, including parting current lineations (p.c.l.), gutter casts, channel margins, clast imbrications, flutes, and axial trends of troughs exposed in three dimensions. In many instances, a uni-directional indicator (e.g., clast imbrication) was used to constrain the flow direction of a co-stratal, bi-directional indicator (e.g., gutter cast, p.c.l.); the two measurements then were treated as a single, uni-directional indicator with the azimuth of the bi-directional indicator. In all such cases, the angular difference between the uni- and bi-directional indicator was less than 60°, usually less than 30°.

Figure 9 reveals a dominant east-to-west transport pattern, with a mean vector slightly south of west. This direction is orthogonal to the strike of the detachment, indicating transverse drainage away from the breakaway. Several localities show a subordinate mode indicating flow to the south or southwest. Beds with southward paleocurrent indicators typically carry volcanic clasts, consistent with derivation from the north. Thus, the dominant east to west trend is seen in all members throughout the entire observed basin, with a subordinate north to south trend most clearly seen in epiclastic strata (Fig. 9).

Member IV paleodrainage is relatively poorly documented compared with that of other members. This is because of outcrop and facies characteristics. Although member IV is recognized over a large area (Fig. 2), it tends to crop out poorly, mostly as slope wash with a rare exposed bed. The best data were obtained along Kingston Wash, which has excellent exposures. Moreover, the predominant facies of member IV is boulder conglomerate with mostly equant clasts. As such, the only paleocurrent indicators present are imbricated clasts, which are uncommon and not very reliable. Still, member IV yields what is a predominantly east to west trend, consistent with the other members.

The only major exceptions to the trend come from the Blacksmith Hills, where paleocurrents flow toward the northwest. However, the Blacksmith Hills have undergone approximately 60° of clockwise rotation about a vertical axis (Friedmann et al., 1994b). This is based on both regional strike-line trends (Fowler, 1992) and paleomagnetic studies (Friedmann, 1995). The data from the Blacksmith Hills area in Figure 9 have been restored 60° counterclockwise, consistent with their declination error. Once this rotation is removed, then paleocurrent directions from that region match the regional basin trend (Fig. 9).

Facies relations

Sedimentary facies were mapped at 1:12,000 scale for most of the basin, with details recorded in the stratigraphic sections. The predominant facies assemblages from the basin are alluvial fan (often debris-flow dominated), lacustrine, rock avalanche, gravity-glide block, and uncommon braided-stream deposits. Sedimentological details used in facies determination are presented in Friedmann (1995).

The most significant aspect of Shadow Valley facies distribution is the persistence of lacustrine deposits far from the detachment breakaway at their time of deposition. More than 1,000 m of lacustrine sediments are preserved in members II and III, indicating a relatively stable, long-lived feature. These deposits consist predominantly of mudstones and fine sandstones and contain turbidites, matrix-rich debris-flow deposits, burrowed muds, abundant slumping, and fan deltas. Lacustrine environments in member III contain substantial playa deposits in the form of gypsiferous siltstones. Although many of the sedimentary rocks formed in water deeper than several meters (Friedmann, 1995, 1996), there are many shallow water indicators including oscillation ripples, bird and mammal footprints, and rootlets. Lacustrine facies interfingered with alluvial fan facies both across strike and strike parallel, and fans prograded

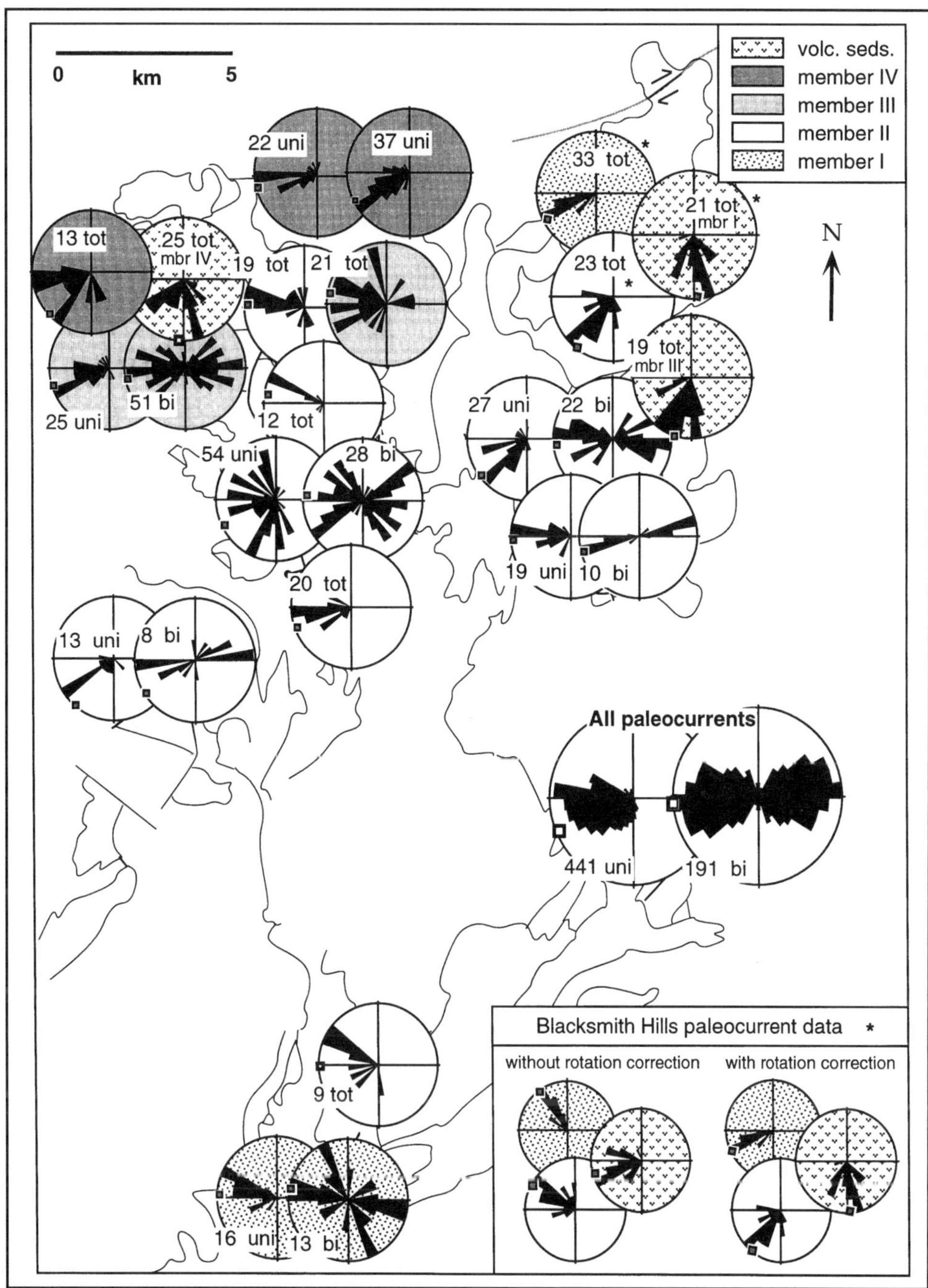

Figure 9. Paleocurrent measurements, both uni- and bi-directional, rosettes and mean vectors. Rosettes are shaded to match stratigraphic member. Volcanic sediments are not member specific.

into the lacustrine environments from the east, north, and south. Note that no units interfinger from the west.

INTERPRETATION OF UNCONFORMITIES

Each unconformity is related to a discrete, short-lived tectonic episode within the Shadow Valley basin. The member I unconformity formed from initation of detachment faulting, suggested by the change in sedimentary style from the older Miocene strata and rapid accumulation (Table 2). Intrusion of the Kingston Peak pluton across the detachment fault in the Kingston Range resulted in initiation of the Kingston Wash accommodation zone, rotation of the Blacksmith Hills block, and the member II unconformity. This is suggested by the age data, which indicate that the Kingston Peak pluton intruded between the top of member I and the base of member II. It is

TABLE 2. STRATIGRAPHIC AGES, ACCUMULATION RATES, AND ERRORS FROM SECTION 292

	Lower Age (Ma)	Lower Uncertainty	Upper Age (Ma)	Upper Uncertainty	Duration (m.y.)	Duration Uncertainty	Thickness* (m)	Thickness Uncertainty	Accumulation Rate* (mm/yr)	Rate Uncertainty
Magnetic Interval										
5r.1r	11.05	0.001	10.95	0.001	0.1	0.0014	[35]	3.5	[0.35]	0.04
5r.1n	11.1	0.001	11.05	0.001	0.05	0.0014	85	8.5	1.70	0.18
5r.2r	11.48	0.001	11.1	0.001	0.38	0.0014	95	9.5	0.25	0.03
5r.2n	11.53	0.001	11.48	0.001	0.05	0.0014	50	5	1.00	0.10
5r.3r	11.94	0.001	11.53	0.001	0.41	0.0014	[115]	11.5	[0.28]	0.03
Average (min)	11.94	0.001	10.95	0.001	0.99	0.0014	[380]	38	[0.38]	0.04
(inside)	11.53	0.001	11.05	0.001	0.48	0.0014	230	23	0.48	0.05
Radiometric Interval										
Member IIc	11.40	0.30	11.05	0.00	0.35	0.3000	130	13	0.37	0.32
Member IIb	11.78	0.24	11.40	0.30	0.38	0.3842	200	20	0.53	0.53
Member IIa	11.94	0.00	11.78	0.24	0.16	0.2400	290	29	1.81	2.72
Member I										
Full section	13.4	0.4†	13.07	0.27	0.33	0.4826	1,000	100	3.03	4.44
(min)	13.8	0	12.8	0	1	0.0000	1,000	100	[1.00]	0.10
Inside	13.12	0.26	13.07	0.27	0.05	0.3748	750	75	15.00	112.46
(min)	13.38	0	12.8	0	0.58	0.0000	750	75	[1.29]	0.13

*Brackets indicate minimum estimage.
†Estimate (error unpublished)

also suggested by the paleomagnetic data, which indicate 60° of clockwise rotation of member I with no rotation of member II (Friedmann et al., 1994a; Friedmann, 1995). The member III unconformity is attributed to growth folding (c.f. Suppe et al., 1992) near the Kingston Wash accommodation zone, suggested by the shallowing of bedding dips across transverse folds above the unconformity.

It is significant that despite the presence of these unconformities, the dips of member I through III beds are essentially the same and do not show fanning or shallowing across the unconformities (Fig. 6; Fowler et al., 1995). This is because the angular discordances across these unconformities are only locally manifested, and in most parts of the basin the unconformities are reflected only by sharp changes in depositional style. This observation suggests that members I through III accumulated during slip along the main detachment, yet before upper plate normal faulting.

Member IV contains evidence for fanning dips, including a decrease of bedding angle upsection, onlap, toplap, and truncation of earlier member IV strata. Whereas members I to III comprise a common dip domain of median 37° and a Gaussian distribution of dips, member IV (Kingston fans) constitutes a separate, shallower dip domain with median 15° and a dip distribution highly skewed to smaller tilts (Fig. 6). Also, faults which cut some member IV beds are truncated by younger member IV beds. These observations lead to the conclusion that deposition of member IV marks the beginning of extension within the Shadow Valley upper plate and deposition of "perched" strata (c.f. Fedo and Miller, 1992). A more detailed explanation of the transition between members I through III and member IV is given by Fowler et al. (1995).

PALEODRAINAGE INTERPRETATION

Thick lacustrine deposits currently lie more than 35 km away from the detachment breakaway, with no contemporaneous lacustrine deposits closer to the detachment fault. In addition, large, long fan bodies characterized by westward paleoflow brought footwall-derived boulders to more than 50 km from their source and show proximal-style deposition more than 35 km from their source.

Extensional transport alone cannot explain the distribution of distally disposed lacustrine and alluvial fan deposits. Estimates of intraplate extension are small (11%–42%) and do not exceed 8 km, whereas upper plate translation is constrained at 5–9 km by pre-Tertiary geologic relationships (see previous discussion). Thus, the maximum extensional translation of the lacustrine deposits is 16–17 km, and could be as little as 8–9 km (Friedmann et al., 1994a, 1994b). Retro-deformation of the *maximum* extension estimate still places the lacustrine depocenter at least 15 km from the active range front. This configuration strongly suggests that long transport distance and transverse drainage away from the footwall are characteristic of members I through III, and inconsistent with popular models for deposition in extensional half-grabens (e.g., Leeder and Gawthorpe, 1987).

ACCUMULATION AND SLIP RATES

The new age controls presented here are sufficient to allow determination of local rock-accumulation rates (Table 2) and extension/slip rates (Table 3). The measured sections chosen for the calculations have the best age control, most continuous outcrop, and greatest proximity to the depocenter (as described previously). The member II subdivisions (IIa, IIb, and IIc) were delineated by dated ashes within the section. Chron boundaries were determined by correlation to the magnetic polarity time scale of Cande and Kent (1995), using the radiometric ages to constrain choice of chrons (Fig. 7B). Choice of correlation substantially affects rates for individual chrons, but does not strongly affect mean rates for member II nor rates for member I. Note that the reference sections used to constrain Table 2 have not been backstripped and decompacted, and as such are only *rock* accumulation rates and *not sediment* accumulation rates. Any compaction associated with progressive burial would result in higher sediment accumulation rates.

Accumulation rates and errors for Tables 2 and 3 were calculated using the formula of Renne et al. (1991), including a 10% stratigraphic uncertainty. "Inside" rates correspond to time spans completely bounded by magnetic reversals, as top and bottom chrons are not completely recorded by strata. Similarly, only minimum thicknesses and rates can be calculated for those stratigraphic intervals without a complete record.

Undecompacted rock accumulation rates are on the order of 0.5–1.0 mm/yr (m/k.y.) and range from >5.0 to <0.2 mm/yr with a mean value close to 0.5 mm/yr. Values are comparable for both radiometric and paleomagnetic age determination (Table 2). The unevenness of rates during deposition can be explained in several ways. First, jumps in accumulation rate may be an artifact of the correlation; however, the radiometric and geologic constraints make other correlations untenable. Second, it has been observed in many places that short-term sediment accumulation rates vary by as much as an order of magnitude across the boundaries of short chrons (e.g., Johnson et al., 1988; Talling and Burbank, 1993). Thus, although unsatisfying, jumps in accumulation rate calculated for section 292 can be attributed to the unevenness inherent in the stratigraphic record.

Although average accumulation rates are higher than those of traditional rift basins (Olsen, 1991; Cohen, 1991; Evans, 1988), Shadow Valley rates are similar to the few published rates from supradetachment basins (Woodburne et al., 1990; Holm et al., 1994; Scrivner, 1984) and consistent with the preponderance of mass-wasting deposits.

Extension rates across the region (Table 3) were calculated similarly, usually with larger uncertainties in both bounding ages and amount of extension. Distance is measured in map view, so extension rate makes no assumption of fault geometry. Above the Halloran Hills segment of the detachment, we know that upper plate tilting commenced after deposition of member III. We assumed that all translation above the detachment occurred before upper plate extension, based on our reconstruction (see previous discussion; Fowler et al., 1995). In the Kingston Range, we lack the temporal resolution necessary to separate an intraplate phase from a translational phase, so all deformation is incorporated into the calculation.

Rates of extension are >2 mm/yr in general and >6 mm/yr in the Kingston Range. These rates are comparable to extension rates from other highly extended terranes (e.g., Foster et al., 1990; Spencer and Reynolds, 1991; Hill and Baldwin, 1993).

DISCUSSION

Timing of upper-plate breakup relative to regional extension

Kingston Range (northern segment). Detachment faulting began between 13.4 Ma and 13.0 Ma, as indicated by the Mesquite Pass sill and the oldest dated bed of member I. No domino-style, tilt-block faulting occurred during this time, and the basin filled as a consequence of upper-plate translation along the detachment. Since the detachment fault in the Kingston Range was intruded by the Kingston Peak pluton between 12.8 Ma and 12.1 Ma, deposition of member I, initiation of tilt-block rotation, and intrusion of the pluton all occurred in less than one million years. By the end of extension, normal faults had cut and rotated all lithologies in the Kingston upper plate, including member I. The above chronometric and geometric relationships require very high slip rates for the Kingston section of the detachment before intrusion (Tables 2 and 3).

Halloran Hills (southern segment). Slip on the Halloran Hills segment of the detachment fault began at the same time as the Kingston segment, but its upper plate did not internally

TABLE 3. EXTENSION/SLIP RATES

Detachment Segment	Lower Age (Ma)	Lower Uncertainty	Upper Age (Ma)	Upper Uncertainty	Duration (m.y.)	Duration Uncertainty	Extension (km)	Extension Uncertainty	Extension Rate (mm/yr)	Rate Uncertainty
Kingston	13.4	0.4	12.5	0.3	0.9	0.50	6	0.5	6.67	0.44
Halloran										
Translation	13.4	0.4	10.5	0.5	2.9	0.64	7	2	2.41	0.18
Intraplate	10.5	0.5	7	1	3.5	1.12	5	2	1.43	0.26

extend until much later. Fowler et al. (1993; 1995) describes two discrete phases of basin extension for Shadow Valley. In the first phase, deformation occurs as pure translation; in the second phase, deformation occurs as intraplate extension and tilt-block rotation. Member IV was deposited during the second phase, such that dips fan toward progressively shallower dips (Fig. 6; Davis et al., 1993; Bishop, 1994; Fowler et al., 1995).

Implications. Assuming that member IV deposition began after 10.5 Ma, shortly after member III (Table 1), and assuming that member IV deposition marks the beginning of the tilt-block deformation, then translational extension (phase one) ran for at least 2.5 m.y. before internal extension of the Halloran Hills upper plate. A similar pattern of deformation is described in the Casteneda Hills/Signal area of west-central Arizona (Lucchitta and Suneson, 1993) and in the Sacramento Mountains of south-eastern California (Fedo and Miller, 1992). In other parts of the Colorado River extensional corridor (Howard and John, 1987), most notably in the Chemehuevi and Whipple Mountains, peak extension and uplift occurred from 23 Ma to 20 Ma based on footwall cooling data (Foster et al., 1990; 1993). However, regional upper-plate tilt-block faulting, with some exceptions, began between 20 Ma and 18 Ma (Nielsen and Beretan, 1990), suggesting a lag of several million years between detachment initiation and internal deformation of the upper plate. If two-phase basin extension is common in highly extended terranes, then estimates of initial extension based on timing of upper plate faulting may be several million years late (Fowler et al., 1995).

The difference in timing between the Halloran Hills segment and Kingston Range segment can be explained in several ways (Fowler et al., 1995). The Kingston Peak pluton may have created a significant thermal and mechanical perturbation during detachment faulting which accelerated deformation in the Kingston Range (Table 3). Perhaps more important, the rocks preserved in the Kingston Range were closer to the breakaway than those exposed in the Halloran Hills region. This relationship implies that the Kingston Range rocks were above a shallower portion of the detachment, which would produce a weaker upper plate, reflected in the close spacing of upper plate normal faults in the Kingston Range (T. K. Fowler, unpublished data).

Sediment transport paths

Paleocurrent and provenance data, combined with mapping of the detachment surface and footwall, suggest that sediment entered the basin primarily from the east along synformal or "negative" corrugations (Figs. 10A and 11). Since the corrugations of the detachment are primary features, synformal parts of the corrugations would be topographic lows and would serve as transport paths from the footwall hinterland. Many clast types, such as the Tapeats Sandstone, the garnetiferous gneisses, and the Delfonte Volcanics, come from sources which lie near the heads of negative corrugations (Fig. 8). Clark Mountain and the Mesquite Mountains are topographic, "antiformal" highs separating these sources and the Shadow Valley basin. Thus, the Mountain Pass and Mesquite Pass lows may have provided major sediment pathways into the basin.

In addition to providing sediment conduits into the basin, negative corrugations may have helped to create large footwall catchments. Finite extension along a low-angle, corrugated surface would produce large scoop-shaped drainage basins (Fig. 11). Nickpoint migration away from the scarp would increase the area of the drainage basin and thus the sediment flux (Hack, 1957; Leopold et al., 1964; Leeder and Jackson, 1993). The relatively gently sloping source region would provide a larger sediment yield than would smaller catchments formed along steeper faults (Hack, 1957). This sediment would then be shunted through the negative corrugations into the large fan systems preserved in the basin (Fig. 11).

Paleotopography

The paleotopography of the hinterland is poorly constrained, but may be reflected in the modern topography. "Positive" or antiformal corrugations are reflected in local topographic highs, such as the Clark and Mesquite Mountains, whereas "negative" or synformal corrugations are reflected in Mountain Pass, Mesquite Pass, and possibly Winters Pass. The presence of abundant debris flows and rock avalanche breccias within the basin fill also constrains paleotopography. Rock avalanches require steep slopes (>25°, usually ~45°) and high topography (>150 m, usually ~1,000 m) to initiate (Keefer, 1984). Debris-flow conglomerates are also typical of tectonically active areas with steep slopes.

This observation is initially at odds with the constraints on the low (~30°) initial slopes of the Kingston/Halloran detachment. Rock avalanche deposits are abundant in many other Miocene terrestrial extensional basins (e.g., Yarnold and Lombard, 1989; Beratan, 1991; Schaller, 1991), many of which are believed to have formed above shallowly dipping faults. One explanation is that these faults formed at steep angles and were subsequently rotated to shallower dips (Wernicke and Axen, 1988; Buck, 1988). In this interpretation, there would be no difficulty in generating the necessary hinterland slopes to supply rock avalanche breccias. Topping (1993) considers the presence of rock avalanche breccias to support the hypothesis of initially steep master faults. However, structural evidence from other locations requires shallow (<30°) initial dips despite associated rock avalanche fill (John and Foster, 1993; Dinter and Royden, 1993; Dokka, 1993).

In the case of several detachment systems, the primary geometry of the detachment surface is curviplanar and corrugated (Howard and John, 1987; Spencer and Reynolds, 1991). In such cases, the *margins* of the corrugations may be quite steep, locally greater than 45°. This also must have been true in Mesquite Pass, where the modern topography has north- and south-dipping slopes greater than 30°, yet a west-dipping slope of 5°–10°. Thus, rock avalanche breccias could be shed from the steep margins of corrugations and then proceed down drainage

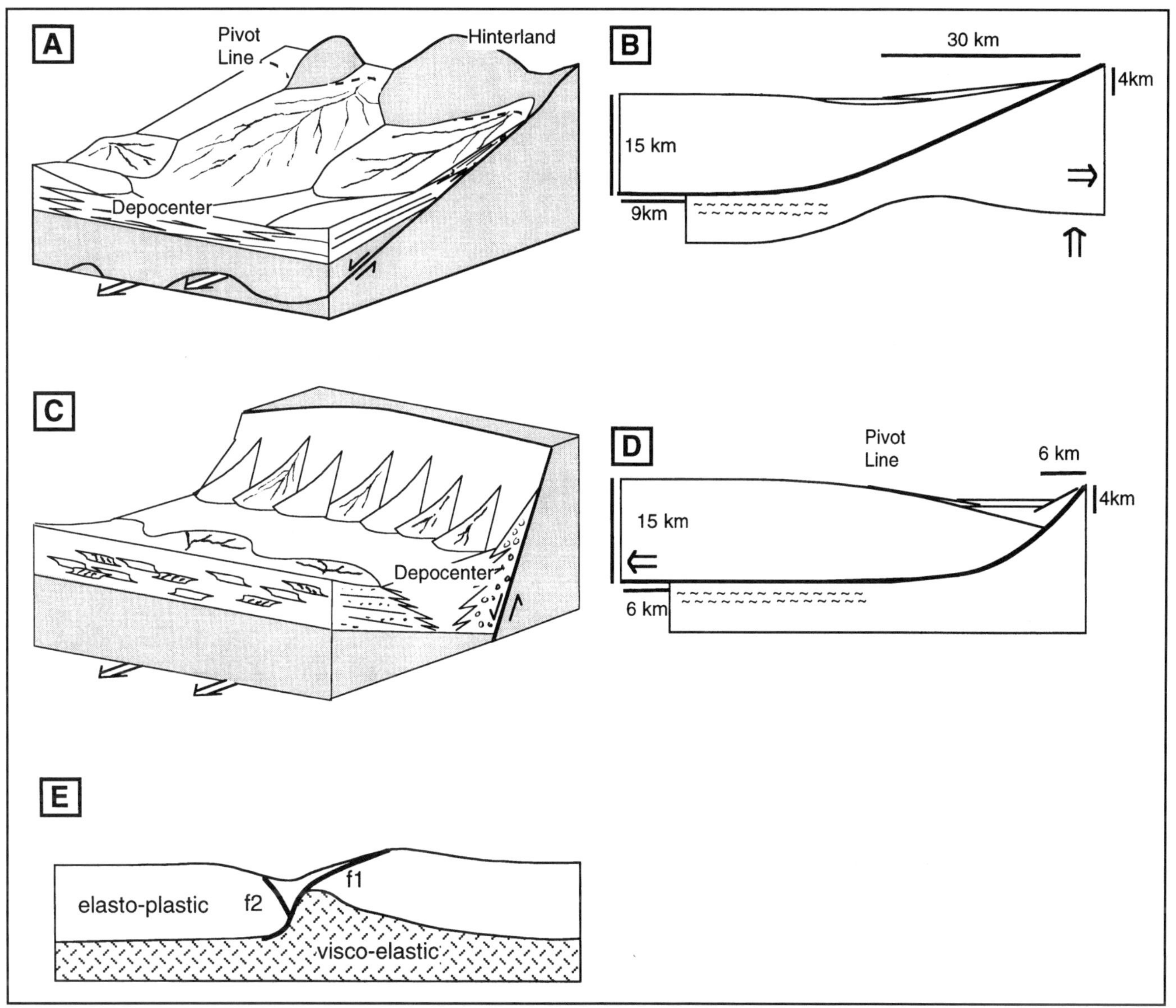

Figure 10. (A) Block diagram of proposed supradetachment basin model; note sediment entry into basin along corrugations and long transverse drainage. (B) Crustal-scale cross section. Note distal depocenter. (C) Asymmetric half-graben model (after Leeder and Gawthorpe, 1987); note predominantly hanging wall source and proximal depocenter. (D) Crustal-scale cross section. (E) Numerical simulation of rapid extension of a brittle plate above a ductile plate (after Schultz-Ela and Bobineau, 1992). Note distal depocenter and footwall highland.

into the basin (Friedmann et al., 1993; Miller and John, 1990). This interpretation reconciles the conflicting structural and sedimentological data and is consistent with the abundance of rock avalanche deposits preserved near the breakaway of synformal or "negative" corrugations (Fig. 4A; Miller and John, 1990).

'Active footwall'

The distant position of the depocenter of Shadow Valley basin with respect to its eastern breakaway and long east-west transverse drainage are not predicted elements of half-graben basin deposition (Fig. 10, C and D; Leeder and Gawthorpe, 1987). In such settings, the depocenter forms very close to the range front, while long drainages with higher sediment yields tend to form within the hanging wall. This is because in extensional settings characterized by steep bounding faults, hanging wall subsidence produced by faulting greatly exceeds associated footwall uplift, usually by a factor of five (Stein and Barrientos, 1985). The half-graben model successfully describes many continental extensional basins (e.g., Cohen, 1991; Olsen, 1991; Roberts and Jackson, 1991; Hutchinson et al., 1992). In contrast to depositional systems described from half-grabens, drainage

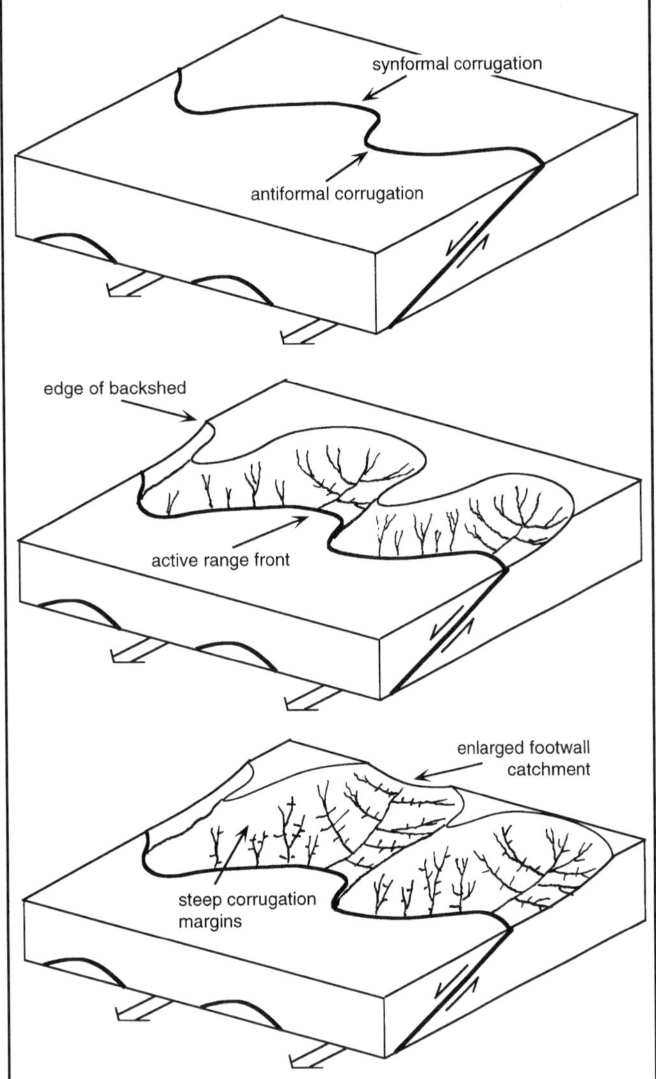

Figure 11. Schematic block diagram of possible topographic evolution of the detachment hinterland.

patterns of the Shadow Valley basin were predominantly transverse across the whole of the basin with a subordinate axial component. Virtually all of the sediment is footwall-derived, rather than hanging wall–derived. Finally, the depocenter lies over 15 km from the breakaway range front to the east rather than close to it.

Other basins found in detachment settings also show long, long-lived, transverse drainages from the footwall (Yarnold, 1994; Lucchitta and Suneson, 1993; Dinter and Royden, 1993). It thus appears that many supradetachment basins do not conform to the continental half-graben model (Friedmann and Burbank, 1995). This discrepancy can be explained if in detachment settings footwall uplift exceeds hanging wall subsidence at the range front (Fig. 10, A and B). Such a pattern of uplift and subsidence might be linked to rapid extension. Numerical models of Schultz-Ela and Bobineau (1992) predict rapid footwall uplift associated with high strain rates (Fig. 10E). During rapid extension, the hanging wall undergoes little subsidence and is essentially passive with respect to a rising footwall. This would result in initial bypassing at the range front and development of a distal depocenter. Bypassing because of footwall uplift might result in basin-wide unconformities followed by continued rapid accumulation, as is observed in the Shadow Valley basin.

Many detachment systems are characterized by very high slip rates along master faults (Foster et al., 1990; Spencer and Reynolds, 1991; Wernicke et al., 1993) and high hinterland topography (Hill et al., 1992; Lee and Lister, 1992). Davis and Lister (1988) argue that the footwall of metamorphic core complexes must rise far more than the hanging wall subsides. This geometry is different from half-graben settings in which subsidence exceeds uplift, yet is consistent with the observed depositional trends in the Shadow Valley basin and other supradetachment settings (Friedmann and Burbank, 1995).

CONCLUSIONS

1. Detachment faulting associated with the Shadow Valley basin began between 13.4 Ma and 13.0 Ma. After intrusion of the Kingston Peak pluton, detachment faulting and related sedimentation ceased in the Kingston Range but continued to the south in the Halloran Hills at 12.0 Ma. Breakup of the Halloran Hills upper plate probably began at 10.5 Ma. Thus, intraplate extension commenced 2–3 m.y. after detachment faulting. Extension and deposition essentially ceased by 7 Ma but may have ended earlier.

2. Unconformities that separate stratigraphic members are a consequence of tectonic episodes. The Member II unconformity relates to the intrusion of the Kingston Peak pluton, the member III unconformity to an episode of transverse growth folding, and member IV to upper plate domino-style block faulting. The structural events before the member IV unconformity are not attributable to internal extension of the upper plate.

3. Sedimentation during detachment faulting was characteristically transverse and derived from the (eastern) footwall. The basin depocenter was distal to the active range front.

4. Rapid accumulation rates calculated from the Shadow Valley basin are largely a consequence of rapid extension. These rates are on the order of 0.5 mm/yr and may be faster.

5. The corrugated geometry of the Kingston Range/Halloran Hills detachment affected sedimentation in the Shadow Valley basin by providing transport conduits for detritus into the basin, a source for rock avalanche breccias, and possibly generating large catchments.

6. Extension resulted in a high footwall and a distal depocenter, most probably due to uplift of the footwall. This may have resulted from unroofing and isostatic reequilibriation. This pattern seems to be characteristic of highly extended or rapidly extended terranes.

ACKNOWLEDGMENTS

This research was conducted under National Science Foundation grants EAR-9118610 (D. W. Burbank) and EAR-9005588 and EAR-9205711 (G. A. Davis). Thanks to Tom Brudos, Kim Bishop, Mary Parke, Clark Burchfiel, Doug Burbank, and Jim Calzia for their work and geological insight, and Jerome Amory for field assistance. Thanks to Andrew Meigs for a very helpful early review, and special thanks to Doug Burbank for discussions concerning magnetic correlations. Reviews from Becky Dorsey, Kathi Beratan, and an anonymous reviewer greatly improved the quality of this manuscript.

REFERENCES CITED

Armstrong, R. L., 1970, Geochronology of the Tertiary igneous rocks, eastern Basin and Range province, western Utah, eastern Nevada, and vicinity, U.S.A: Geological Society of America Bulletin, v. 34, p. 203–232.

Bell, R. E., 1971. Geology of the Shadow Mt. area, eastern San Bernardino County, California [M.S. thesis]: San Diego, California, San Diego State University, 47 p.

Beratan, K. K., 1991. Miocene Synextensional patterns, Whipple Mountains, southeastern California: Implications for the geometry of the Whipple detachment system, Journal of Geophysical Research, v. 96, p. 12,425–12,442.

Bishop, K. M., 1994, Mesozoic and Cenozoic extensional tectonics in the Halloran and Silurian Hills, eastern San Bernardino County, California [Ph. D. dissert.]: Los Angeles, University of Southern California, 181 p.

Boyer, S. E., and Hossack, J. R., 1992, Structural features and emplacement of surficial gravity-slide sheets, northern Idaho–Wyoming thrust belt, *in* Link, p. K., Kuntz, M. A., and Platt, L. B., eds., Regional geology of eastern Idaho and western Wyoming: Boulder, Colorado, Geological Society of America Memoir 179, p. 197–213.

Buck, R. W., 1988, Flexural rotation of normal faults: Tectonics, v. 7, p. 959–973.

Burchfiel, B. C., and Davis, G. A., 1975, Nature and controls of Cordilleran orogenesis, extensions of an earlier synthesis: American Journal of Science, v. 275-a, p. 363–396.

Burchfiel, B. C., and Davis, G. A., 1988, Mesozoic thrust faults and Cenozoic low-angle normal faults, eastern Spring Mountains, Nevada, and Clark Mountains thrust complex, California, *in* Weide, D. L., and Faber, M. L., eds., This extended land: Geological excursions in the southern Basin and Range: Las Vegas, University of Nevada, Las Vegas, p. 87–106.

Burchfiel, B. C., Walker, D., Davis, G. A., and Wernicke, B., 1983, Kingston Range and related detachment faults—A major "breakaway" zone in the southern Great Basin: Geological Society of America Abstracts with Program, v. 15, p. 536.

Calzia, J. P., 1991, Geology of the Kingston Range, southern Death Valley, California, *in* Reynolds, R. E., ed., Crossing the borders: Quarternary studies in eastern California and southwest Nevada: San Bernardino County Museum Mojave Desert Quaternary Research Center Field Guidebook, p. 176–188.

Cande, S. C., and Kent, D. V., 1995, A new geomagnetic polarity time scale for the late Cretaceous and Cenozoic: Journal of Geophysical Research, v. 100, p. 6093–6095.

Cohen, A. S., 1991, Tectono-stratigraphic model for sedimentation in Lake Tanganyika, Africa, *in* Katz, B., ed., Lacustrine basin exploration: Case studies and modern analogues: Tucson, Arizona, American Association of Petroleum Geologists Memoir 50, p. 137–150.

Davis, G.A., and Burchfiel, B.C., 1993, Tectonic problems revisited: The eastern terminus of the Miocene Garlock Fault and the amount of slip on the southern Death Valley fault zone: Geological Society of America Cordilleran Section, Abstracts with Programs, Reno, v. 25, p. 28.

Davis, G. A., and Lister, L. S., 1988, Detachment faulting in continental extension: Perspectives from the southwest U.S. Cordillera: Boulder, Colorado, Geological Society of America Special Paper 218, p. 133–159.

Davis, G. A., Fowler, T. K., Bishop, K. M., Brudos, T. C., Friedmann, S. J., Burbank, D. W., Parke, M. A., and Burchfiel, B. C., 1993, Pluton pinning of an active Miocene detachment fault system, eastern Mojave Desert, California: Geology, v. 21, p. 267–630.

Dinter, D. A., and Royden, L. H., 1993, Late Cenozoic extension in northern Greece: Strymon Valley detachment system and Rhodope metamorphic core complex: Geology, v. 21, p. 45–49.

Dokka, R. K., 1989, The Mojave extensional belt of California: Tectonics, v. 8, p. 363–390.

Dokka, R. K., 1993, Original dip and subsequent modification of a Cordilleran detachment fault, Mojave extensional belt, California: Geology, v. 21, p. 711–714.

Evans, A. L., 1988, Neogene tectonic and stratigraphic events in the Gulf of Suez rift area, Egypt: Tectonophysics, v. 153, p. 235–247.

Fedo, C. M., and Miller, J. M. G., 1992, Evolution of a Miocene half-graben basin, Colorodo River extensional corridor, southeastern California: Geological Society of America Bulletin, v. 104, p. 481–493.

Foster, D. A., Harrison, T. M., Miller, C. F., and Howards, K. A., 1990, The $^{40}Ar/^{39}Ar$ thermochronology of the eastern Mojave Desert California, and adjacent western Arizona with implications for the evolution of metamorphic core complexes: Journal of Geophysical Research, v. 95, p. 20,0005–20,024.

Foster, D. A., Gleadow, A. J. W., Reynolds, S. J., and Fitzgerald, p. G., 1993, The denudation of metamorphic core complexes and the reconstruction of the Transition Zone, west-central Arizona: Thermochronologic constraints: Journal of Geophysical Research, v. 98, p. 2167–2186.

Fowler, T. K., 1992, Geology of Shadow Mountain and the Shadow Valley basin: Implications for Tertiary tectonics of the eastern Mojave desert [M.S. thesis]: Los Angeles, University of Southern California.

Fowler, T. K., Friedmann, S. J., and Davis, G. A., 1993, Two-phase evolution as a record of footwall uplift during extensional detachment faulting: Geological Society of America Annual Meeting, Abstracts with Program, Boston, v. 25, p. A352.

Fowler, T. K., Friedmann, S. J., Davis, G. A., and Bishop, K., 1995, Two-phase evolution of the Shadow Valley basin, southeastern California: A record of footwall uplift during extensional detachment faulting, Basin Research, v. 7, p. 165–179.

Friedmann, S. J., 1993, High-precision radiometric dates from the Miocene Shadow Valley basin: Geological Society of America Cordilleran Section, Abstracts with Program, Reno, v. 25, A39.

Friedmann, S. J., 1995, The Shadow Valley basin, eastern Mojave Desert, California: The structural and stratigraphic evolution of a supradetachment basin, and its implications for extensional tectonics [Ph.D. Thesis]: Los Angeles, University of Southern California, 452 p.

Friedmann, S. J., and Burbank, D. W., 1992, Active footwall uplift recorded in a supradetachment basin: The Miocene Shadow Valley Basin: San Francisco, American Geophysical Union, p. 549.

Friedmann, S. J., and Burbank, D. W., 1994, Chronostratigraphic constraints on paleodrainage, depositional systems, and mass wasting, Miocene Shadow Valley basin, eastern California: Geological Society of America Cordilleran Section, Abstracts with program, San Bernardino, Geological Society of America, 26, p. 53.

Friedmann, S. J., and Burbank, D. W., 1995, Rift basins and Supradetachment basins: Intracontinental extensional endmembers: Basin Research, v. 7, p. 109–127.

Friedmann, S. J., and Davis, G. A., 1994, Characteristics of rock avalanche breccias and gravity-driven glide blocks in the Shadow Valley basin, SE California: Geological Society of America Annual Meeting, Abstracts with Program, Seattle, v. 26.

Friedmann, S. J., Davis, G. A., Burbank, D. W., and Brudos, T. C., 1993, The effect of a corrugated breakaway on drainage configuration: The Miocene Shadow Valley Supradetachment Basin: Geological Society of America Cordilleran section, Abstracts with Program, Reno, v. 25, p. 39.

Friedmann, S. J., Davis, G. A., Fowler, T. K., Brudos, T., Burbank, D. W., and Burchfiel, B. C., 1994a, Stratigraphy and gravity-glide elements of a Miocene supradetachment basin, Shadow Valley, eastern Mojave desert, *in* McGill, S. F., and Ross, T. M., eds., Geological investigations of an active margin: Geological Society of America Cordilleran Section Guidebook: San Bernardino, California, San Bernardino County Museum, p. 302–320.

Friedmann, S. J., Davis, G. A., Fowler, T. K., 1994b, Tectonic and stratigraphic evolution of the Shadow Valley basin, eastern California: Geological Society of America Cordilleran Section, Abstracts with Program, San Bernardino, v. 26, p. 52.

Hack, J. T., 1957, Studies of longitudinal stream profiles in Virginia and Maryland: U.S. Geological Survey Professional Paper 294B.

Hewett, D. F., 1956, Geology and mineral resources of the Ivanpah Quadrangle, California and Nevada: U.S. Geological Survey Professional Paper, 172 p.

Hill, E. J., and Baldwin, S. L., 1993, Exhumation of high-pressure metamorphic rocks during crustal extension in the D'Entrecasteaux region, Papua New Guinea: Journal of Metamorphic Geology, v. 11, p. 261–277.

Hill, E. J., Baldwin, S. L., and Lister, G. S., 1992, Unroofing of active metamorphic core complexes in the D'Entrecasteaux Inslands, Papua New Guinea: Geology, v. 20, p. 907–910.

Holm, D. K., Pavlis, T. K., and Topping, D. J., 1994, Black Mountains crustal section, Death valley extended terrane, California, *in* McGill, S.F., and Ross, T.M., eds., Geological investigations of an active margin: Geological Society of America Cordilleran Section Guidebook: San Bernardino, California, San Bernardino County Museum, p. 31–54.

Howard, K. A., and John, B. E., 1987, Crustal extension along a rooted system of low angle faults, Colorodo River extensional corridor, California and Arizona, *in* Coward, M. P., Dewey, J. F., and Hancock, P. L., eds., Continental extensional tectonics: London, Geological Society of London Special Publication 28, p. 299–311.

Hutchinson, D. R., Golmshtok, A. J., Zonenshain, L. P., Moore, T. C., Scholtz, C. A., and Klitgord, K. D., 1992, Depositional and tectonic framework of the rift basins of Lake Baikal from multi-channel seismic data: Geology, v. 20, p. 589–592.

John, B. E., 1987, Geometry and evolution of a mid-crustal level extensional fault system, Chemehuevi Mountains, southeastern California, *in* Coward, M. P., Dewey, J. F., and Hancock, P. L., eds., Continental extensional tectonics: London, Geological Society of London Special Publication 28, p. 313–335.

John, B. E., and Foster, D. A., 1993, Structural and thermal constraints on the initiation angle of detachment faulting in the southern Basin and Range: The Chemehuevi Mountains case study: Geological Society of America Bulletin, v. 105, p. 1091–1108.

Johnson, N. M., Sheikh, K. A., Dawson-Saunders, E., and McRae, L. E., 1988, The use of magnetic-reversal time lines in stratigraphic analysis: A case study in measuring variability in sedimentation rates, *in* Kleinspehn, K. L., and Paola, C., New perspectives in basin analysis: New York, Springer-Verlag, p. 189–199.

Jones, C. 1983, Paleomagnetism of the Kingston Range, San Bernardino County, California: A tectonic interpretation: Eos, v. 64, p. 686.

Keefer, D. K., 1984, Rock avalanches caused by earthquakes: source characteristics: Science, v. 223, p. 1288–1290.

Lee, J., and Lister, G. S., 1992, Late Miocene ductile extension and detachment faulting, Mykonos, Greece: Geology, v. 20, p. 121–124.

Leeder, M. R., and Gawthorpe, R. L., 1987, Sedimentary models for extensional tilt-block/half-graben basins, *in* Coward, M. P., Dewey, J. F., and Hanckock, P. L., eds., Continental extensional tectonics: Geological Society of London Special Publication 28, p. 139–152.

Leeder, M. R., and Jackson, J. A., 1993, The interaction between normal faulting and drainage in active extensional basins, with examples from the western United States and central Greece: Basin Research, v. 5, p. 79–102.

Leopold, L. B., Wolman, M. G., and Miller, J. P., 1964, Fluvial Processes in Geomorphology: San Fransisco, Freeman.

Lucchitta, I., and Suneson, N. H., 1993, Dips and extension: Geological Society of America Bulletin, v. 105, p. 1346–1356.

McMackin, M. C., 1992, Tectonic Evolution of the Kingston Range, Death Valley, California [Ph.D. dissert.]: University Park, Pennsylvania State University, 158 p.

Miller, J. M. G., and John, B. E., 1988, Detached strata in a Tertiary low-angle normal fault terrane, southeatern California: A sedimentary record of unroofing, breaching, and continued slip: Geology, v. 16, p. 645–649.

Miller, J. M. G., and John, B. E., 1990, Sedimentation patterns near an undulating detachment fault during Cenozoic continental extension, Chemehuevi Mountains, eastern California: International Sedimentological Congress, Nottingham, England, v. 13, p. 359.

Nielson, J. E., and Beratan, K. K., 1990, Tertiary basin development and tectonic implications, Whipple detachment system, Colorado River extensional corridor, California and Arizona: Journal of Geophysical Research, v. 95, p. 571–580.

Olsen, P. E., 1991, Tectonic, climatic, and biotic modulation of lacustrine ecosystems—Examples from Newark Supergroup of eastern North America, *in* Katz, B., ed., Lacustrine basin exploration: Case studies and modern analogues: Tucson, Arizona, American Association of Petroleum Geologists Memoir 50, p. 209–224.

Parke, M., and Davis, G. A., 1990, Gravity gliding in the eastern Mojave Desert? Only the shadows know. . . : Geological Society of America Cordilleran Section, Abstracts with Programs, v. 22, p. 74.

Renne, P. R., Fulford, M. M., Busby-Spera, C., 1991, High resolution $^{40}Ar/^{39}Ar$ chronostratigraphy of the late Cretaceous El Gallo Formation, Baja California del Norte, Mexico: Geophysical Research Letters, v. 18, p. 459–462.

Reynolds, R. E., 1990, Erosion, deposition, and detachment: The Halloran Hills sequence, *in* Reynolds, J., ed., At the end of the Mojave: Quaternary studies in the eastern Mojave Desert: Redlands, California, San Bernardino County Museum, p. 101–104.

Reynolds, R. E., and McMackin, M. R., 1988, Cenozoic Tectonics in the Halloran Hills and Kingston Range: Field trip road log, *in* Weide, D. L., and Faber, M. L., eds., This extended land: Geological excursions in the southern Basin and Range: Las Vegas, University of Nevada, p. 201–206.

Roberts, S., and Jackson, J., 1991, Active normal faulting in central Greece: An overview, *in* Roberts, A. M., Yielding, G., and Freeman, B., eds., The geometry of normal faults: Geological Society of London Special Publications, p. 125–142.

Rosendahl, B. R., 1987, Architecture of continental rifts with special reference to East Africa: Annual Review of Earth and Planetary Science, v. 15, p. 445–503.

Schaller, P. J., 1991, Analysis and implications of large Martian and terrestial landslides [Ph.D. dissert.]: Pasadena, California Institute of Technology, 451 p.

Schultz-Ela, D. D., and Bobineau, J. P., 1992, Extension of a brittle layer over a ductile layer: Eos, v. 73, no. 43, p. 561–562.

Scrivner, P. J., 1984, Stratigraphy, sedimentology, and vertebrate ichnology of the Copper Canyon Formation (Neogene), Death Valley National Monument [M.S. thesis]: Los Angeles, University of Southern California, 486 p.

Skirvin, T. M., 1989, Quarternary to Miocene deformation and sedimentation in the Old Dad Mountains. [M.S. thesis]: University of New Mexico, 128 p.

Spencer, J. E., and Reynolds, S. J., 1991, Tectonics of Mid-Tertiary extension along a transect through west-central Arizona: Tectonics, v. 10, p. 1204–1221.

Stein, R. S., and Barrientos, S. E., 1985, Planar high-angle faulting in the Basin and Range: Geodetic analysis of the 1983 Borah Peak, Idaho, earthquake: Journal of Geophysical Research, v. 90, p. 11,355–11,366.

Suppe, J., Chou, G. T., and Hook, S. C., 1992, Rates of folding and faulting determined from growth strata, *in* McKlay, K. R., ed., Thrust tectonics: London, Chapman and Hall, p. 105–121.

Talling, P. J., and Burbank, D. W., 1993, Assessment of uncertainties in magnetostratigraphic dating of sedimentary strata, *in* Aïssaoui, D. M., McNeill, D. F., and Hurley, N. F., ed., Application of paleomagnetism to sedimentary geology, SEPM Special Publication 49, p. 41–59.

Topping, D. A., 1993, Paleogeographic reconstruction of the Death Valley extended region: Evidence for Miocene large rock-avalanche deposits in the Amaragosa Chaos basin, California: Geological Society of America Bulletin, v. 105, p. 1190–1213.

Turrin, B. D., Dohrenwend, J. C., Drake, R. E., and Curtis, G. H., 1985, K-Ar ages from the Cima volcanic field, eastern Mojave Desert, California: Isochron/West, v. 44, p. 9–16.

Varnes, D. J., 1978, Slope movement types and process, *in* Schuster, R. L., and Krinzek, R. J., eds., Landslides, analysis, and control: Washington, D.C., Transportation Board, National Academy of Sciences, Specal Report 176, p. 11–33.

Wernicke, B. P., and Axen, G. J., 1988, On the role of isostacy in the evolution of normal fault systems: Geology, v. 16, p. 848–851.

Wernicke, B. R., Axen, G. J., and Snow, J. K., 1988, Basin and Range extensional tectonics at the latitude of Las Vegas, Nevada: Geological Society of America Bulletin, v. 100, p. 1738–1757.

Wernicke, B. P. , Snow, J. K. , Hodges, K. V., and Walker, J. D., 1993, Structural constraints on Neogene tectonism in the southern Great Basin, *in* Lahren, M. M., Trexler, J. H., and Spinosa, C., eds., Crustal evolution of the Great Basin and the Sierra Nevada: Reno, Nevada, Mackay School of Mines, p. 453–480.

Woodburne, M. O., Tedford, R. T., and Swisher, C. C., 1990, Lithostratigraphy, biostratigraphy, and geochronology of the Barstow Formation, Mojave Desert, southern California: Geological Society of America Bulletin, v. 102, p. 459–477.

Wright, L. A., and Troxel, B. W., 1967, Limitations on right-lateral, strike-slip displacement, Death Valley and Furnace Creek fault zone, California: Geological Society of America Bulletin, v. 78, p. 933–950.

Yarnold, J. C., 1993, Rock-avalanche characteristics in dry climates and the effect of flow into lakes: Insights from the mid-Tertiary sedimentary breccias near Artillery Peak, Arizona: Geological Society of America Bulletin, v. 105, p. 345–360.

Yarnold, J. C., 1994, Tertiary sedimentary rocks associated with the Harcuvar core complex in Arizona (U.S.A.): Insights into paleogeographic evolution during displacement along a major detachment fault system: Sedimentary Geology, v. 89, p. 43–63.

Yarnold, J. C., and Lombard, J. P., 1989, A facies model for rock-avalanche deposits formed in dry climates, *in* Colburn, I. P., Abbot, P. L., and Minch, J., eds., Conglomerates in basin analysis: A symposium dedicated to A. O. Woodford: Society of Economic Paleontologists and Mineralogists, Pacific Section, v. 62, p. 9–31.

Manuscript Accepted by the Society April 21, 1995

Printed in U.S.A.

Geological Society of America
Special Paper 303
1996

Evolution of a supradetachment extensional basin: The Lower Miocene Pickhandle basin, central Mojave Desert, California

Robert P. Fillmore* and J. Douglas Walker
Department of Geology, University of Kansas, Lawrence, Kansas 66045

ABSTRACT

The Lower Miocene Pickhandle Formation was deposited in an elongate, northwest–trending supradetachment basin that formed in response to northeast-directed extension on the central Mojave metamorphic core complex. Regional paleogeographic reconstruction suggests that the basin was bounded to the southwest by the detachment breakaway, to the northeast by the hanging wall, and to the southeast by the sidewall of a right-lateral transfer fault; the northwest boundary or continuation of the basin is uncertain.

Silicic to intermediate volcanism initiated the basin at ~23.7 Ma ($^{40}Ar/^{39}Ar$, Gravel Hills). Primary lava flows and mudflow deposits in the Calico Mountains, Waterman Hills, and Mud Hills grade to the northwest, in the Gravel Hills, to epiclastic fluvial deposits. Axial fluvial strata in the Gravel Hills intertongue with coarse alluvial-fan deposits that had a source in the hanging wall to the northeast. During this period, volcanic deposition was also occurring in the southern part of the basin, in the Lead Mountain and Elephant Mountain areas.

During the interval from 21.7 Ma to 21.4 Ma ($^{40}Ar/^{39}Ar$, Waterman Hills and Mud Hills) volcanism waned, and fluvial and rock avalanche deposition became dominant. Clast imbrication and detrital composition in fluvial deposits in the Mud Hills indicate a granitic source to the south. Overlying rock avalanche breccias of identical granitic composition strongly suggest the same source terrane. This source is interpreted to be the breakaway escarpment. During this period, the southern part of the basin (Waterman Hills, Lead Mountain, Elephant Mountain) became partitioned from the northern part (Mud Hills, Gravel Hills). In the southern sub-basin, debris-flow deposits in the Waterman Hills grade southward, at Lead Mountain, into fluvial conglomerate. These north-derived fluvial deposits intertongue with lacustrine carbonate and siltstone. At Elephant Mountain 5 km to the south, southern-derived alluvial-fan deposits grade abruptly northward into lacustrine deposits. Conglomerate at Waterman Hills and Lead Mountain likely were derived from highlands associated with an intrabasinal transfer fault that separated the less-extended southern sub-basin from the more highly extended area to the north. Fan deposits at Elephant Mountain are interpreted to have been derived from the sidewall escarpment of a northeast-trending, right-lateral transfer fault that marked the southern termination of the central Mojave metamorphic core complex extensional system.

*Present address: Geology Department, Northern Arizona University, Flagstaff, Arizona 86011-4099.

Fillmore, R. P., and Walker, J. D., 1996, Evolution of a supradetachment extensional basin: The Lower Miocene Pickhandle basin, central Mojave Desert, California, *in* Beratan, K. K., ed., Reconstructing the History of Basin and Range Extension Using Sedimentology and Stratigraphy: Boulder, Colorado, Geological Society of America Special Paper 303.

INTRODUCTION

The study of strata associated with metamorphic core complex–style extensional systems allows insight into processes of upper crustal extension that is not otherwise available. Such studies are useful in evaluating the tectonic controls on basin evolution, determining the basin geometry, and refining the timing of sedimentologic and tectonic events.

The purpose of this paper is twofold: to document the stratigraphic and structural evolution of a supradetachment basin during the early stage of formation, and to determine, using the sedimentary record, the amount of extension that has occurred in the central Mojave Desert. Additionally, sedimentologic and chronostratigraphic evidence from the Lower Miocene Pickhandle Formation is presented that indicates a temporal and spatial relationship with development of the central Mojave metamorphic core complex (CMMCC).

The Pickhandle Formation is exposed in a discontinuous, elongate, northwest-southeast–trending outcrop belt in the Gravel Hills, Mud Hills, Calico Mountains, Waterman Hills, Lead Mountain, and Elephant Mountain areas (Fig. 1). New $^{40}Ar/^{39}Ar$ dates from the Pickhandle (Walker and Fillmore, 1993; Walker et al., 1995) indicate that deposition was synchronous with northeast-directed extension on the CMMCC. Paleogeographic reconstruction based on provenance studies and sediment dispersal patterns suggests that the basin was bounded to the southwest by the detachment breakaway zone and to the northeast by the hanging wall of the detachment system.

Evidence for large magnitude, Lower Miocene extension in the central Mojave Desert has grown in the past few years (e.g., Dokka and Glazner, 1982; Dokka, 1986, 1989; Dokka and Woodburne, 1986; Glazner et al., 1989; Walker et al., 1990). The Waterman Hills detachment fault (WHDF), which has been recognized in the Waterman Hills, Mitchel Range, and Mt. General area (Fig. 1), separates a mylonitized footwall complex of igneous and metamorphic rocks from steeply southwest-dipping to overturned, volcanic and sedimentary rocks of the Pickhandle Formation. These exposures comprise the central Mojave metamorphic core complex (Fig. 1). Sense of shear indicators show a top-to-the-northeast transport direction (Dokka and Woodburne, 1986; Glazner et al., 1988; Bartley et al., 1990).

Geochronologic data from the footwall of the CMMCC indicate that mylonitization and cooling occurred from ca. 24–19 Ma. Walker et al. (1990) obtained an age of 23.0 ± 0.9 Ma (U-Pb zircon) from a synmylonitic dacite intrusion in the Waterman Hills. An age of 22 to 23 Ma from zircon discordia was reported from the variably mylonitized Waterman Hills granodiorite and $^{40}Ar/^{39}Ar$ analyses on biotite yielded cooling ages of 20.1 Ma and 20.9 Ma (Glazner et al., 1992). Apatite fission track data from the footwall in the Mitchel Range indicate cooling below 70 °C by 19 Ma (Dokka and Baksi, 1988).

Age data from the Pickhandle Formation indicate that basin formation began at ca. 23.7 Ma, and Pickhandle sedimentation was over by ca. 19.3–18.9 Ma. New age data from Waterman Hills and Lead Mountain strata indicate that they are correlative with the Pickhandle Formation elsewhere in the study area and, based on lithologic similarities, are interpreted to be part of the Pickhandle Formation. The bracketing of Pickhandle sedimentation between 23.7 Ma and 18.9 Ma indicates that basin formation was synchronous with regional crustal extension on the CMMCC, which is constrained between 23.0 Ma and 19 Ma.

Recent paleomagnetic studies in the Mud Hills, Gravel Hills, and Waterman Hills suggest that the Pickhandle Formation has experienced 23° of counterclockwise rotation (Valentine et al., 1993). Data from the Oligo-Miocene Lane Mountain Quartz Latite and the Jackhammer Formation suggest about 55° of net clockwise rotation (Valentine et al., 1993). Data from the overlying Barstow Formation and younger rocks show no rotation, suggesting that Pickhandle rotation was related to hanging wall movement during extensional deformation (Bartley and Glazner, 1991; Valentine et al., 1993). Without data regarding which specific intervals of the Pickhandle Formation were analyzed, it is difficult to use these data conclusively in the paleogeographic reconstruction of this study. Although 23° of counterclockwise rotation changes paleocurrent vectors, it is not enough to significantly alter our interpretations or reconstruction.

METHODS

Eight complete sections of the Pickhandle Formation were measured in detail; partial sections were measured where lateral variations warranted additional data. Lateral and vertical facies variations were examined to determine depositional environments. Where structural complications obscured these relationships, detailed facies mapping was conducted.

Provenance of clastic deposits was constrained by clast counts on conglomeratic units using a grid system on the outcrop that varied in dimension with the average clast size. A minimum of 200 clasts per station were counted. Paleocurrent data were obtained primarily from clast imbrication; cross-stratification and bidirectional data from trough axes were also measured where present. Data were corrected for structural tilt before plotting rose diagrams and calculating the mean vector.

STRATIGRAPHY

Pickhandle Formation

In the Calico Mountains, Lead Mountain, and locally in the Mud Hills, the Pickhandle Formation conformably overlies the Jackhammer Formation (McCulloh, 1952). The Jackhammer Formation at its type locality comprises about 200 m of, in ascending order, sandstone, conglomerate, tuff, and basalt (McCulloh, 1952); elsewhere the Jackhammer is much thinner or missing. The age of the Jackhammer Formation is unknown, but the apparently conformable relationship with the Pickhandle Formation suggests a late Oligocene or early Miocene age. The Jackhammer likely represents the incipient stage of extension,

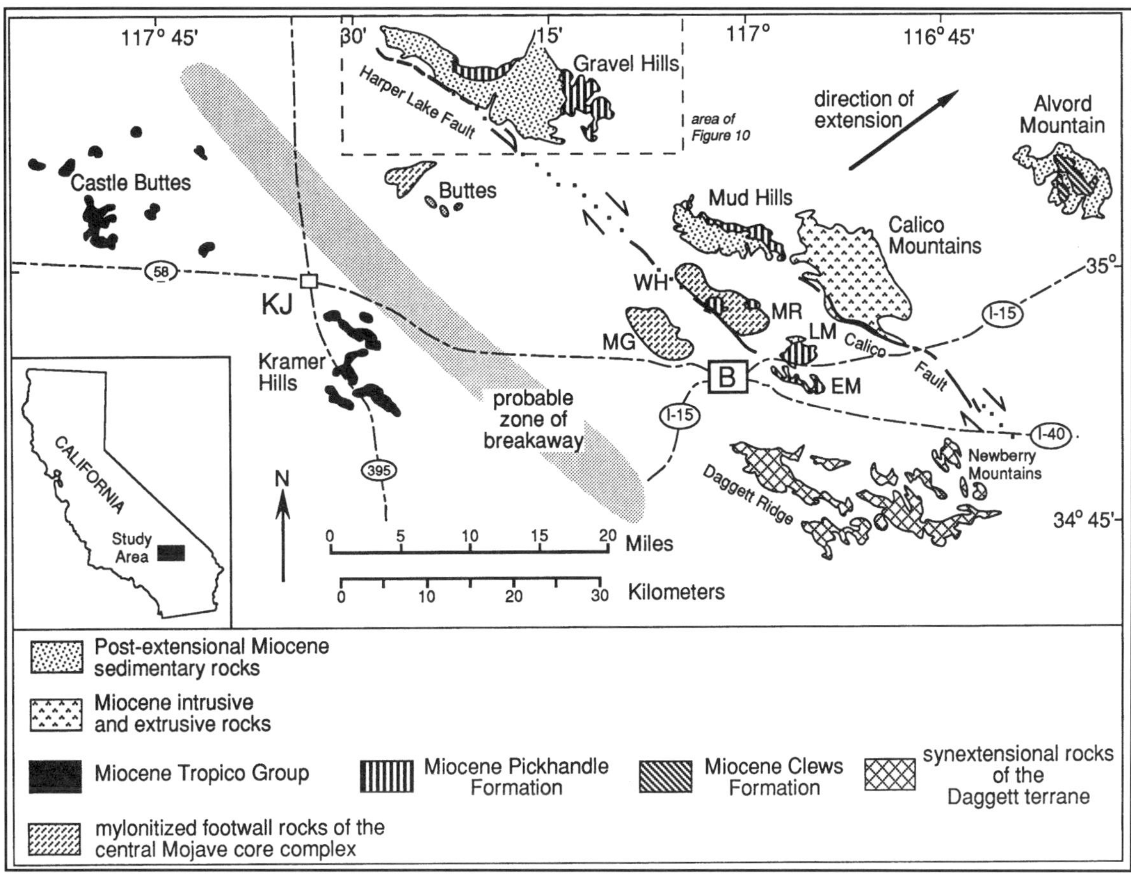

Figure 1. Generalized geologic map of Miocene rocks in the central Mojave Desert, California, showing distribution of Miocene rocks, the central Mojave metamorphic core complex, and the localities discussed in this chapter. KJ, town of Kramer Junction; WH, Waterman Hills; MR, Mitchel Range; MG, Mt. General; LM, Lead Mountain; EM, Elephant Mountain; B, town of Barstow.

prior to large-scale faulting. Where the Jackhammer Formation is absent, the Pickhandle Formation rests nonconformably on plutonic basement rock.

The Pickhandle Formation everywhere is overlain by the Barstow Formation (Merriam, 1915), with the contact marked by a slight angular unconformity. The Barstow Formation ranges to more than 1,000 m in thickness. In the Mud Hills it consists of a basal fluvial conglomerate and sandstone (Owl Canyon Conglomerate Member) and an upper succession of lacustrine claystone, sandstone, and carbonate (Dibblee, 1968). The Barstow Formation in the Gravel Hills grades southeastward from a thick sequence of coarse conglomerate of alluvial fan origin to interbedded lacustrine sandstone, siltstone, and claystone.

The Pickhandle Formation comprises as much as 1,000 m of volcanic and clastic sedimentary rock. Over most of its exposure the formation consists of a lower succession of primary and epiclastic volcanic deposits (Unit 1) and an upper succession of lacustrine-, fluvial- and rock-avalanche–derived deposits (Unit 2). In general, the transition between the lower volcanic and upper clastic successions is gradational over an interval of as much as ~100 m, although in the Mud Hills it occurs abruptly over a 5 m interval. The abrupt transition in the Mud Hills led Dokka et al. (1991) to redefine the upper clastic succession as the "Mud Hills Formation." This terminology was also used by Ingersoll et al. (1993; this volume) in a study of the synextensional strata in the Mud Hills. This redefinition is rejected here for several reasons. First, the "Mud Hills Formation" is only applicable in the Mud Hills. The transition is gradational or ill-defined at other Pickhandle Formation localities. Furthermore, epiclastic volcanic strata in the Gravel Hills are intercalated with coarse fluvial and alluvial fan deposits and no upper and lower successions can be readily discerned. The redefinition of Dokka et al. (1991) is based on the examination of a single locality of the regionally extensive Pickhandle Formation and cannot be applied at other localities.

Age data—Pickhandle Formation

Miocene stratigraphy of the study area and age data pertinent to the Pickhandle Formation are summarized in Figure 2. The age of the Pickhandle Formation is based on data from volcanic rocks (Fig. 2). In the Gravel Hills, Burke et al. (1982) reported an age of 18.9 ± 1.3 (whole rock, K-Ar) from a quartz latite welded tuff at the top of the Pickhandle Formation. An $^{40}Ar/^{39}Ar$ age of 23.7 ± 0.5 was obtained from a tuff located 60 m above the base of the formation in the western Gravel Hills (plagioclase; Walker and Fillmore, 1993; Walker et al., 1995).

Several dates have been obtained from the Pickhandle Formation in the Mud Hills (Fig. 2). A dacite breccia in the lower part of Unit 1 at Solomon Canyon has yielded an $^{40}Ar/^{39}Ar$ age of 22.4 ± 0.5 Ma (plagioclase; Walker and Fillmore, 1993; Walker et al., 1995). In the middle of the formation, at the transition between the lower volcanic sequence of Unit 1 and the overlying clastic succession (Unit 2 of this paper; Mud Hills Formation of Dokka et al., 1991), two tuffs have been dated (lower and upper, respectively) at 21.7 ± 0.5 Ma and 21.4 ± 0.5 Ma (biotite; Walker and Fillmore, 1993; Walker et al., 1995). A minimum age for the Pickhandle Formation in the Mud Hills is provided by an $^{40}Ar/^{39}Ar$ age of 19.3 ± 0.02 Ma from the Red Tuff, located at the base of the Owl Canyon Conglomerate Member of the Barstow Formation (Fig. 2; Woodburne et al., 1990). The Owl Canyon Conglomerate unconformably overlies the Pickhandle Formation.

An $^{40}Ar/^{39}Ar$ date of 21.7 ± 0.5 Ma was obtained from a tuff in the Pickhandle Formation in the Waterman Hills (biotite; Walker and Fillmore, 1993; Walker et al., 1995). Like similar-age tuffs in the Mud Hills, the dated unit occurs just below the boundary between Unit 1 and Unit 2 (Fig. 2). In the Lead Mountain area, two $^{40}Ar/^{39}Ar$ dates have been obtained (biotite; Fig. 2; Walker and Fillmore, 1993; Walker et al., 1995). A tuff bed in the lower part of the section yielded an age of 23.0 ± 0.5 Ma. In the upper part of the section, a tuff unit interbedded with carbonate and siltstone yields an age of 21.3 ± 0.5 Ma.

Regional correlations

Newly acquired age data from the Pickhandle Formation allow for more accurate correlation with other Miocene strata in the central Mojave Desert. The Tropico Group (Dibblee, 1958), exposed west of the Pickhandle outcrop belt (Fig. 1), is also of Lower Miocene age. Tropico Group strata are divided into Lower and Upper units (Dibblee, 1958) that are separated by an angular unconformity. Two distinct basalt flows have been identified at the top of the Lower Tropico Group. Whole rock K-Ar analyses of the basalts have yielded ages of 21.2 ± 0.5 Ma (Red Buttes Basalt; Dokka and Baksi, 1989), and 19.8 ± 0.7 Ma (Saddleback Basalt; Armstrong and Higgins, 1973). These data indicate that the Tropico Group is largely time correlative to the Pickhandle Formation, although we will show that they formed in different basins.

East of the Pickhandle outcrop belt, at Alvord Mountain (Fig. 1), the Clews Formation (Byers, 1960; Fillmore, 1993; Fillmore et al., 1994) lies nonconformably on basement rock

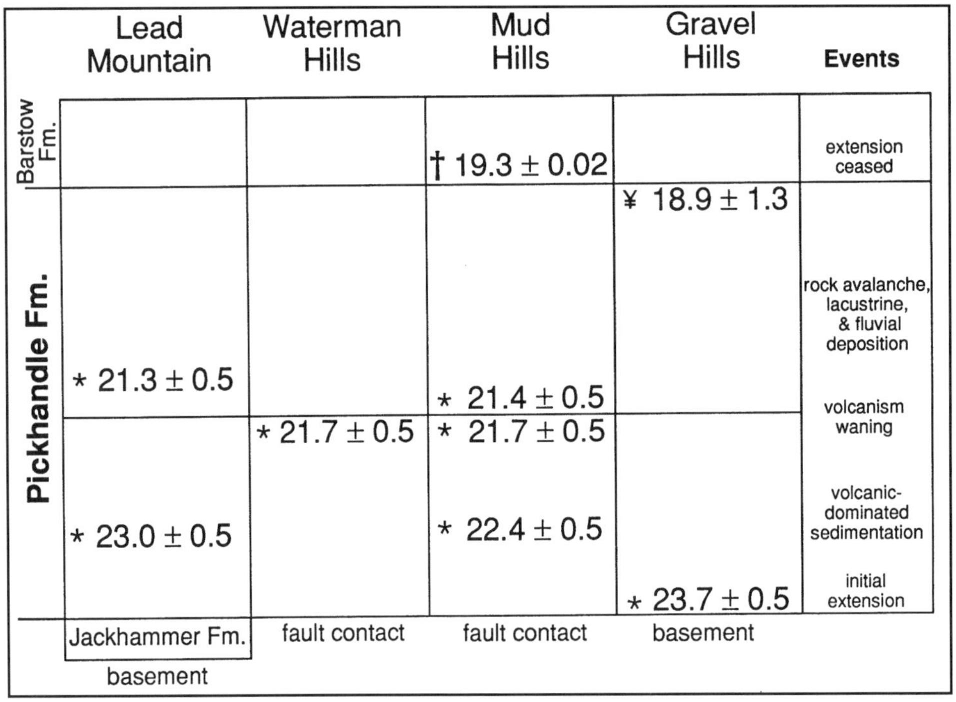

	Lead Mountain	Waterman Hills	Mud Hills	Gravel Hills	Events
Barstow Fm.			† 19.3 ± 0.02		extension ceased
Pickhandle Fm.				¥ 18.9 ± 1.3	rock avalanche, lacustrine, & fluvial deposition
	* 21.3 ± 0.5		* 21.4 ± 0.5		volcanism waning
		* 21.7 ± 0.5	* 21.7 ± 0.5		
	* 23.0 ± 0.5		* 22.4 ± 0.5		volcanic-dominated sedimentation
				* 23.7 ± 0.5	initial extension
	Jackhammer Fm. / basement	fault contact	fault contact	basement	

Figure 2. Stratigraphic and age data for Pickhandle Formation and associated strata in the central Mojave Desert. The symbol * refers to $^{40}Ar/^{39}Ar$ data from mineral separates from this study (Walker et al., 1995) ; † refers to $^{40}Ar/^{39}Ar$ data from Woodburne et al., 1990; ¥ refers to K-Ar whole rock analysis of quartz latite from Burke et al., 1982.

and is interpreted to be correlative to the Pickhandle Formation. The maximum age is unknown, but a minimum age is provided by the conformably overlying Peach Springs Tuff (18.5 ± 0.2 Ma; $^{40}Ar/^{39}Ar$; Nielson et al., 1990).

The Tropico, Pickhandle, and Clews basins are interpreted as a tripartite set of extensional basins associated with deformation on the CMMCC (Fig. 3), each representative of a different structural setting in the extensional system (Fillmore et al., 1994). In a west to east direction, the Tropico basin is interpreted to be located on the backside of the detachment breakaway, and was bounded to the east by the isostatically uplifted zone of the footwall (Fig. 3). The Pickhandle basin formed in a supradetachment setting with the west margin defined by the breakaway zone, and the east margin bounded by the hanging wall of the detachment (Fig. 3). The Clews basin formed well within the hanging wall of the extensional system and marks the eastern limit of recognizable crustal deformation in the system (Fig. 3; Fillmore, 1993; Fillmore et al., 1994).

SEDIMENTOLOGY

In this section, lithofacies in the Pickhandle Formation are described and interpreted. The section is divided into nonvolcanic and volcaniclastic-dominated lithofacies. For the sake of clarity the individual lithofacies are discussed in a proximal to distal order. Lithofacies assemblages and interpretations of depositional environments are shown in Table 1. Spatial rela-

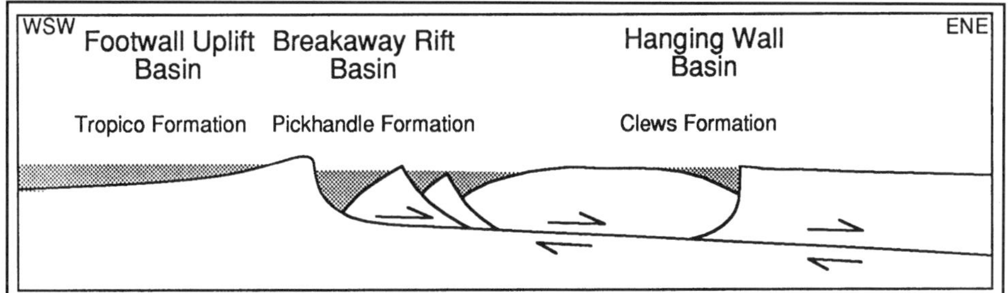

Figure 3. Schematic cross-section of basins formed early in the development of the central Mojave extensional system and related tectonic elements.

TABLE 1. LITHOFACIES ASSEMBLAGES IN THE PICKHANDLE FORMATION, CENTRAL MOJAVE DESERT, CALIFORNIA*

NONVOLCANIC ASSEMBLAGES (UNIT 2)									
Assemblage 1: Rock Avalanche Deposits		Assemblage 2: Proximal/Medial Alluvial Fan		Assemblage 3: Medial/Distal Alluvial Fan		Assemblage 4: Braided Stream		Assemblage 5: Lacustrine	
Facies 1	Monolithologic breccia	Facies 2	Structureless conglomeratic sandstone	Facies 4	Sheetlike sandstone and conglomerate	Facies 5	Lenticular sandstone and conglomerate	Facies 6	Horizontal bedded pebbly sandstone
		Facies 3	Clast-supported conglomerate					Facies 7	Carbonate rock
								Facies 8	Siltstone
VOLCANIC ASSEMBLAGES (UNIT 1)									
		Assemblage 6: Proximal Debris Apron		Assemblage 7: Sheetflood-Dominated Plain		Assemblage 8			
		Facies 9	Conglomeratic lithic tuff	Facies 10	Clast-supported conglomerate	Facies 13	Well-bedded pumiceous tuff		
		Facies 10	Clast-supported conglomerate	Facies 11	Tuffaceous siltstone	Nonsedimentary rhyolite lava and basalt flows			
		Facies 11	Tuffaceous siltstone	Facies 12	Sheetlike sandstone and conglomerate				

*Assemblages are arranged from proximal to distal.

tions between the assemblages are discussed so that depositional settings can be reconstructed on a basinwide scale. These reconstructions are combined with sediment dispersal and provenance data to evaluate controls on sedimentation throughout the depositional history of the basin.

Non-volcanic lithofacies

Facies 1—Monolithologic breccia. Monolithologic breccia deposits range up to 77 m thick and where outcrop permits, may be traced up to 6 km laterally. Breccia consists of angular rock fragments that range up to several meters in length and is composed dominantly of plutonic rock, but may locally consist of felsite or contain felsite lenses. Breccia units contain a typical vertical sequence that begins with a 1–2 m interval of deformed substrate overlain by 1–2 m of mixed substrate and breccia. This mixed zone consists of breccia lenses up to 205 as much as mm thick bordered by thin (as much as 0.5–25 mm), subhorizontal, undulatory shear zones. The base may be planar or undulatory. Thin (as much as 1–35 mm) clastic dikes composed of substrate sediment commonly extend vertically from the base of the mixed zone and terminate abruptly within a shear zone or at the base of large clasts. Matrix-poor breccia overlies the mixed zone. Mixed zone thickness varies from body to body but is usually at least 2 m. This subfacies typically coarsens upward and grades to matrix- and clast-supported breccia in which shattered blocks up to several meters in length are common. The tops of some breccia units contain a red-stained weathered zone ~1 m thick. Monolithologic breccia is a common lithofacies in Unit 2 of the Pickhandle Formation.

Monolithologic breccias are interpreted as the products of large rock avalanches that originated on nearby fault escarpments. Similar deposits are common in Tertiary extensional basins throughout the Basin and Range (e.g., Longwell, 1951; Burchfiel, 1966; Krieger, 1977; Kerr, 1984; Yarnold and Lombard, 1989; Nielson and Beratan, 1990; Fillmore, 1993; Yarnold, 1993). The basal mixed zone represents a horizon of intense shear that formed during avalanche movement, as the body "rode" on a shearing basal layer. Rapid upward coarsening within the breccia body is interpreted to reflect an upward decrease in the degree of fragmentation as a result of a vertical decrease in the effective normal stress within the rock mass as the overburden thickness of overlying breccia body decreases upward (Fillmore, 1993).

Facies 2—Structureless conglomeratic sandstone. Structureless beds of sandstone may exhibit crude normal grading or are ungraded. Depositional units are lenticular or tabular and range to as much as 4 m thick. A notable feature is the common presence of isolated boulders (as much as 1 m in diameter) in the middle or upper part of a depositional unit (Fig. 4a). Facies 2 is common in Unit 2 in the Waterman Hills (Fig. 5) and the Gravel Hills (Fig. 6).

Structureless conglomeratic sandstone units are interpreted to be debris flow deposits. Lenticular geometries are attributed

Figure 4. (A) Discrete boulder supported in tuffaceous sandstone matrix from channelized debris-flow deposit (Assemblage 1) in middle part of the Black Canyon–Gravel Hills section. Staff is marked in feet. (B) Stacked lenticular conglomerate and sandstone facies association (Assemblage 1) showing basal conglomerate in scour hollow, overlain by low-angle cross-stratified and horizontal stratified pebbly sandstone. Hammer for scale.

to flow within earlier-formed channels. Tabular units represent unconfined debris flow. A high strength for all units of facies 2 is indicated by the presence of large (>1 m) matrix-supported boulders high in the successions (Fig. 4a). This is attributed to a large component of fine-grained ash and pumice in the matrix (e.g. Waresback and Turbeville, 1990; Turbeville, 1991).

Facies 3—Clast-Supported Conglomerate. Clast-supported boulder conglomerate successions are laterally continuous and form thick sequences with individual depositional units ranging >8 m thick. Imbrication in the form of a(t)b(i) is common throughout facies 3. Intercalations of conglomeratic sandstone are present but rarely are more than 0.5 m thick. Clasts range up to boulder size (up to 1.5 m) and are angular to subrounded. Matrix is composed of tuffaceous sandstone. Facies 3 is a common component of the Pickhandle Formation in Unit 2 at Elephant Mountain (Fig. 5), and in Unit 1 in the Gravel Hills-Black Canyon section (Fig. 6).

Facies 3 deposits represent deposition on the proximal to medial part of an alluvial fan (Table 1). The paucity of matrix-supported debris flow deposits that are typical of proximal deposits may indicate a more medial location on the fan. In such a setting, proximal matrix-supported deposits would be reworked during flood events and transported downfan.

Facies 4—Sheetlike stratified sandstone and conglomerate. Laterally extensive bodies of interstratified sandstone and conglomerate range to as much as 5 m thick and are traceable laterally as far as outcrop permits, as much as 2 km. Depositional units may display normal or inverse grading, or they may be ungraded. Horizontal stratification is defined by centimeter-scale alternations of sandstone and conglomerate. Sandstone is conglomeratic and commonly grades laterally and/or vertically into clast-supported conglomerate with imbrication, a(t)b(i). Deposits are poorly sorted with clasts ranging from pebble to boulder size, and sand ranging from fine- to coarse-grained. Facies 4 is common in Unit 2 of the Pickhandle Formation at Elephant Mountain, the Waterman Hills, and the Mud Hills (Figs. 5 and 6).

Poorly sorted sheetlike sandstone and conglomerate bodies were deposited during unconfined, traction-dominated sheet-flood flow in the medial to distal reaches of an alluvial fan (Table 1). The lateral continuity of depositional units and paucity of internal erosion surfaces suggest rapid, episodic sedimentation in an unconfined setting. Well-developed imbrication in clast-supported conglomerate suggests bedload traction transport (Rust, 1972).

Facies 5—Lenticular stratified sandstone and conglomerate. Lenticular sandstone and conglomerate bodies occur in stacked, overlapping successions. Individual bodies range to as much as 12 m thick and scour underlying strata with as much as 1 m of relief (Fig. 4B). Units fine upward, typically beginning with clast-supported conglomerate filling the basal scour. Conglomerate passes upward into a succession of low-angle cross-stratified pebbly sandstone that may be capped by horizontally bedded sandstone (Fig. 4B). Imbrication [a(t)b(i)] is common in clast-supported conglomerate. Facies 5 deposits are poorly to moderately sorted with clasts up to boulder size in the basal parts and pebbles dominating the upper parts. Sandstone ranges from fine- to coarse-grained and typically is tuffaceous. Facies 5 is abundant in Unit 1 in the Gravel Hills–Black Canyon section (Fig. 6).

Lenticular sandstone and conglomerate bodies represent channels of shallow, braided-stream systems (Table 1). The paucity of well-developed cross-stratification, a common component of sandy braided streams (Cant and Walker, 1978), is attributed to an elevated concentration of fine sediment in the system that can inhibit the development of bedforms that produce cross-stratification (Allen and Leeder, 1980). Abundant sediment apparently was available in the form of large amounts of unconsolidated ash and pumice.

Facies 6—Horizontal bedded pebbly sandstone. Horizontal bedded pebbly sandstone forms thick successions of depositional units that range 3–15 cm thick. Bedding is defined by planar contacts and vertical grain-size changes. Units typically display normal grading but may be structureless. Beds are laterally extensive and may grade vertically or laterally into laminated siltstone. Siltstone is also present as thin (3–5 cm) lenses. Sandstone ranges from coarse- to fine-grained. Pebbles and granule-sized grains are dominated by pumice fragments. Facies 6 is common in Unit 2 in the Waterman Hills and at Elephant Mountain (Fig. 5).

Deposits of facies 6 represent a nearshore lacustrine environment (Table 1). This interpretation is based on graded bedding and lateral and vertical facies relations. The coarse grain size indicates a proximity to a fluvial system, and the sporadic presence of coarser sediment is attributed to episodic fluvial progradation into the lake during periods of high discharge.

Facies 7—Carbonate rock. Beds of gray to black carbonate rock range from <10 cm to 3 m thick and may be traced laterally for as much as 1 km. Massive and laminated beds were observed and in some cases both occur within a single succession. Irregular, wavy laminations are present locally and are sometimes associated with oncoids. Thicker carbonate units contain thin interbeds of orange ripple laminated siltstone. Carbonate beds also occur within clastic-dominated successions of conglomerate and sandstone at Elephant Mountain (Fig. 5). Carbonate composition ranges from limestone to dolomite. Black chert is common as small nodules (<5 cm) and thin beds (2–4 cm). Carbonate beds occur throughout the sections at Lead Mountain and Elephant Mountain (Fig. 5).

Interbedded carbonate rock and siltstone are interpreted to represent a shallow lacustrine setting (Table 1). Wavy laminae and oncoids are of algal origin and suggest clear shallow water and sporadic biogenic activity. Carbonate beds within coarse clastic successions suggest a temporary cessation of clastic influx.

Facies 8—Siltstone. Siltstone commonly envelops carbonate beds and occurs as thin (3–5 cm) lenses within carbonate and horizontal bedded sandstone successions. Units may be horizontal or ripple laminated. Siltstone typically is calcareous and

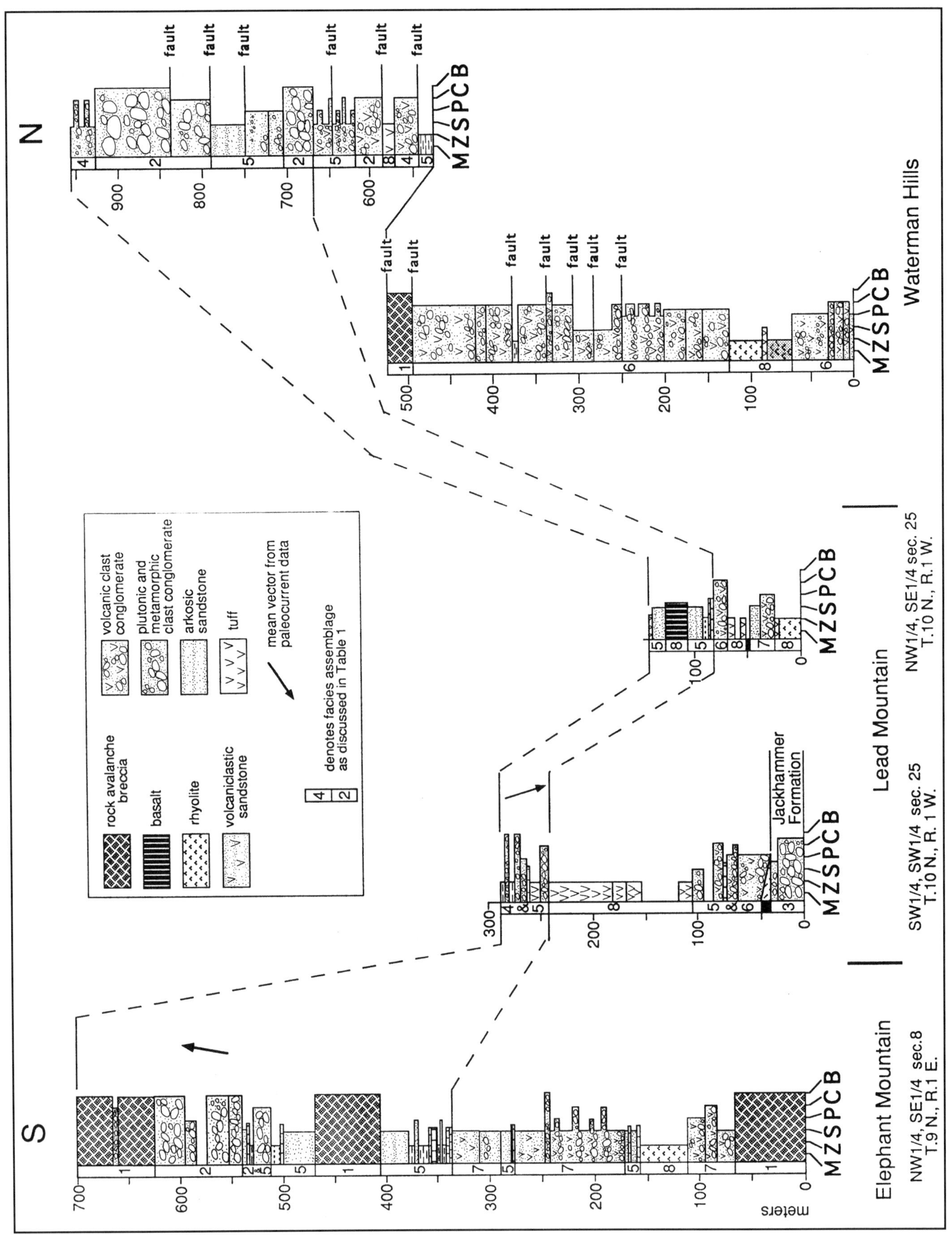
S
N
rock avalanche breccia
basalt
rhyolite
volcaniclastic sandstone
volcanic clast conglomerate
plutonic and metamorphic clast conglomerate
arkosic sandstone
tuff
mean vector from paleocurrent data
denotes facies assemblage as discussed in Table 1
meters
fault
MZSPCB
Jackhammer Formation
Elephant Mountain
NW1/4, SE1/4 sec.8 T.9 N., R.1 E.
Lead Mountain
SW1/4, SW1/4 sec. 25 T.10 N., R. 1 W.
NW1/4, SE1/4 sec. 25 T.10 N., R.1 W.
Waterman Hills

may grade to coarse pebbly sandstone. In Unit 2 in the Lead and Elephant Mountain areas, siltstone successions are traceable laterally for as much as 2 km (Fig. 5).

Siltstone at Lead and Elephant Mountain represent lacustrine deposition during periods of clastic influx from adjacent fluvial systems (Table 1). Thicker, laterally continuous successions that grade from horizontal bedded sandstone represent deeper water lacustrine deposition.

Volcaniclastic lithofacies

Facies 9—Conglomeratic lithic tuff. Conglomeratic lithic tuff forms thick successions of matrix-supported volcanic conglomerate. Depositional units range to as much as 18 m thick and are laterally extensive. Units are structureless and no grading is evident. Angular clasts range from pebble to cobble size and are composed of red rhyolite and gray andesite. Matrix is dominated by tuff and pumice fragments. Conglomeratic lithic tuff beds are an abundant component of Unit 1 in the Waterman Hills and Mud Hills (Figs. 5 and 6).

Thick units of poorly sorted, matrix-supported, conglomeratic lithic tuff are interpreted to be the deposits of volcaniclastic debris flows. These deposits represent a proximal volcanic apron and likely were derived from the remobilization of loose pyroclastic material on the flanks of an adjacent volcanic complex.

Facies 10—Clast-supported conglomerate. Thin, clast-supported volcaniclastic conglomerate units typically overlie debris flow deposits of facies 9. Units are 1–2 m thick and grade rapidly from underlying matrix-supported conglomerate. Clasts and matrix are the same as underlying deposits of facies 9.

Clast-supported conglomerate represents water reworking of the tops of underlying debris-flow deposits during periods of debris-flow inactivity. These deposits provide important markers for separating thick depositional units of facies 9.

Facies 11—Tuffaceous siltstone. In Unit 1 in the Mud Hills, tuffaceous siltstone units cap thick conglomeratic lithic tuff beds of facies 9. Siltstone units are lenticular and range from 10 cm–1.3 m thick. Siltstone is finely laminated. Sand-sized material in the form of biotite and feldspar crystals and lithic fragments is common.

Laminated tuffaceous siltstone units that overlie debris flow deposits likely accumulated in topographic lows that formed between or on top of debris flow lobes. Tuff may have been derived from muddy ponded water after cessation of debris flow, or may represent small fallout events.

Facies 12—Sheetlike volcaniclastic sandstone and conglomerate. Facies 12 forms thick successions (as much as 75 m thick) in Unit 1 at Elephant Mountain (Fig. 5). Individual depositional units range to as much as 3 m thick. Horizontal bedding is defined by pebble and cobble stringers. Boulders are present sporadically. Matrix is dominated by tuffaceous sediment with arkosic sand a subordinate component.

Facies 12 represents sheetflood deposits with sediment derived from the reworking of coarser debris flow and mudflow deposits. Interbedded lacustrine carbonates and tuffaceous siltstones indicate a setting transitional between proximal debris aprons and a distal lacustrine system.

Facies 13—Well-bedded pumiceous tuff. Well-bedded successions of pumiceous tuff are common in Unit 1 in both Gravel Hills localities (Fig. 6) and range to as much as 28 m thick. Individual bed thickness ranges from 2–150 cm, but most beds are 10–20 cm thick. Varicolored silicic to intermediate volcanic fragments up to granule size are common.

Well-bedded tuff is interpreted to represent the fallout of pyroclastic ejecta during explosive volcanic eruption. Tuff sequences separated by other unrelated strata indicate multiple discrete eruptions, although all were apparently confined to the period of Unit 1 deposition.

Nonsedimentary rock units

Rhyolite flows and hypabyssal intrusives. Rhyolite bodies of variable thickness (5–30 m) are present in the Pickhandle Formation at most localities and are interpreted as the product of viscous lava flow. In the Gravel Hills, some bodies intruded and deformed previously deposited Unit 1 strata. Evidence for this is a radial dip pattern in strata surrounding these intrusions and local faulting of the host sedimentary rocks.

Basalt flows. Basalt flows in the Pickhandle Formation were observed only in the Lead Mountain area (Fig. 5) where flow units range to as much as 7 m thick. Near the top of the formation three stacked flow units are discerned by the presence of brecciated bases and vesicular tops. A short distance to the west (<1 km) basalt flows interfinger with lacustrine siltstone and carbonate rock.

Lithofacies assemblages

Facies are grouped into assemblages based on observed field relations. Each assemblage represents a specific depositional setting within the Pickhandle basin. Facies assemblages are shown in Table 1 and are arranged in a proximal to distal setting. Table 1 is further divided into volcanic and nonvolcanic systems that coincide with Unit 1 and Unit 2, respectively. Additionally, facies assemblages are shown on measured sections in Figures 5 and 6.

◄———

Figure 5. Stratigraphic sections of the Pickhandle Formation from the northern sub-basin localities. Dominant grain size for each unit denoted at the base of the sections: M, mud; Z, silt; S, sand; P, pebbles; C, cobbles; B, boulders. Numbers to the left of the columns represent the facies assemblages in Table 1: (1) rock avalanche deposits; (2) proximal/medial alluvial fan deposits; (3) medial/distal alluvial-fan deposits; (4) braided-stream deposits; (5) lacustrine deposits; (6) proximal volcaniclastic-debris apron; (7) volcaniclastic sheetflood-dominated plain; (8) fallout tuff deposits and primary lava flows. Vertical scale is in meters.

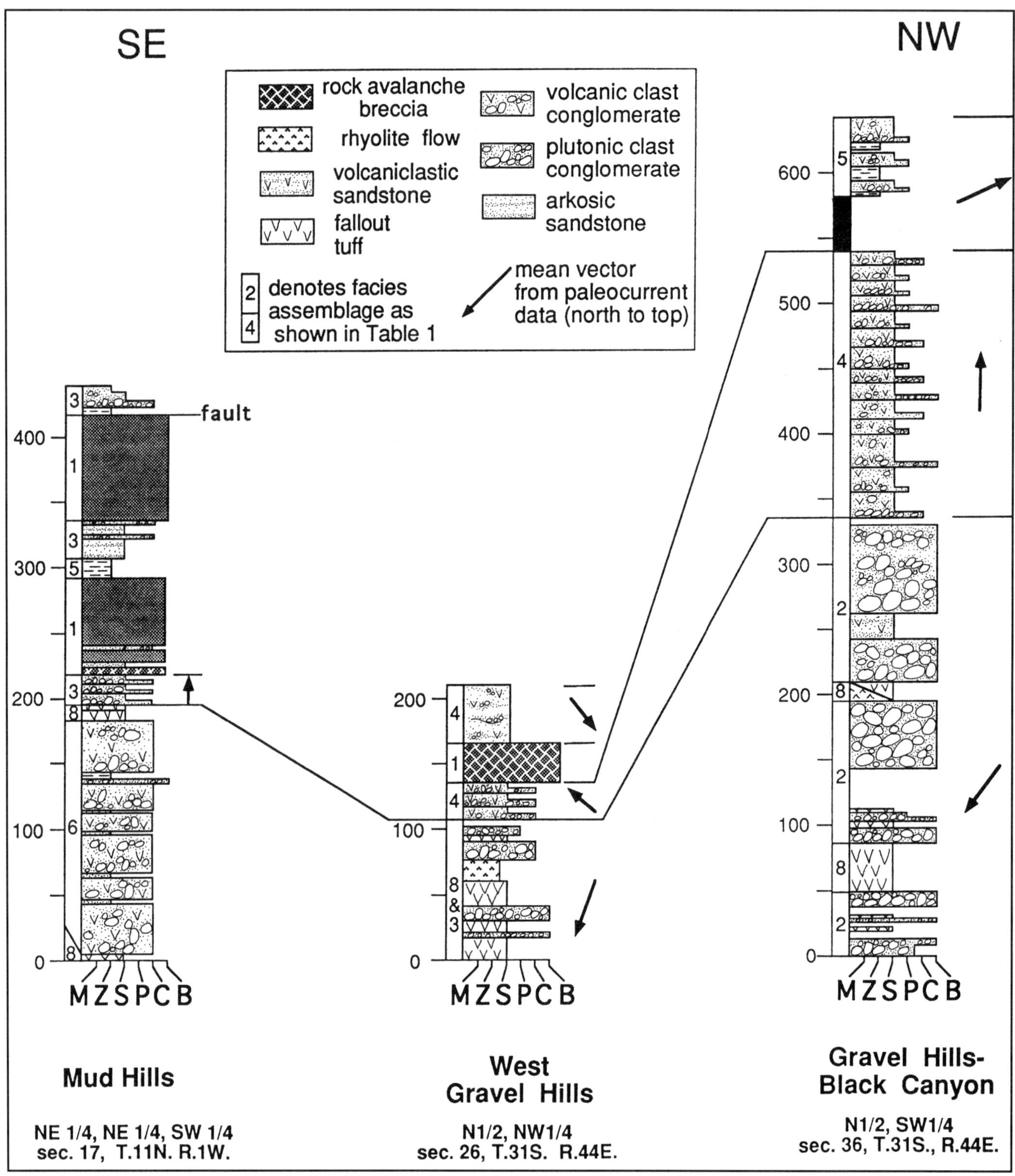

Figure 6. Stratigraphic sections of the Pickhandle Formation from the northern sub-basin localities. Dominant grain size for each unit denoted at the base of the sections: M, mud; Z, silt; S, sand; P, pebbles; C, cobbles; B, boulders. Numbers to the left of the columns represent the facies assemblages in Table 1. See Figure 5 for explanation of numbers. Vertical scale is in meters.

DEPOSITIONAL SETTING AND PALEOGEOGRAPHIC RECONSTRUCTIONS

The Pickhandle Formation has been interpreted as the fill of an elongate, northwest-southeast–trending basin based on outcrop patterns of the Pickhandle Formation and facies patterns in the overlying Barstow Formation (Dibblee, 1968). Age data, stratigraphic and sedimentologic studies, and paleogeographic reconstructions suggest that the elongate, northwest-trending, Pickhandle basin was partitioned prior to deposition of Unit 2 into two sub-basins. The northern basin encompassed the Gravel Hills and Mud Hills (Figs. 1 and 6). The southern basin includes the Waterman Hills, and the Lead and Elephant Mountain areas (Figs. 1 and 5). Strata of the southern sub-basin are discussed first; the following discussion of the northern sub-basin is divided into a lower volcanic-dominated succession (Unit 1) and an upper clastic succession (Unit 2).

Facies distribution and provenance—Southern sub-basin

Strata of the southern sub-basin, exposed at Elephant Mountain, Lead Mountain, and the Waterman Hills, grade vertically from lower volcanic and volcaniclastic deposits (Unit 1) to nonvolcanic deposits of Unit 2 (Fig. 5).

Waterman Hills. The Pickhandle Formation in the Waterman Hills is separated from the underlying footwall of the CMMCC by the Waterman Hills detachment fault (Walker et al., 1990). Because the formation is entirely fault-bounded, the basal depositional contact is not present. This incomplete section is the thickest at ~960 m (Fig. 5). This apparent thickness is the result of numerous normal faults within the section that thicken or repeat units (Fig. 5). As a result, only the overall lithologic succession from this locality is useful in basin reconstruction, and interpretations based on thickness must be used with care.

Unit 1 in the Waterman Hills is a heterogeneous sequence of primary and epiclastic volcanic deposits (Fig. 5). These strata are dominated by conglomeratic lithic tuff of debris-flow origin (facies 9) overlain by thin units of fluvial sandstone and conglomerate (facies 10). Laminated siltstone (facies 11), rarely displaying mudcracks, recur throughout the succession and indicate local ponding. Two thick rhyolite and flow-breccia units also occur in this succession (assemblage 8; Fig. 5). These flows, coupled with coarse volcanic debris flow deposits, suggest deposition proximal to the volcanic source.

Unit 2 in the Waterman Hills displays a crude upward coarsening pattern (Fig. 5). Lower strata comprise lenticular sandstone and conglomerate of fluvial origin (facies 5); these pass upward into sheetlike, graded pebbly sandstone beds that represent a lacustrine margin (facies 6). Rock avalanche breccia (facies 1) and debris-flow deposits (facies 2) form thin, discontinuous units. Uppermost strata are dominated by coarse, matrix-supported conglomerate (facies 2) intercalated with lacustrine shoreline deposits of facies 6. Unit 2 in the Waterman Hills represents a fluctuating lacustrine shoreline bounded by an alluvial-fan system that was fed by nearby highlands.

Debris flow conglomerates in Unit 2 contain a clast assemblage that cannot be traced to any presently exposed adjacent localities (Walker et al., 1990). This assemblage includes biotite diorite, black hornfels, porphyritic granitoids with K-feldspar phenocrysts as much as 35 cm long, coarse-grained marble, pelitic gneiss, and gneissic granitoids. Based on comparison with rocks in surrounding ranges, Walker et al. (1990) identified a source on the western flank of the Alvord Mountain basement complex located 35 km to the northeast (Figs. 1 and 7), where all of these rock types are exposed. They proposed that Alvord Mountain bounded the Pickhandle basin during deposition of these conglomerate units and then was transported to the northeast along the Waterman Hills detachment fault (Fig. 7). It is our interpretation that Alvord Mountain constitutes a separate hanging-wall block that was decoupled from the rest of the hanging wall by extensional faulting and restores to bound the southern part of the northeast Pickhandle basin margin (Fig. 7).

Paleocurrent data were not collected from the Waterman Hills section because conglomerate occurs as debris-flow deposits with no imbrication. Although it is difficult to determine precisely where Alvord Mountain was located during extension, the very coarse and disorganized nature of conglomerate in the Waterman Hills suggests that the source terrane was close.

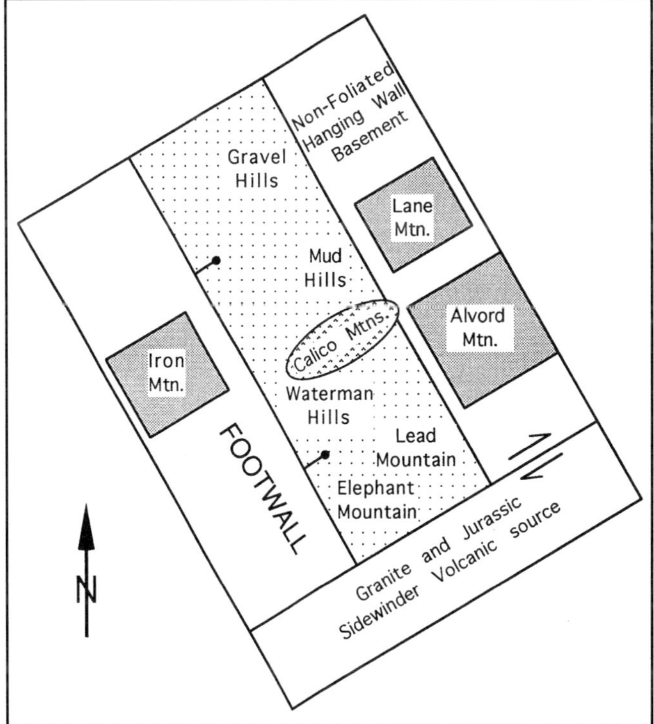

Figure 7. Reconstruction of various sediment sources in the footwall and hanging wall, and within the basin for the Pickhandle Formation and their spatial relations with the studied localities within the basin.

Lead Mountain. The Pickhandle Formation at Lead Mountain is a heterogeneous sequence of volcanic, fluvial, and lacustrine deposits that overlies conglomerate of the Jackhammer Formation. Unit 1 is dominated by volcaniclastic debris flow deposits (facies 9; Fig. 5). Intercalated with debris-flow deposits are basalt and rhyolite flows, thin lacustrine carbonate beds (facies 7), and volcaniclastic fluvial deposits (facies 12 ; Fig. 5).

Unit 2 consists of lacustrine limestone and laminated siltstone (facies 7 and 8), interbedded with fluvial sandstone and conglomerate (facies 4 and 5). Coarse fluvial deposits associated with lacustrine strata suggest a lake margin bounded by a fluvial braidplain environment. These deposits grade southward to lacustrine siltstone and carbonate. To the east, a 20-m-thick succession of basalt flows is intercalated with Unit 2 (Fig. 5).

Two conglomerate beds in Unit 2 at Lead Mountain contain a clast assemblage that is similar to that found in Unit 2 in the Waterman Hills, to the north. Clasts in the lower conglomerate are composed of 45% porphyritic biotite granite with K-feldspar megacrysts, 35% banded micaceous marble, and 4% quartzofeldspathic gneiss, with the remainder made up of various granitoid types. Gneissic clasts in the overlying conglomerate increase to ~15% of the population. Paleocurrent data from imbrication show a southward paleoflow (Fig. 8B), suggesting a source to the north. Conglomerate units at Lead Mountain are interpreted as the distal equivalent of debris-flow deposits in Unit 2 in the Waterman Hills section based on similar clast composition and stratigraphic position (Fig. 5). Additionally, facies changes and paleocurrent data suggest a north-to-south proximal-to-distal setting.

Elephant Mountain. The section at Elephant Mountain begins with a 68 m sequence of tuff clast avalanche breccia (facies 1; Fig. 5), probably derived from an escarpment associated with extensional faulting. Thick units of volcaniclastic sandstone and conglomerate (facies 9 and 12) that overlie breccia represent sheetflood and debris-flow activity, and attest to continued reworking of primary volcanic material. Several thick (as much as 27 m) rhyolite flows also occur in this lower interval (Fig. 5). Fluvial deposits in the upper part of Unit 1 alternate between volcanic and granitic compositions. Thin lacustrine limestone beds (facies 7) occur sporadically in these fluvial deposits.

No paleocurrent data were obtained from Unit 1; however, the coarse angular clasts indicate a proximal source. Based on lithologic similarities, it is likely that these deposits shared a source with Unit 1 in the Waterman Hills and Lead Mountain.

Unit 2 begins with intercalated limestone, pebbly siltstone, and calcareous conglomerate deposits (assemblage 5; Fig. 5, at ~340 m). These strata are interpreted to represent a nearshore lacustrine environment bounded by a coarse-grained fluvial system. Shoreline lacustrine deposits grade upward into very coarse conglomerate and rock avalanche breccia deposits representing a prograding alluvial fan setting (assemblages 1 and 2; Fig. 5, 380–700 m). A rapid transition northward, from conglomerate to laminated siltstone and limestone, indicates that the lacustrine system was located immediately north of the alluvial fan and that the fan prograded from south to north.

Clast composition in alluvial fan deposits at Elephant Mountain is dominated by distinctive metavolcanic rocks of the Middle Jurassic Sidewinder series and equigranular granitoids. Imbrication shows a northerly paleoflow direction (Fig. 8A), suggesting a provenance immediately to the south. Cox and Wilshire (1993) report basement composed of Sidewinder metavolcanic rocks in the north part of the Elephant Mountain area. However, these rocks are overlain by the Pickhandle Formation and occur to the north of the southerly derived Sidewinder-bearing conglomerates and could not have been the source. The probable source for granitic and metavolcanic rocks is in the Stoddard Mountain area, presently 24 km to the west-southwest. It is our interpretation that the conglomerates were deposited along the southern basin margin and were subsequently translated east-northeast ~20 km from their source, on the upper plate of the CMMCC detachment fault.

Facies distribution and provenance—Northern Sub-basin

Mud Hills. Unit 1 in the Mud Hills consists almost entirely of thick volcanic-debris-flow deposits of facies 9, capped by thin beds of clast-supported conglomerate (facies 10) indicative of fluvial reworking (Fig. 6, 0–195 m). Debris-flow deposits in the Mud Hills, although compositionally similar to those in the Waterman Hills, contain higher proportions of matrix and smaller average clasts. Thin, tuffaceous siltstone lenses (facies 11) reflect local ponding on the surface of the debris-flow deposits.

Strata of the lower succession fine to the northwest. Thickly bedded debris-flow deposits (facies 9) and silicic lava flows in the northern Calico Mountains grade northward, in the southern Mud Hills, to debris-flow deposits with common but thin units of fluvial conglomerate (assemblage 6; Fig. 6). Farther northwest, in the northern Mud Hills, debris-flow deposits become finer grained and contain a higher percentage of tuff matrix. This suggests a source to the south, probably within the Calico Mountains (Fig. 1).

The base of the nonvolcanic Unit 2 in the Mud Hills contains sheetflood sandstone and conglomerate deposits (facies 4), representing deposition in a distal alluvial fan setting (Table 1). These are overlain by several thick bodies (as much as 77 m thick) of rock avalanche breccia (facies 1) that are separated by a variety of facies assemblages (facies 4, 5, and 8; Fig. 6). The successive catastrophic emplacement of rock avalanche breccia suggests the presence of a nearby escarpment. Intercalated debris-flow deposits are of the same composition as the breccia units and may be the reworked product of the rock avalanche lobes or, alternatively, may represent material shed off the same highland source.

Fluvial deposits at the base of the upper succession in the Mud Hills are composed mostly of quartz monzonite detritus. Imbrication shows a northern paleoflow direction (Fig. 8C),

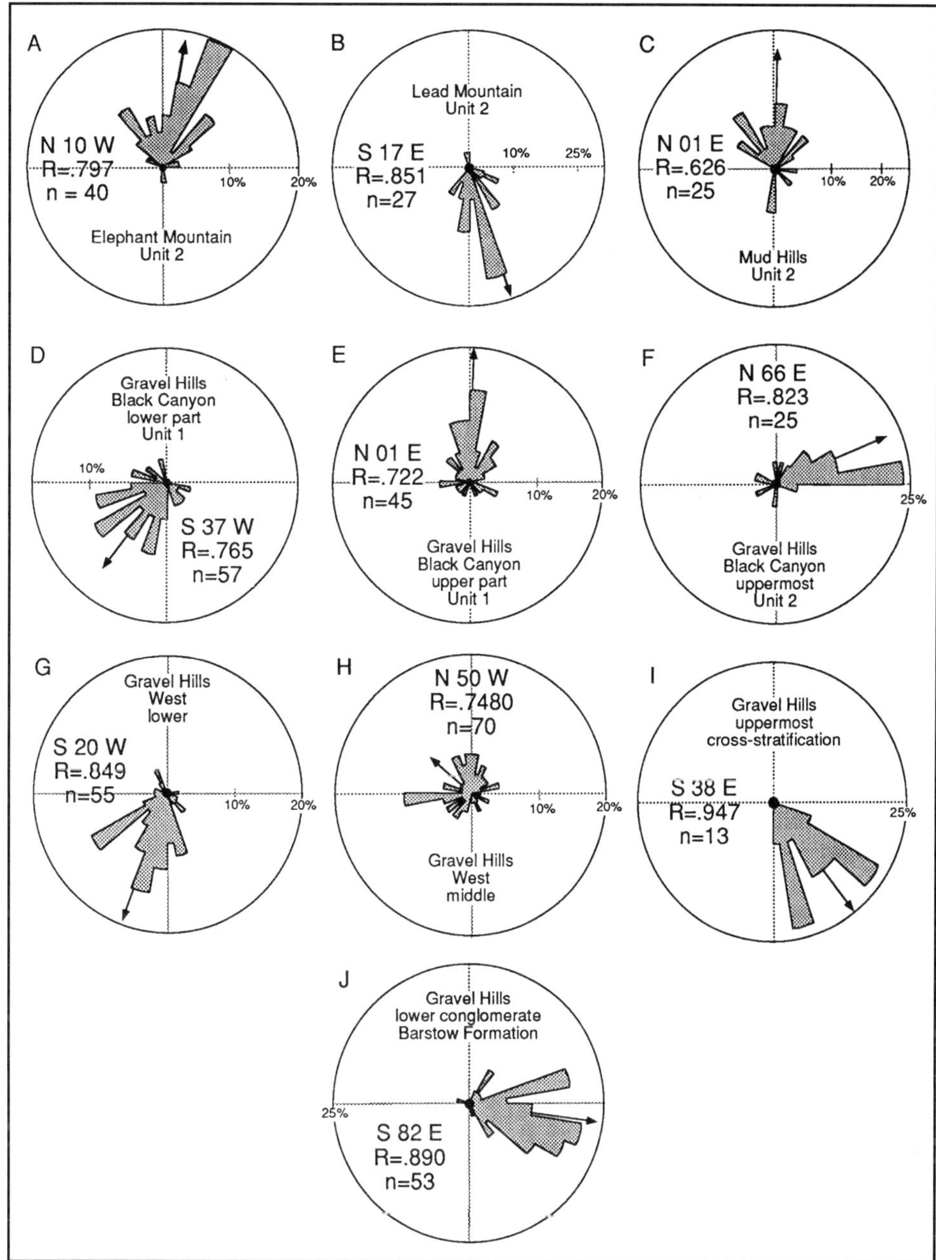

Figure 8. Paleocurrent data from all Pickhandle localities plotted on rose diagrams showing distribution of data and mean vector.

indicating a source to the south. These deposits are overlain by a thick sequence of rock avalanche breccia (Fig. 6) also composed of quartz monzonite with zones of white felsite. The quartz monzonite composition suggests that they also were derived from the south. Rock avalanche deposits suggest an oversteepened slope; the southern source is interpreted to have been an escarpment that was related to the breakaway zone of the detachment fault.

Gravel Hills—Black Canyon. The use of lithologic criteria to distinguish Units 1 and 2 in the two Gravel Hills localities is difficult because of the lack of distinctive compositional changes. It is possible, however, to correlate a shift in paleocurrent directions in the Black Canyon and west Gravel Hills sections (Fig. 6). Thus, stratigraphic boundaries in these localities are based primarily on changes in paleocurrent directions, although subtle changes in clast compositions are also used.

In the Gravel Hills, the base of the Black Canyon section consists of a 333 m thick sequence of clast- and matrix-supported boulder conglomerate (assemblage 2), representing a proximal alluvial fan setting (Table 1; Fig. 6). This succession differs from Unit 1 at other localities in that clasts consist of plutonic basement rock, although the matrix contains a significant tuffaceous component. The basal part of this sequence contains several thick bedded tuff units (as much as 36 m) that are interpreted to represent syneruption ash fallout (facies 13; Fig. 6).

Unit 2 is not easily defined in the Black Canyon section. Basal alluvial fan deposits (assemblage 2) of Unit 1 are overlain by a 230-m-thick middle sequence of stacked, lenticular sandstone and conglomerate bodies (facies 5; Fig. 6; 330–560 m) that represents a transition between Unit 1 and Unit 2. These fluvial braidplain deposits contain boulders of plutonic basement, and abundant volcanic pebbles and cobbles. Sandstone is dominated by volcanic fragments, chiefly tuff and lapilli. This is overlain by 264 m of poorly exposed sandstone and laminated tuffaceous siltstone of Unit 2 (assemblage 5; Fig. 6). These fine-grained strata are interpreted to reflect low-energy, fluvio-lacustrine conditions. Possible causes for the change in depositional setting are discussed below.

The boulder conglomerate at the base of Unit 1 at Black Canyon is dominated by quartz monzonite and hornblende diorite clasts with imbrication showing flow to the southwest (Fig. 8D). These data indicate a source in plutonic basement rocks exposed immediately to the northeast in the hanging wall.

Paleoflow directions for the middle transitional succession and for Unit 2 of the Black Canyon section show abrupt changes (Fig. 8, E and F). Paleocurrent indicators in the middle succession (335–540 m; Fig. 6) display a shift to the north (Fig. 8E); conglomerate clasts are composed of hornblende granitoids, quartz monzonite, and Tertiary volcanic rocks. Sandstone is arkosic with a large tuff component.

Paleoflow indicators in uppermost deposits at Black Canyon (580 m to top of section, Fig. 6) show a shift to the east. Some of these deposits exhibit evidence of soil development. Less than 5 km to the east, uppermost Pickhandle strata include a thin limestone bed interpreted to have formed in a lacustrine setting. It is likely that the low energy fluvial system represented by uppermost deposits at Black Canyon drained into this lacustrine system.

Gravel Hills, West. The base of the section in the western Gravel Hills is dominated by primary fallout tuff and fluvially reworked tuffaceous deposits (facies 10, 12, and 13; Fig. 6). Reworked units are characterized by poor sorting and plutonic rock fragments whereas fallout tuffs are typified by normal grading and contain only rhyolite lithic fragments.

Rhyolite lava flows are common in the western Gravel Hills with at least two episodes of emplacement evident. The earliest episode was extrusive and was synchronous with deposition of lower tuffaceous deposits. This is indicated by flow-banded rhyolite that changes laterally to cross-stratified, monomictic rhyolite conglomerate. Over a distance of 0.5 km this conglomerate grades to tuffaceous fluvial deposits with the rhyolite component diluted by plutonic rock fragments. It is likely that this rhyolite flow produced relief within the basin. Subsequent erosion of the flow margins provided rhyolite fragments for the monomictic conglomerate. Evidence for a later intrusive episode consists of rhyolite lava bodies that are bounded on all sides by tuffaceous strata that dip radially away from the body, suggesting deformation by intrusion of the viscous lava.

Unit 2 is dominated by epiclastic volcanic sandstone and conglomerate (facies 5, 10, and 12; Fig. 6, 102–210 m). Lower strata of Unit 2 comprise lenticular fluvial sandstone bodies with several intercalated, compositionally distinctive conglomerate horizons. These strata are overlain by several quartz monzonite breccia bodies of rock avalanche origin (facies 1). Overlying the weathered tops of rock avalanche deposits are lenticular fluvial sandstone bodies composed of volcanic and granitic detritus. Dibblee (1968) assigned these breccia deposits and overlying volcaniclastic fluvial deposits to the Barstow Formation; however, based on detailed facies mapping and correlations with the Pickhandle Formation in the Mud Hills, we instead assign these strata to the Pickhandle Formation. Thus, the top of the Pickhandle Formation is defined as the uppermost white, lenticular conglomeratic sandstone unit of dominantly volcanic composition. These strata overlie granitic rock avalanche breccia deposits. The base of the Barstow Formation is defined as the first nonvolcanic boulder conglomerate bed overlying these volcaniclastic deposits in angular unconformity. Basal Barstow strata are composed of granitic boulders in an arkosic matrix.

Unit 1 in the west Gravel Hills section, located 13 km west of the Black Canyon locality, is much thinner than the Black Canyon locality and contains less conglomerate (Fig. 6), but clasts are of the same composition and both exhibit a southwest flow direction (Fig. 8G). Similarly, data from Unit 1 in the west Gravel Hills conglomerates indicate a source immediately to the northeast.

Fluvial deposits in Unit 2 of the west Gravel Hills section are composed dominantly of volcanic detritus. Two conglomerate bodies, however, that occur in the section from ~110–125 m (Fig. 6) contain a clast assemblage of foliated hornblende granitoids, quartzite, rare marble, and silicic volcanic rocks. Imbrication in Unit 2 shows a northwest paleoflow direction (Fig. 8H) that is interpreted to indicate the development of an axial fluvial system for the northwest-trending basin. The source for volcanic clasts is interpreted to be older volcanic deposits and/or contemporaneous volcanism within the basin. Foliated granitoid and metasedimentary clasts likely were introduced to the basin somewhere to the southeast and transported northwestward in the axial system. Hornblende granitoids and Paleozoic metasedimentary rocks are currently exposed in the Lane Mountain area, northeast of the basin in the hanging wall, and in the footwall southwest of the basin at Iron Mountain (Fig. 7). It is possible that one or both of these areas acted as sources for sediment in the basin. A northwest paleoflow direction coupled with a clast assemblage that likely entered the basin somewhere to the southeast suggests a northwest-trending axial drainage system. Thick deposits of quartz monzonite rock avalanche breccia near the top of the west Gravel Hills section (Fig. 6), like those in the

Mud Hills, are interpreted to have been derived from the detachment breakaway escarpment that probably bounded the basin along its southwest margin.

Paleocurrent data from cross-stratification in uppermost sandstone in the west Gravel Hills show an abrupt change to an east-southeast paleoflow direction (Fig. 8I). This drainage pattern continues into the overlying Barstow Formation (Fig. 8J) which consists of coarse granitic conglomerate of alluvial-fan origin.

Paleogeographic and tectonic reconstruction

During Unit 1 deposition (23.7–21.7 Ma) the Pickhandle was an extensive, elongate northwest-trending basin in which sediment was derived primarily from an intrabasinal volcanic center. Tectonic partitioning did not divide the basin until Unit 2 deposition (post 21.7 Ma). Because of this, the following discussion for Unit 1 deposition includes the entire basin, whereas discussion of Unit 2 is divided into southern and northern sub-basins.

Unit 1. Compositional similarities between Unit 1 volcanic deposits throughout the Pickhandle basin suggest that the southern and northern sub-basins shared a volcanic source. A fining of volcanogenic deposits from the Waterman Hills southward and a fining to the northwest from the northern Calico Mountains to the northwest Mud Hills and the Gravel Hills suggest the volcanic source was located in the Calico Mountains. The Calico Mountains are composed of highly altered silicic to intermediate volcanic rocks of unknown age overlain by lacustrine deposits of the Barstow Formation(?).

Coarse volcanic debris flows and primary lava flows in the Waterman Hills grade south to debris-flow, fluvial, and lacustrine deposits at Lead Mountain. Limestone and laminated tuffaceous siltstone interbedded with volcaniclastic deposits at Lead Mountain and Elephant Mountain suggest at least periodic lacustrine conditions and a closed drainage system (Fig. 9). During most of Unit 1 deposition in the southern part of the basin, a volcanic-debris apron extended southward from Calico Mountain and included the Waterman Hills area. This apron fined southward through Lead Mountain where debris-flow deposits are interbedded with volcaniclastic sheetflood deposits. At the southernmost locality at Elephant Mountain, sheetflood deposits intertongue with lacustrine deposits, indicating a lacustrine system in the far southern part of the basin (Fig. 9).

Basal Pickhandle strata in the northern Calico Mountains consist of primary lava flows, block and ash, and debris-flow deposits that represent the proximal deposits of the volcanic center. Unit 1 in the southern Mud Hills is dominated by coarse debris flow deposits of a proximal volcanic apron (Fig. 9). To the northwest, in the central and northwest part of the Mud Hills, Unit 1 fines as clasts become smaller and tuff matrix becomes a dominant component of the debris flows. These deposits represent a mudflow-dominated volcanic apron and grade distally into a lapilli and ash-choked fluvial braidplain system in the western Gravel Hills (Fig. 9). Fluvial deposits in the Gravel Hills interfinger with northeast-derived arkosic conglomerates that represent a source in the hanging wall (Fig. 9). The outlet for the northwest-flowing axial drainage system remains unknown because of the lack of exposure of the Pickhandle Formation north or west of the Gravel Hills. The absence of low-energy lacustrine deposits in lower strata, however, suggests an externally drained basin. It is possible that tuffaceous, lacustrine deposits of the Tropico Group to the southwest represent this low-energy facies of the Pickhandle Formation, but the lack of evidence for an outlet to the southwest makes this speculative.

In the Gravel Hills, the thick basal succession (~330 m) of plutonic boulder conglomerate at Black Canyon, compared with the relatively thin succession in the west Gravel Hills section (Fig. 5), is attributed to faulting in the hanging wall source immediately to the northeast. According to Dibblee (1968), the most recent movement on the northwest-trending Blackwater fault, which bounds the Gravel Hills to the northeast, has a scissorlike motion: the segment immediately northwest of Black Canyon has a down-to-the-northeast sense of movement, whereas the segment adjacent to Black Canyon has a down-to-the-southwest sense of movement (Fig. 10; Dibblee, 1968). The latest movement on the Blackwater fault is constrained by associated deformation of the Black Mountain Basalt which has been dated at 2.55 ± 0.58 Ma (whole rock K-Ar, Burke et al.,

Figure 9. Paleogeographic reconstruction of the Pickhandle basin during deposition of the lower volcanic-dominated succession. The right-lateral transfer fault at the southern margin is the extensional system-bounding fault. There is no evidence for the development of the intrabasinal transfer zone at this time, but it is believed to have developed in the approximate location of the volcanic center. The volcanic center likely partitioned the basin during this period.

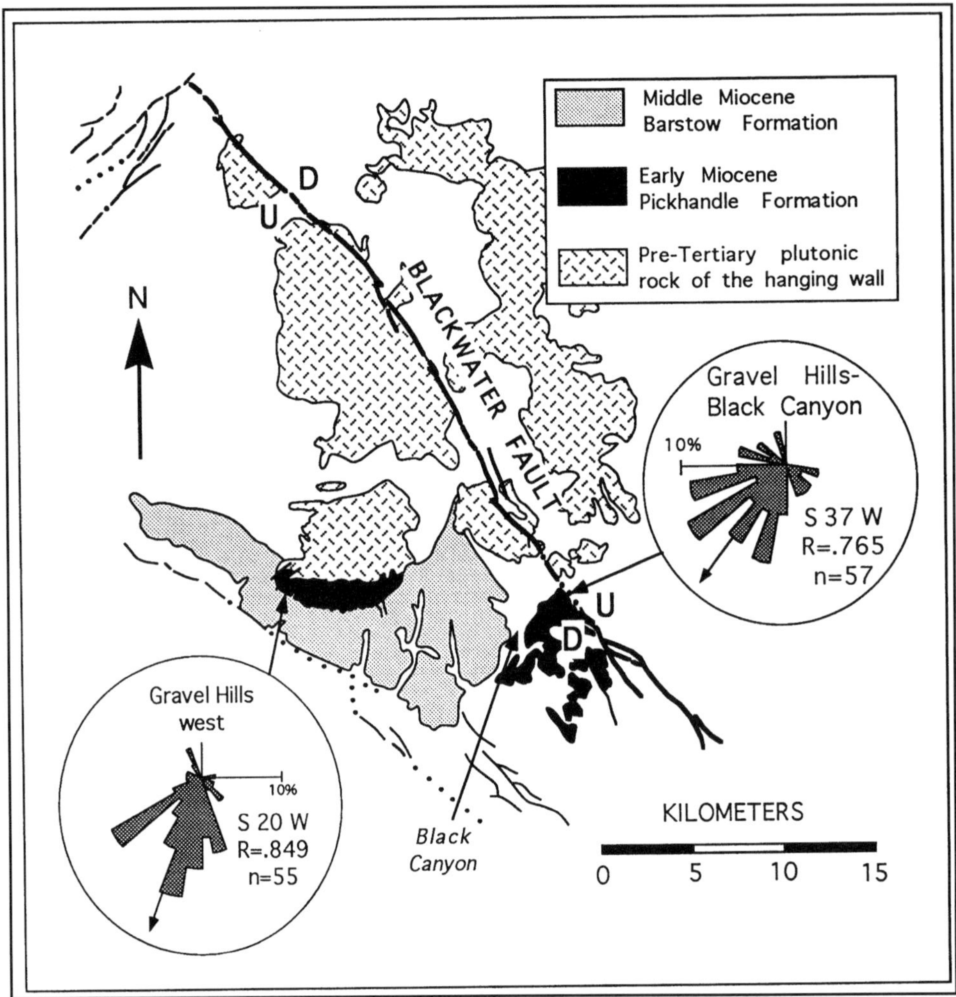

Figure 10. Detailed geologic map of the Gravel Hills area showing the location of the Blackwater fault and paleocurrent data from the lower strata in the Gravel Hills west and Black Canyon localities (modified from Dibblee, 1968).

1982). It is our interpretation that this fault was active with the same sense of movement during deposition of the lower succession of the Pickhandle Formation and represents the brittle breakup of the trailing edge of the hanging wall. The thick, coarse nature of the Black Canyon section is attributed to deposition on the downthrown southwest side of this fault; lower strata in the west Gravel Hills are thinner and finer-grained because of deposition on the upthrown, but still southwest-sloping, side of the fault (Fig. 10). The Blackwater fault terminates to the northwest at a northeast-trending fracture zone (Fig. 10) within a post-Miocene(?) volcanic eruptive center. These fractures may mark the transfer zone that forms the northwest terminus of the Pickhandle basin.

Unit 2: Southern basin. During deposition of Unit 2 in the southern sub-basin, several sediment sources became active (Fig. 11). Highlands that supplied sediment are associated with faulting at the sub-basin margins and partitioning of the earlier elongate basin.

At the southernmost locality of the formation, at Elephant Mountain, the great thickness of alluvial-fan deposits and north-directed paleoflow suggest a fault-bounded basin margin immediately to the south. Alluvial-fan deposits intertongue with, and grade northward, into fine-grained lacustrine deposits that represent a perennial lake that occupied the basin depocenter (Fig. 11). Alluvial-fan deposits have a clast assemblage of equigranular granite and Jurassic metavolcanic rocks. The nearest possible source terrane of this composition is located 24 km to the west-southwest, in the Stoddard Mountain area. The offset of these deposits from this source suggests that a right-lateral transfer fault formed the southern margin of the Pickhandle basin, accommodating a southward decrease in extension between the central Mojave extensional system to the north and the Daggett "terrane" of Dokka (1989) to the south.

The upper succession at Lead Mountain is thinner and finer grained than at Elephant Mountain. These strata represent lacustrine conditions, except for several coarse conglomerate units with south-directed paleocurrent indicators that record periods of southward fluvial progradation from the north.

Upper strata in the Waterman Hills contain proximal alluvial-fan deposits that mark the north margin of the southern sub-basin. Interbedded debris-flow deposits and lacustrine sandstone with graded bedding suggest the presence of an alluvial-fan–lacustrine margin during Unit 2 deposition. These strata also mark the northern extent of the lacustrine system (Fig. 11). The "exotic" clast conglomerate that fines from debris-flow deposits in the Waterman Hills to fluvial deposits at Lead Mountain likely came from highlands that developed during partitioning. These highlands, which must have been located north of the Waterman Hills, are interpreted to represent an intrabasinal transfer zone. This transfer zone likely developed as the Alvord Mountain block decoupled from the main hanging-wall block during differential extension. Highlands that supplied coarse sediment to the southern sub-basin are interpreted to be related to local transpression along this postulated transfer zone.

The southern sub-basin was bounded to the northwest and southeast by right-lateral transfer zones that probably formed as a result of an overall stepwise decrease in extension toward the southeast. Alluvial fans prograded south and north from these margins into a closed basin that was occupied by an extensive lake (Fig. 11). The west margin remains unknown and preserved strata do not record any influx of sediment from that direction. It is possible that the lacustrine system abutted the footwall escarpment to the southwest and very little sediment was preserved. The northeast basin margin likely was a gentle westerly slope composed of the trailing edge of the hanging wall of the extensional system. The relatively thin nature of the Lead Mountain section is attributed to deposition high on this slope. Granitic basement exposed at Lead Mountain represents the undeformed hanging wall.

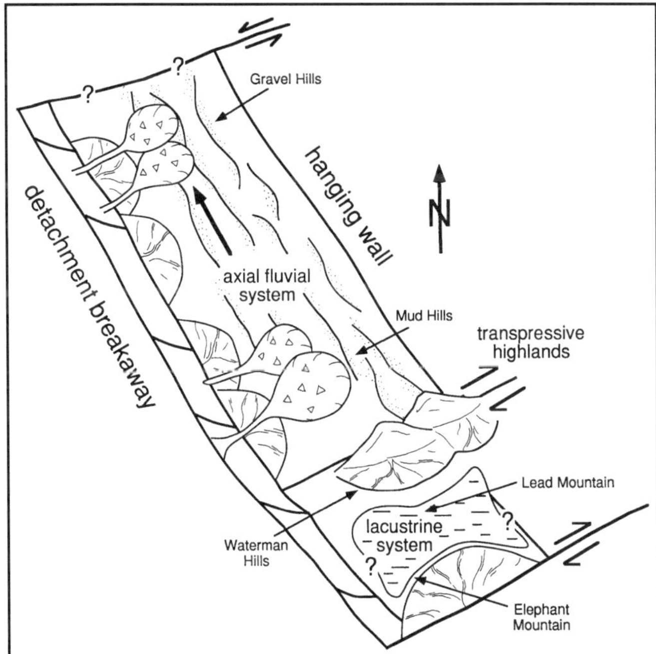

Figure 11. Paleogeographic reconstruction of the Pickhandle basin during deposition of the upper clastic succession. At Elephant Mountain alluvial fans prograded north from the right-lateral fault that bounded the basin to the south. Highlands associated with a right-lateral transfer zone partitioned the southern and northern sub-basins. In the northern sub-basin, alluvial fans and rock avalanche breccia lobes indicate a source to the southwest along the breakaway zone.

Unit 2: Northern sub-basin. Exposed strata of the northern sub-basin are believed to represent the northeast part of the northwest-trending Pickhandle basin. Sediment dispersal patterns, detrital composition, facies distributions, and regional stratigraphic relations suggest that this basin was bounded on its southwest margin by highlands of the detachment breakaway zone and along its northeast margin by the hanging wall that was a southwest-dipping slope (Fig. 11).

Lower Unit 2 fluvial deposits in the Mud Hills indicate a source terrane to the south composed dominantly of quartz monzonite. The thick succession of overlying quartz monzonite avalanche breccia suggests that this source was an escarpment. The source is interpreted to be related to the detachment breakaway zone.

Several thick quartz monzonite breccia deposits are also present near the top of the western Gravel Hills section to the northwest. These are also interpreted to be derived from the breakaway zone to the southwest and suggest that the breakaway formed a northwest-trending escarpment that extended over the entire length of the basin, forming the southwest basin margin (Fig. 11).

The occurrence of rock avalanche deposits along the northeast basin margin that are inferred to have a source on the southwest margin suggests a relatively narrow basin. Large rock avalanches have been shown to travel as much as 15 km from their source (Yarnold and Lombard, 1989). The interpretation that these deposits were derived from the breakaway escarpment to the southwest and traveled across the basin, near the northeast margin, suggests that the basin was probably <10–15 km wide.

In the west Gravel Hills, lower Unit 2 strata indicate a northwest-flowing axial stream system that had an outlet to the northwest. This system was disrupted by previously mentioned rock avalanche deposits from the southwest. Overlying fluvial deposits show a complete drainage reversal with flow directed to the northeast. This reversal is corroborated by a similar pattern in the Black Canyon area in which flow also shifted to the northeast (Fig. 6). A short distance east of the Black Canyon area (< 5 km), uppermost Pickhandle strata consist of algal carbonate of lacustrine origin; east-flowing fluvial systems represented by uppermost strata in both Gravel Hills localities likely drained into this lacustrine system.

Evidence for the cause of the sudden drainage reversal is sparse, but several consecutive depositional episodes can be linked to hypothetically reconstruct the events that led to fluvial derangement. A synchronous basinwide episode of rock avalanche deposition demonstrated by breccia in the Mud and

Gravel Hills at approximately the same stratigraphic interval may signal an increased rate of fault movement on the detachment breakaway. A corresponding increase in the rate of extension had to be accommodated laterally at the boundaries of the extensional system. The absence of synextensional deposits or extension-related structures northwest of the Gravel Hills supports the interpretation that the basin and the central Mojave extensional system terminated in this area. The subsequent drainage reversal in the Gravel Hills may be related to the transfer of extensional strain across this boundary into transpression, causing uplift and damming of the earlier drainage outlet. Continued uplift eroded northernmost Pickhandle deposits and directed flow eastward into the newly developed lacustrine depocenter. As transpression at the northern basin margin continued, highlands developed and shed coarse sediment of the Barstow Formation southeastward into the basin. The unconformity between the Pickhandle and Barstow Formations probably does not represent a long interval but simply reflects an episode of rapid extensional deformation.

BASIN DEVELOPMENT

The southeastern margin of the entire Pickhandle basin complex (southern and northern sub-basins) and the central Mojave extensional system is delineated by the north-draining alluvial-fan complex at Elephant Mountain. This margin is interpreted as a northeast-trending, right-lateral transfer fault that separated the highly extended central Mojave system from the less extended "Daggett terrane" of Dokka (1986) to the southeast (Figs. 11 and 12). This fault coincides with the northeast-trending Mojave Valley fault that was hypothesized by Martin and others (1993) to have right-lateral movement based on several different lines of evidence. Right-lateral displacement is consistent with offset of the Elephant Mountain alluvial-fan system from its Mesozoic volcanic and plutonic source terrane, now located a minimum of 24 km to the west-southwest.

Highlands north of Lead Mountain and the Waterman Hills that were a source for conglomerate are interpreted to have been related to the development of an intrabasinal transfer zone. This right-lateral transfer zone is interpreted to have caused partitioning and formed the northwest boundary of the southern sub-basin, between the Waterman Hills and the Mud Hills. This fault is interpreted to be a response to a southward decrease in extension (Fig. 12).

There is no sedimentary evidence for the presence of the intrabasinal transfer zone during Unit 1 deposition. Recently, however, Ebinger et al. (1993) reported that in the northern Afar Depression, Neogene to Holocene volcanic centers occur preferentially along oblique slip transfer zones that separate depositional basins. Our basin reconstruction suggests that a similar relationship may have existed in the Pickhandle basin. The location of the Calico Mountains volcanic source prior to differential intrabasinal extension was coincident with the postulated location of the transfer zone. However, a genetic relationship between transfer faulting and volcanism in the Pickhandle basin is speculative.

The interpretation that Lower Miocene strata at Lead Mountain and Elephant Mountain are part of a sub-basin of the Pickhandle basin conflicts with interpretations by Dokka and Woodburne (1986), Lambert (1987), and Dokka et al. (1991). Dokka et al. (1991) believed that the area was a nested caldera complex based on their interpretation of intercalated volcanic and lacustrine deposits as intracaldera fill and of shallow-level intrusive rocks as "ring intrusives." They further suggested that the caldera formed in an area unaffected by extension, possibly pre-dating extensional deformation.

Based on similarities with well-established Pickhandle successions in the Mud Hills, $^{40}Ar/^{39}Ar$ dates from Lead Mountain strata, and regional paleogeographic reconstructions, we argue that this area simply represents an extensional sub-basin of the central Mojave extensional system. Although shallow intrusive rocks are present rarely, Cox and Wilshire (1993) found that most of the felsite units in the basal part of the Tertiary section at Elephant Mountain are flow units. Furthermore, upper clastic strata that are used in this study to reconstruct the regional paleogeographic and tectonic setting appear to have been ignored by Dokka and co-workers. There is no evidence that we could discern to suggest the presence of a caldera in this region. Cox and Wilshire (1993) likewise dismissed the caldera interpretation and attributed most of the deformation in the area to extension.

Reconstructions using provenance data and offset across transfer faults indicate that extension of at least 60 km is distrib-

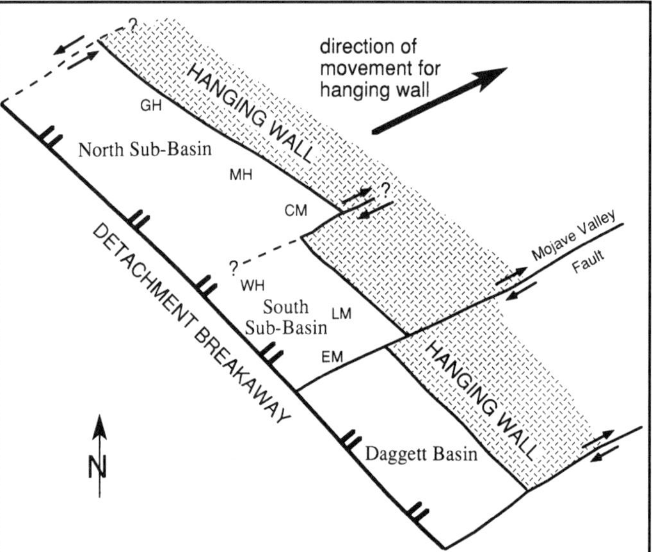

Figure 12. Reconstruction of extension-related faults in the central Mojave Desert during the Lower Miocene showing location of main detachment fault and associated transfer faults. Localities within the basin are denoted by GH, Gravel Hills; MH, Mud Hills; CM, Calico Mountains; WH, Waterman Hills; LM, Lead Mountain; EM, Elephant Mountain.

uted in a southwest-northeast direction in the central Mojave Desert. Alluvial-fan deposits in the upper succession at Elephant Mountain restore to a Mesozoic volcanic and granitic source at Stoddard Mountain, 24 km to the southwest, along a vector of 237°. Right-lateral offset was along the southeastern basin-bounding transfer fault (Fig. 12; Mojave Valley fault of Martin et al., 1993). Additionally, Alvord Mountain has experienced ~35 km of northeast-directed translation relative to debris flow deposits in the Waterman Hills that are interpreted to have a source in western Alvord basement rocks (Walker et al., 1990; this study). There is no evidence for Lower Miocene extensional deformation east of the Alvord Mountain area, and it is likely that the Clews basin east of Alvord Mountain represents the eastern limit of Lower Miocene extension for this system (Fillmore, 1993). The ~60 km of extension from this study is in good agreement with estimations by Martin et al. (1993) of ~60–70 km of right lateral offset on the Mojave Valley fault. Their estimates came from the offset of pre-Tertiary tectonic elements so that constraints for the age of the fault were poor. They did, however, suggest that it was associated with Lower Miocene extension.

Transfer zone development is attributed to a stepwise, southeastward decrease in extension. The transfer zone that partitioned the southern and northern sub-basins is one such element (Fig. 12). The transfer fault that bounds the Pickhandle basin on its southeast margin also represents a southward decrease in extension. This fault separates the central Mojave extensional system from the weakly extended "Daggett terrane" of Dokka (1986; Fig. 12). There is no evidence for extensional deformation south of the Daggett system, so that on an even larger scale there is a regional stepwise decrease in total extension to the south.

CONCLUSIONS

This study demonstrates the utility of synextensional strata in reconstructing large-magnitude extensional terranes. Paleogeographic reconstructions of the Pickhandle basin suggest that the southwest basin margin was formed by the detachment breakaway zone (Fig. 1). Based on vastly different types of basin-fill in the Pickhandle and Tropico basins to the southwest, the breakaway zone likely formed a drainage divide between the two basins (Fig. 3; Fillmore et al., 1994). The southeast basin margin was formed by the west-southwest–trending, right-lateral Mojave Valley fault (Martin et al., 1993); this is based on the offset of southernmost Pickhandle strata from their probable source area 24 km to the west-southwest.

Partitioning of the Pickhandle basin by an intrabasinal transfer fault, also believed to be a southwest-trending fault with right-lateral movement, provides insight into how large-magnitude extensional systems terminate along strike. Termination in the central Mojave extensional system was by a stepwise southeastward decrease in extension rather than an abrupt termination.

Finally, provenance determination has allowed for the restoration of depositional systems with their source areas. This restoration suggests that a minimum of 60 km of extension has occurred on central Mojave extensional system, an estimation that is in good agreement with earlier suggestions that were based on the offset of Paleozoic and Mesozoic elements (Martin et al., 1993).

ACKNOWLEDGMENTS

This work was supported through National Science Foundation grants EAR-8816628 and EAR-8916802 to J. D. Walker and by student research grants to R. P. Fillmore by the Geological Society of America, the University of Kansas Geology Associates Fund, and a Shell Oil Student Fellowship. Discussions with Allen Glazner, John Fletcher, Mark Martin, Jonathon Miller, and Drew Coleman were particularly helpful.

REFERENCES CITED

Allen, J. R. L., and Leeder, M. R., 1980, Criteria for the instability of upper-stage plane beds: Sedimentology, v. 27, p. 209–217.

Armstrong, R. L., and Higgins, R., 1973, K-Ar dating of the beginning of Tertiary volcanism in the Mojave Desert, California: Geological Society of America Bulletin, v. 84, p. 1095–1100.

Bartley, J. M., and Glazner, A. F., 1991, En echelon Miocene rifting in the southwestern United States and model for vertical-axis rotation in continental extension: Geology, v. 19, p. 1165–1168.

Bartley, J. M., Fletcher, J. M., and Glazner, A. F., 1990, Tertiary extension and contraction of lower-plate rocks in the central Mojave metamorphic core complex, southern California: Tectonics, v. 9, p. 521–534.

Burchfiel, B. C., 1966, Tin Mountain landslide, southeastern California, and the origin of megabreccia: Bulletin of the Geological Society of America, v. 77, p. 95–100.

Burke, D. B., Hillhouse, J. W., McKee, E. H., Miller, S. T., and Morton, J. L., 1982, Cenozoic rocks in the Barstow basin area of southern California—stratigraphic relations, radiometric ages, and paleomagnetism: U.S. Geological Survey Bulletin 1529-E, 16 p.

Byers, F. M., Jr., 1960, Geology of the Alvord Mountain quadrangle, San Bernardino County, California: U.S. Geological Survey Bulletin 1089-A, 71 p.

Cant, D. J., and Walker, R. G., 1978, Fluvial processes and facies sequences in the sandy braided South Saskatchewan River, Canada: Sedimentology, v. 25, p. 625–648.

Cox, B. F., and Wilshire, H. G., 1993, Geologic map of the area around the Nebo Annex, Marine Corps Logistic Base, Barstow California: U.S. Geological Survey Open-File Report 93-568, 36 p.

Dibblee, T. W., Jr., 1958, Tertiary stratigraphic units of western Mojave Desert: American Association of Petroleum Geologists Bulletin, v. 42, p. 135–144.

Dibblee, T. W., Jr., 1968, Geology of the Fremont Peak and Opal Mountain quadrangles, California: California Division of Mines and Geology, Bulletin 188, 64 p.

Dokka, R. K., 1986, Patterns and modes of early Miocene crustal extension, central Mojave Desert, California, *in* Mayer, L., ed., Extensional tectonics of the southwestern United States: A perspective on processes and kinematics: Boulder, Colorado, Geological Society of America Special Paper 208, p. 75–95.

Dokka, R. K., 1989, The Mojave extensional belt of southern California: Tectonics, v. 8, p. 363–390.

Dokka, R. K., and Baksi, A. K., 1988, Thermochronology and geochronology of detachment faults of the Mojave extensional belt, California: Geological Society of America Abstracts with Programs, v. 20, no. 7, p. A16.

Dokka, R. K., and Baksi, A. K., 1989, Age and significance of the Red Buttes Andesite, Kramer Hills, Mojave Desert, California, *in* Reynolds, R. E., ed., The west-central Mojave Desert: Quaternary studies between Kramer and Afton Canyon: San Bernardino, California, San Bernardino County Museum Association, p. 51.

Dokka, R. K., and Glazner, A. F., 1982, Aspects of early Miocene extension in the central Mojave Desert, Geological Excursions in the California Desert: Anaheim, California, Geological Society of America, Cordilleran Section Fieldtrip Guidebook, p. 31–46.

Dokka, R. K., and Woodburne, M. O., 1986, Mid-Tertiary extensional tectonics and sedimentation, central Mojave Desert, California: Baton Rouge, Louisiana State University Publications in Geology and Geophysics 1, 55 p.

Dokka, R. K., and 10 others, 1991, Aspects of the Mesozoic and Cenozoic geologic evolution of the Mojave Desert: Geological Excursions in Southern California and Mexico: Geological Society of America, Annual Meeting Guidebook: San Diego, California, San Diego State University, Department of Geological Sciences, p. 1–43.

Ebinger, C., Hayward, N., and Cann, J., 1993, Rift kinematics and along-axis segmentation in N. Afar: Geological Society of America Abstracts with Program, v. 25, no. 6, p. A198.

Fillmore, R. P., 1993, Sedimentation and extensional basin evolution in a Miocene metamorphic core complex setting, Alvord Mountain, central Mojave Desert, California, USA: Sedimentology, v. 40, p. 721–742.

Fillmore, R. P., Walker, J. D., Bartley, J. M., and Glazner, A. F., 1994, Development of three genetically related basins associated with detachment-style faulting: Predicted characteristics and an example from the central Mojave Desert: Geology, v. 22, p. 1087–1090.

Glazner, A. F., Bartley, J. M., and Walker, J. D., 1988, Geology of the Waterman Hills detachment fault, central Mojave Desert, California, *in* Weide, D. L., and Faber, M. L., eds., This extended land—Geological journeys in the southern Basin and Range: Las Vegas, University of Nevada, Special Publication 2, p. 225–237.

Glazner, A. F., Bartley, J. M., and Walker, J. D., 1989, Magnitude and significance of Miocene crustal extension in the central Mojave Desert, California: Geology, v. 17, p. 50–53.

Glazner, A. F., Fletcher, J. M., Bartley, J. M., Walker, J. D., Martin, M. W., Miller, J. S., Taylor, W. J., and Coleman, D. S., 1992, Widespread Miocene plutonism and ductile deformation in the central Mojave metamorphic core complex, California [abs.]: Eos (American Geophysical Union Transactions), v. 73, p. 548.

Ingersoll, R. V., and 8 others, 1993, The Mud Hills, Mojave Desert, California: Structure, stratigraphy and sedimentology of a rapidly extending terrane: Geological Society of America Abstracts with Program, v. 25, no. 5, p. 56.

Kerr, D. R., 1984, Early Neogene continental sedimentation in the Vallecito-Fish Creek Mountains, western Salton Trough, California: Sedimentary Geology, v. 38, p. 217–246.

Krieger, M. H., 1977, Large landslides, composed of megabreccia, interbedded in Miocene basin deposits, southeastern Arizona: U.S. Geological Survey Professional Paper 1008, 25 p.

Lambert, J. R., 1987, Middle Tertiary structure and stratigraphy of the southern Mitchell Range, San Bernardino County, California [M.S. thesis]: Baton Rouge, Lousiana State University, 119 p.

Longwell, C. R., 1951, Megabreccia developed downslope from large faults: American Journal of Science, v. 249, p. 343–355.

Martin, M. W., Glazner, A. F., Walker, J. D., and Schermer, E. R., 1993, Evidence for right-lateral transfer faulting accommodating en echelon Miocene extension, Mojave Desert, California: Geology, v. 21, p. 355–358.

McCulloh, T. H., 1952, Geology of the southern half of the Lane Mountain quadrangle, California [Ph.D. thesis]: Los Angeles, University of California, 182 p.

Merriam, J. C., 1915, Extinct faunas of the Mohave Desert, their significance in a study of the origin and evolution of life in America: Popular Science Monthly, v. 86, p. 245–264.

Nielson, J. E., and Beratan, K. K., 1990, Tertiary basin development and tectonic implications, Whipple detachment system, Colorado extensional corridor, California and Arizona: Journal of Geophysical Research, v. 95, p. 571–580.

Nielson, J. E., Lux, D. R., Dalrymple, G. B., and Glazner, A. F., 1990, Age of the Peach Springs Tuff, southeastern California and western Arizona: Journal of Geophysical Research, v. 95, p. 571–580.

Rust, B. R., 1972, Pebble orientation in fluvial sediments: Journal of Sedimentary Petrology, v. 42, p. 384–388.

Turbeville, B. N., 1991, The influence of ephemeral processes on pyroclastic sedimentation in a rift-basin, volcaniclastic-alluvial sequence, Española basin, New Mexico, *in* Cas, R., and Busby-Spera, C. eds., Volcaniclastic sedimentation: Sedimentary Geology, v. 74, p. 139–155.

Valentine, M. J., Brown, L. L., and Golombek, M. P., 1993, Cenozoic crustal rotations in the Mojave Desert from paleomagnetic studies around Barstow, California: Tectonics, v. 12, p. 666–677.

Walker, J. D., and Fillmore, R. P., 1993, Age and depositional characteristics of the synextensional Pickhandle Formation, central Mojave Desert, California: Eos (American Geophysical Union Transactions), v. 74, p. 588.

Walker, J. D., Bartley, J. M., and Glazner, A. F., 1990, Large-magnitude Miocene extension in the central Mojave Desert: Implications for Paleozoic to Tertiary paleogeography and tectonics: Journal of Geophysical Research, v. 95, p. 557–569.

Walker, J. D., Fletcher, J. M., Fillmore, R. P., Martin, M. W., Taylor, W. J., Glazner, A. F., and Bartley, J. M., 1995, Connection between igneous activity and extension in the central Mojave metamorphic core complex: Journal of Geophysical Research, v. 100, no. B6, p. 10,477–10,495.

Waresback, D. B., and Turbeville, B. N., 1990, Evolution of a Plio-Pleistocene volcanogenic-alluvial fan: The Puye Formation, Jemez Mountains, New Mexico: Geological Society of America Bulletin, v. 102, p. 298–314.

Woodburne, M. O., Tedford, R. H., and Swisher, C. C., III, 1990, Lithostratigraphy, biostratigraphy, and geochronology of the Barstow Formation, Mojave Desert, southern California: Geological Society of America Bulletin, v. 102, p. 459–477.

Yarnold, J. C., 1993, Rock-avalanche characteristics in dry climates and the effect of flow into lakes: Insights from mid-Tertiary sedimentary breccias near Artillery Peak, Arizona: Geological Society of America Bulletin, v. 105, p. 345–360.

Yarnold, J. C., and Lombard, J. P., 1989, A facies model for large rock-avalanche deposits formed in dry climates, *in* Colburn, I. P., Abbott, P. L., and Minch, J., eds., Conglomerates in basin analysis: A symposium dedicated to A. O. Woodford: Society of Economic Paleontologists and Mineralogists, Pacific Section, v. 62, p. 9–31.

Manuscript Accepted by the Society April 21, 1995

Printed in U.S.A.

Geological Society of America
Special Paper 303
1996

Evolution of the Miocene north Whipple basin in the Aubrey Hills, western Arizona, upper plate of the Whipple detachment fault

Rebecca J. Dorsey and Patricia Roberts*
Department of Geology, Northern Arizona University, Flagstaff, Arizona 86011-4099

ABSTRACT

A thick sequence of Miocene sedimentary and minor volcanic rocks in the northwestern Aubrey Hills, western Arizona, was deposited in part of the tectonically active north Whipple basin during regional extension and related basin formation in the upper plate of the Whipple detachment fault. The studied sequence (sequence III of Nielson, 1986) overlies the 18.5 Ma Peach Springs Tuff and older Miocene volcanic rocks along an angular unconformity. Sedimentary lithofacies of sequence III consist of (1) red sandy conglomerate, which accumulated by sheetflood and minor debris-flow deposition in a middle alluvial-fan environment; (2) sandstone with minor conglomerate and limestone, which record channel migration and deposition in a low-gradient, distal alluvial-fan or braidplain setting; (3) lacustrine and lacustrine-margin mudstone, algal limestone, and sandstone; and (4) megabreccia, including matrix-rich, jigsaw, and crackle breccia, which was emplaced catastrophically as a large rock avalanche derived from the lower plate of the Whipple detachment fault. The megabreccia sits high in the sedimentary sequence along a quasi-conformable contact with underlying sandstone and conglomerate. Vertical trends in conglomerate clast composition, combined with lithologic information from the megabreccia, provide a step-like record of uplift and unroofing in the Whipple Mountains.

Southward thickening and coarsening of alluvial-fan facies and other aspects of stratigraphic architecture provide evidence for deposition of sequence III in an alluvial-fan system that was controlled by asymmetric subsidence and tilting of the basin toward the south. Sediments were shed to the north-northeast off the proto–Whipple Mountains, which were topographically rising and rapidly eroding during this time. Of several hypotheses for basin subsidence in the Aubrey Hills portion of the north Whipple basin, two seem reasonable: (1) oblique, right-lateral and down-on-the north displacement along an east-trending transfer fault; and (2) syndepositional growth of an asymmetric, extension-parallel syncline that folded the active Whipple detachment fault. We favor the second interpretation because it explains numerous features observed in the basin, and it is consistent with a recent study (Yin and Dunn, 1992) that demonstrates near-surface folding of the Whipple detachment about the same set of extension-parallel fold axes shortly after deposition of the Aubrey Hills sequence.

*Present address: Souder, Miller, and Associates, 1201 Parkway Drive, Suite C, Santa Fe, New Mexico 87501.

Dorsey, R. J., and Roberts, P., 1996, Evolution of the Miocene north Whipple basin in the Aubrey Hills, western Arizona, upper plate of the Whipple detachment fault, *in* Beratan, K. K., ed., Reconstructing the History of Basin and Range Extension Using Sedimentology and Stratigraphy: Boulder, Colorado, Geological Society of America Special Paper 303.

INTRODUCTION

The Whipple detachment fault is located in the Colorado River extensional corridor (Howard and John, 1987), a well-studied zone of early to middle Miocene continental extension and low-angle detachment faulting (Figs. 1 and 2; Shackelford, 1976; Davis and Coney, 1979; Davis, 1980, 1983, 1988; Davis et al., 1980, 1986; John, 1987; Howard and John, 1987; Davis and Lister, 1988; Spencer and Reynolds, 1989, 1991). Detachment faulting in this region was accompanied by strong crustal thinning of the upper plate, isostatic rebound accompanied by ductile-plastic flow of mid-crustal rocks, and exhumation of mylonitic lower-plate rocks from depths of 12–15 km to the surface (e.g., Wernicke, 1981, 1985, 1992; Spencer, 1984; Buck, 1988; Wernicke and Axen, 1988; Block and Royden, 1990; Kruse et al., 1991). At shallow crustal levels, upper-plate deformation occurred by high-angle normal faulting and related development of sedimentary basins. Recent studies of Miocene basin evolution have produced a wealth of new information about syn-extensional basin subsidence, sedimentation, and source-area evolution in the Rawhide Mountains (Spencer et al., 1989; Yarnold, 1994), Casteneda Hills (Lucchitta and Suneson, 1988a, 1988b, 1993), Chemehuevi Mountains (Miller and John, 1988, 1993), Sacramento Mountains (Fedo, 1990; Fedo and Miller, 1992), and the Whipple Mountains area (Beratan, 1990, 1991; Nielson and Beratan, 1990; Yin and Dunn, 1992; Roberts, 1992; Dorsey and Roberts, 1993).

From the studies just listed, it is known that upper-plate basins in the Colorado River extensional corridor formed by asymmetric subsidence and tilting in fault-bounded half grabens, with the possible influence of slip on extension-parallel transfer faults. North- to northwest-trending high-angle normal faults that bound these basins sole into and are commonly truncated by major low-angle detachments. In spite of this new information, much remains unknown about the details of timing, style, and kinematics of upper-plate deformation and basin development. Detailed stratigraphic, paleocurrent, and facies analysis of well-exposed basin-fill sequences provides a powerful approach to a better understanding of the processes of brittle deformation and dismemberment of detachment upper plates. For example, the orientation of major axes of block rotation can be determined from the reconstructed geometry and sediment dispersal patterns of stratigraphic packages. Ultimately, a complete understanding of Miocene detachment faulting in the lower Colorado River region will depend on understanding the timing and style of upper-plate deformation and basin development.

This chapter presents the results of detailed sedimentologic and stratigraphic analysis of Miocene sedimentary rocks exposed in the Aubrey Hills, western Arizona (Fig. 2). These rocks were deposited in part of the north Whipple basin, a synextensional nonmarine basin that formed and subsided in the upper plate of the active Whipple detachment fault on the north flank of the rising Whipple Mountains. We devote much of this chapter to sedimentologic, stratigraphic, petrologic, and paleocurrent analysis of the sedimentary sequence. We then evaluate different possible basin models and develop an interpretation of the major structures that controlled basin subsidence. We show that a secondary but important component of upper-plate deformation and basin development occurred by growth of large extension-parallel synforms and antiforms in the active Whipple detachment fault.

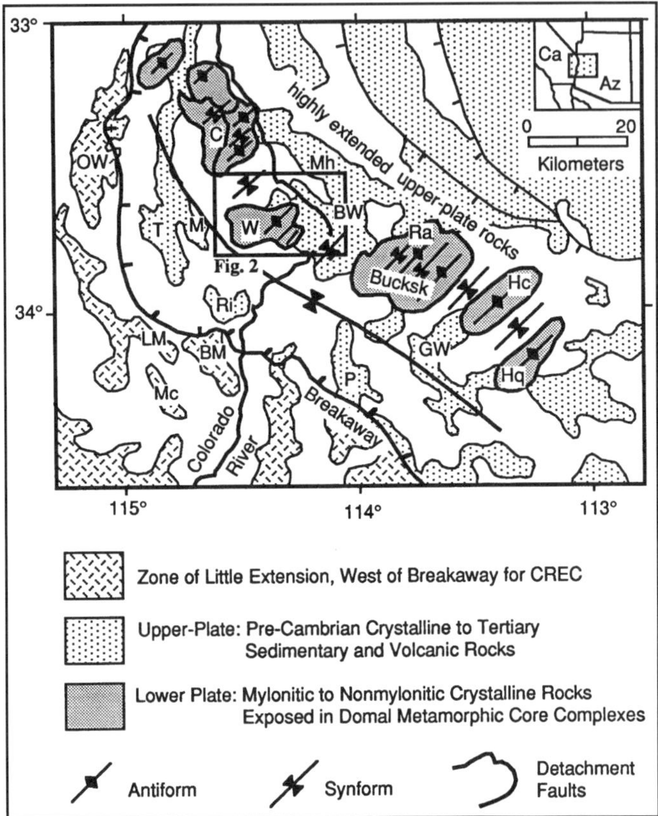

Figure 1. Simplified tectonic map of the Colorado River extensional corridor, adapted from Yin (1991) and Spencer and Reynolds (1991). BM = Big Maria Mts; BW = Bill William Mts; Bucksk = Buckskin Mts; C = Chemehuevi Mts; GW = Granite Wash Mts; Hc = Harcuvar Mts; Hq = Harquahala Mts; LM = Little Maria Mts; M = Mopah Range; Mc = McCoy Mts; Mh = Mohave Mts; OW = Old Woman Mts; P = Plomosa Mts; Ra = Rawhide Mts; Ri = Riverside Mts; T = Turtle Mts; W = Whipple Mts. Box shows location of Figure 2.

TECTONIC SETTING AND GENERAL STRATIGRAPHY

The Colorado River extensional corridor is a highly deformed detachment-fault terrane in which rocks of the upper plate were tectonically transported to the east-northeast relative to the lower plate during early to middle Miocene regional extension (Fig. 1; Howard and John, 1987; Davis and Lister, 1988; Spencer and Reynolds, 1989, 1991). This detachment-fault system is bounded on the west and southwest by a steeply east- to northeast-dipping breakaway fault (Fig. 1). A prominent

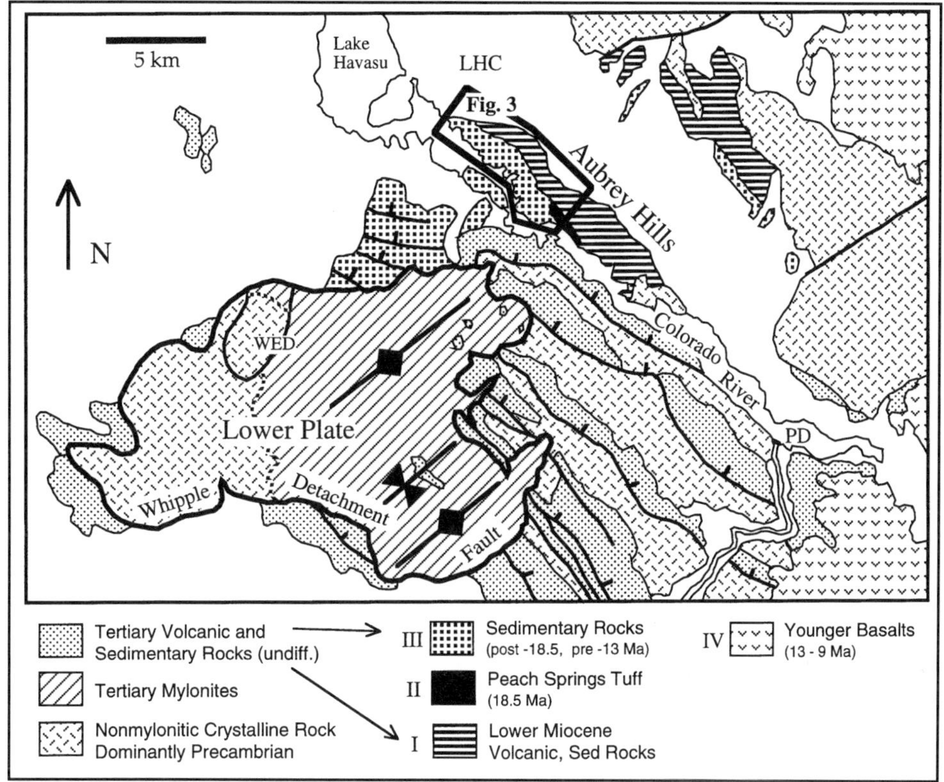

Figure 2. Geologic map of the Whipple Mountains and adjacent areas to the east and northeast, modified from Davis et al. (1980), Dunn (1986), Howard et al. (1990), Davis and Anderson (1991), and Yin and Dunn (1992). Roman numerals represent sequences as defined by Nielson (1986). WED = War Eagle detachment fault; LHC = Lake Havasu City; PD = Parker Dam. Box shows location of Figure 3.

northwest-trending synform in the detachment fault lies between the breakaway and a centrally located belt of domal lower-plate upwarps, or metamorphic core complexes (Fig. 1). Northeast of the core complexes, a zone of highly faulted and extended upper-plate rocks (extensional allochthon of Spencer and Reynolds, 1989) rests on the down-dip projection of the detachment faults (Fig. 1). The Aubrey Hills are located northeast of the Whipple Mountains within this zone of highly extended upper-plate rocks (Figs. 2 and 3). The breakaway zone, synform, antiformal belt of core complexes, and numerous high-angle normal faults that cut the upper plate are all north- to northwest-trending, extension-perpendicular structures that formed in response to northeast-southwest extension. The domal geometry of metamorphic core complexes (Fig. 2) results from antiforms and synforms (corrugations) that are oriented parallel to the direction of tectonic transport (Davis and Lister, 1988; Spencer and Reynolds, 1991; Yin and Dunn, 1992). The significance and origin of extension-parallel folds are discussed later in this chapter.

Rocks in the upper plate of the Whipple detachment fault are highly extended and segmented into a series of southwest-tilted half graben that are bounded by northwest-trending, high-angle normal faults (Fig. 2). The upper plate consists of nonmylonitic Precambrian crystalline rocks that are overlain by syn- to postextensional Miocene volcanic and sedimentary rocks. Miocene rocks can be subdivided into four genetically related sequences having distinctive lithologic and structural characteristics (Nielson, 1986; Nielson and Beratan, 1990). Sequence I ranges in age from about 23 Ma to 19 Ma and consists mainly of voluminous mafic to intermediate volcanic rocks and locally abundant clastic sedimentary rocks. Sequence II consists of the widespread Peach Springs Tuff, dated at 18.5 ± 0.2 Ma by Nielson et al. (1990). In some areas the Peach Springs Tuff is missing and an erosional unconformity exists between rocks belonging to sequence I and III. Sequence III, the focus of this chapter (which includes the Copper Basin Formation of the Whipple Mountains), consists mainly of clastic sedimentary rocks, megabreccia, basalt, and limestone. Sequence IV unconformably overlies older rock sequences. It is a bimodal suite of mostly flat-lying basaltic and rhyolitic flows, tuffs, breccias, and plugs, in places dominated by basalt flows, which range in age from about 13 Ma to 9 Ma (Nielson and Beratan, 1990; Spencer and Reynolds, 1989).

The geologic map of the northwestern Aubrey Hills reveals a southwest-dipping package of rocks belonging to sequences I, II, and III, all bounded by unconformities (Fig. 3). The contact

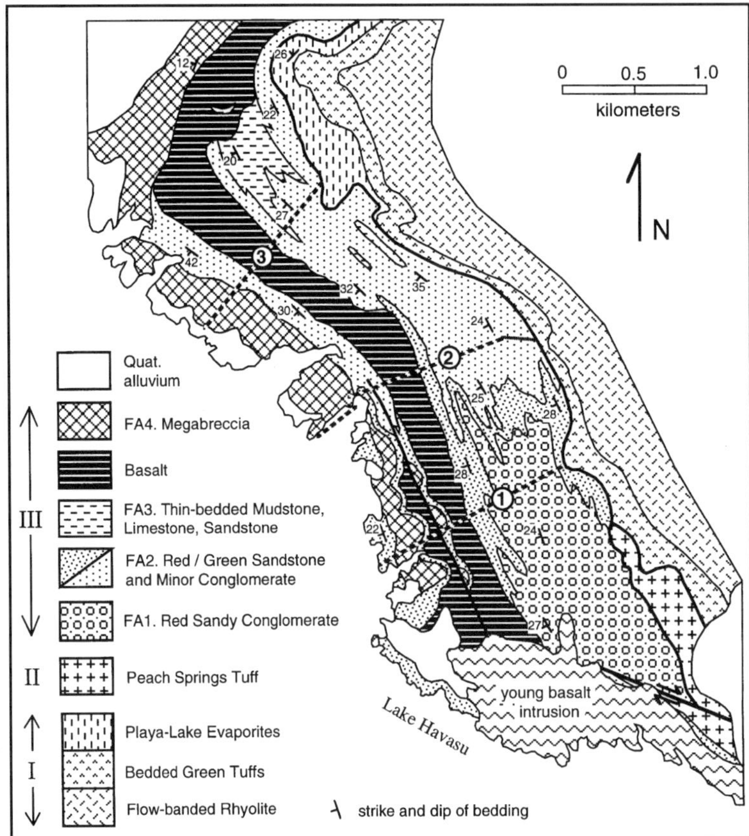

Figure 3. Geologic map of the study area in Aubrey Hills, showing distribution of major lithofacies and location of measured sections. Roman numerals I, II, and III identify unconformity-bounded sequences of Nielson (1986). FA1, FA2, etc. refer to facies associations described in text. Dense-stipple pattern represents red variant of facies association 2; open-stipple pattern represents green variant. Circled numbers indicate location of measured sections in Figure 9.

between sequence I and the overlying Peach Springs Tuff (sequence II) is interpreted to be an unconformity based on the highly irregular (erosional) nature of the contact and truncation of sequence I stratigraphy. The basal contact of sequence III is a low-angle erosional unconformity that truncates stratigraphy in older rocks of sequence I and II and is characterized by local soil development in underlying tuffs (Figs. 3 and 4). Sequence III deposits in the Aubrey Hills include a lithologically variable package of coarse- to fine-grained sedimentary rocks that show a pronounced lateral change in lithofacies from red sandy conglomerate in the south to mudstone, limestone, and sandstone in the north (Fig. 3). This heterogeneous sedimentary package is overlain by a basalt flow unit along a contact that is conformable in the south but locally is an angular unconformity in the north, near a northwest-plunging anticline (Fig. 3). The basalt is overlain by a thin interval of sandstone and minor conglomerate that is capped by megabreccia. Megabreccia contains nonmylonitic to variably mylonitized quartz monzonite and gabbro derived from the lower plate of the Whipple detachment fault. It can be correlated to the southwest, across Lake Havasu (Fig. 2), with a large breccia sheet that is widely exposed in a similar stratigraphic position in the northeastern Whipple Mountains (Dunn, 1986; Yin and Dunn, 1992; Dorsey et al., 1993; Dorsey and Becker, 1995).

LITHOFACIES ANALYSIS

Sedimentary lithofacies of sequence III in the Aubrey Hills are grouped into four facies associations that correspond to map units in Figure 3: (1) red sandy conglomerate; (2) sandstone and minor conglomerate; (3) mudstone, limestone and sandstone; and (4) megabreccia (Table 1).

Facies association 1: Red sandy conglomerate

Red sandy conglomerate is divided into two main lithofacies types: (1a) clast-supported conglomerate and pebbly sandstone, which comprise the majority of facies association 1 (80% to 90%) and (1b) interbedded matrix-supported conglomerate.

Description. *Facies 1a* consists of poorly sorted, massive to

Figure 4. Outcrop photographs of unconformity separating rocks of sequence I from overlying sequence III. (A) Silicified limestone of sequence I overlain by thin-bedded sandstone of sequence III along a low-angle erosional truncation surface at the base of section 3 (Fig. 3). (B) Argillic paleosol in uppermost tuff of sequence I (section 1, Fig. 3), showing a progressive upward increase in pedogenic alteration, sharp planar contact, and overlying less resistant pebbly sandstone of sequence III. Hammer is 32.5 cm long. Peach Springs Tuff is missing at the unconformity in both A and B as a result of post 18.5 Ma erosion.

crudely planar and trough cross-stratified sand-rich pebble to pebble-cobble conglomerate and pebbly sandstone (Fig. 5A). Cross-bedding is sporadically developed and occurs only at low angles. This facies typically shows little or no vertical organization in grain size or stratification features, and sharp bases of beds are either unchannelized or exhibit very shallow (<20 cm deep), broad (1–3 m wide) geometries. Conglomerate resembles facies Gm and Gt of Miall (1978) and Rust (1978), and thin pebbly sandstone interbeds are equivalent to facies Sh and St. In the southern part of the study area we measure stratigraphic intervals in this facies >100 m thick that show no appreciable vertical change in grain size. Because of the subequant clast shape and abundance of sandy matrix in conglomerate, pebble- and cobble-sized clasts are rarely imbricated, making collection of reliable paleocurrent data difficult. Facies 1a ranges from deep to light red to orange-red in color.

Facies 1b consists of matrix-supported, pebble-cobble to pebble-cobble-boulder conglomerate in which clasts are randomly oriented and grading of clasts is typically absent. Matrix-supported conglomerate is internally structureless, and beds range from about 0.5–1.5 m thick; inverse and normal grading is rarely present at the base of conglomerate beds. Matrix consists of very poorly sorted muddy pebbly sandstone. This facies is equivalent to facies Gms of Rust and Koster (1984).

Interpretation. We interpret red sandy conglomerate and pebbly sandstone (facies 1a) to be the product of rapid, unchannelized, flood-flow deposition under conditions of episodic discharge. Traction sedimentation is indicated by clast-supported fabric and uncommon planar to low-angle cross-stratification. Rapid deposition with little or no sediment reworking is indicated by the overall poor sorting and lack of well-developed stratification or bedding features. We infer that rapid deposition took place on a broad, flat surface with rare shallow channels. Facies 1b (matrix-supported conglomerate) was deposited en masse by mud-bearing debris flows, in which clast support was accomplished by a combination of cohesive matrix support and clast buoyancy (Nemec and Steel, 1984).

The close association of facies 1a and 1b is indicative of deposition in a subaerial alluvial-fan environment (Miall, 1978; Rust, 1978; Rust and Koster, 1984). Facies 1a records episodic, rapid deposition by sheetfloods in a rapidly aggrading, unchannelized portion of the alluvial fan (Bull, 1972; Collinson, 1986). The systematic lack of channel geometries indicates a middle alluvial-fan depositional setting located downstream of the fan's intersection point and upstream of the distal alluvial braidplain (facies association 2). Gravel-bearing flood flows emanated from fan-head channels and expanded laterally to form unconfined sand- and gravel-bearing sheetfloods. Sheetflood deposition alternated with less common deposition of debris flows derived from the same system of upper-fan channels. Based on the southward coarsening of age-equivalent deposits, presence of interbedded debris-flow deposits, and north- to northeast-directed paleocurrents, we infer that the entrenched upper fan, not observed in this study, was located south of the study area.

TABLE 1. SUMMARY DESCRIPTIONS AND INTERPRETATIONS OF LITHOFACIES ASSOCIATIONS, SEQUENCE III, AUBREY HILLS

Lithofacies Association	Summary Description	Interpretation
1. Red Sandy Conglomerate	1a. Clast-supported pebble-cobble conglomerate and pebbly sandstone, crude horizontal bedding, massive to weakly planar and trough cross-stratified. Gm, Gt, St.	Sheet-flood deposition in medial alluvial fan.
	1b. Matrix-supported pebble-cobble-boulder conglomerate, internally massive and ungraded. Gms.	Debris-flow deposition in medial alluvial fan.
2. Sandstone and Minor Conglomerate	2a. Planar and trough cross-stratified sandstone and pebbly sandstone, thin- to medium-bedded. Sh, St.	Channel filling in low-gradient distal alluvial fan or braidplain.
	2b. Clast-supported pebble-cobble conglomerate, massive to cross-bedded, commonly at base of fining-up sequences. Gm, Gt.	Channel-lag gravel, longitudinal and transverse bars, distal alluvial fan or braidplain.
	2c. Massive nodular calcareous sandstone with rhizocretions, mottling, and calcareous platy peds. P.	Paleosols formed in interchannel areas, distal fan or braidplain.
	2d. Massive to algal-laminated limestone, tabular, thin-bedded.	Organic precipitation of carbonate in shallow inter-channel ponds.
3. Mudstone, Limestone, Sandstone	3a. Interbedded mudstone, claystone, limestone, and minor sandstone. Planar- to ripple-laminated, thin-bedded.	Quiet-water deposition in lacustrine and lacustrine-margin setting.
4. Megabreccia	4a. Matrix-rich breccia. Matrix-supported, chaotic mix of sedimentary matrix and War Eagle Pluton-derived clasts.	Basal and lateral margins of large rock avalanche, or landslide (sturzstrom).
	4b. Jigsaw breccia. Clast-supported, tightly packed, matrix = pulverized crystalline rock.	Also at margins of breccia sheet, partial brecciation of crystalline rock.
	4c. Crackle Breccia. Internally intact but highly fractured crystalline rock (War Eagle Pluton).	Large rock-avalanche breccia sheet, from lower plate of Whipple detachment fault.

Facies association 2: Sandstone and minor conglomerate

Facies association 2 consists of (2a) planar to cross-stratified sandstone and pebbly sandstone, (2b) massive to weakly stratified conglomerate, (2c) massive nodular calcareous sandstone, and (2d) algal-laminated limestone.

Description. *Facies 2a* (planar to cross-stratified sandstone and pebbly sandstone) is volumetrically the most abundant (about 60–70%) of facies association 2. It consists of weakly bedded, planar to low-angle trough cross-stratified sandstone interbedded and interstratified with thin lenticular bodies of planar and trough cross-stratified pebbly sandstone (Fig. 5B). Sandstone and pebbly sandstone occur in tabular beds that commonly show amalgamation. This facies resembles facies Sh and St of Miall (1978). Sandstone and pebbly sandstone facies are found in the upper parts of fining-up intervals, 2 to 8 m thick,

Figure 5. Outcrop photographs of facies associations 1, 2, and 3. (A) Poorly sorted, weakly stratified, red sandy conglomerate and pebbly sandstone (facies 1a). Clipboard is 31.6 cm long.

Figure 5 (B) Thin- to medium-bedded sandstone, pebbly sandstone and conglomerate (facies 2a and 2b). Channel-lag conglomerate can be seen in lower-right corner and upper ledge; base of upper conglomerate shows shallow erosional scours.

Figure 5 (C) Rhizocretions and nodular calcite in massive calcareous sandstone (facies 2c). White part of pen is 8.5 cm long.

Figure 5 (D) Rhythmically interbedded limestone and mudstone of facies association 3. Person on left for scale.

that have channelized conglomerate (facies 2b) at their base (Fig. 5B).

Facies 2b makes up roughly 20–25% of facies association 2. It consists of poorly sorted, massive to weakly trough cross-stratified, clast-supported pebble and pebble-cobble conglomerate. Conglomerate beds, ranging from 30–100 cm thick, commonly are scoured and channelized at their bases and grade upward into sandstone of facies 2a. Thinner conglomerate beds commonly display a lenticular, discontinuous geometry. Locally, facies 2b consists of laterally discontinuous single-clast horizons of large pebbles to cobbles that are abruptly overlain by sandstone or pebbly sandstone beds (facies 2a).

Facies 2c consists of massive, weakly bedded fine- to medium-grained nodular calcareous sandstone that in places contains calcite- and iron-oxide-filled root casts, or rhizocretions (Fig. 5C). Thin beds of nodular limestone locally are interbedded with calcareous sandstone. Distinctive reddish and greenish mottling in massive sandstone is commonly associated with rhizocretions, and gypsum veins locally are present. These deposits commonly are interbedded with planar stratified sandstone (facies 2a) and thin limestone beds (facies 2d).

Facies 2d consists of tabular to broadly lenticular limestone, occurring in beds 10–60 cm thick, which commonly contains wavy to irregular algal laminations and locally shows silica replacement. Some outcrops of limestone release a fetid odor when broken. Limestone facies occur in association with either the planar-stratified sandstone facies (facies 2a) and/or the massive, nodular calcareous sandstone (facies 2c). Facies 2c and 2d together make up approximately 10% of facies association 2.

Color in facies association 2 shows a systematic distribution in the geologic map, with a laterally restricted, dominantly red variant in the south passing abruptly northward into a green and tan variant (Fig. 3). In terms of lithofacies and sedimentary structures, these two variations are basically identical and are therefore grouped together in the same facies association.

Interpretation. The close association of widely varying grain sizes, lithologies, and sedimentary structures in facies association 2 provides evidence for deposition under a wide range of physical and chemical conditions in a low-gradient, distal alluvial-fan or braidplain setting. Gradational fining-up intervals (facies 2b to 2a) represent channel-fill sequences in which the energy of stream flows waned through time, probably in response to lateral shifting and abandonment of shallow channels. The majority of limestones in the Aubrey Hills (facies 2d) formed by organic precipitation of calcareous algae (e.g., Dean and Fouch, 1984). The common occurrence of algal-laminated limestone, although minor in total volume, provides evidence that shallow, standing bodies of water were sometimes established in interchannel ponds or swamps. This indicates that the distal part of the alluvial-fan system had a low topographic gradient. This inference is supported by the abrupt northward lateral transition from facies association 1 to green and tan deposits of facies association 2 over a horizontal distance of 0.3–0.5 km (Fig. 3). An abrupt facies transition provides evidence for an abrupt downstream decrease in topographic gradient.

Rhizocretions, calcite nodules, and mottling structures in facies 2c are common features of paleosols; they record intermittent soil development in an arid or semi-arid climate under variable, oxidizing and reducing conditions. Nodular limestone in facies 2c probably consists of inorganically precipitated calcrete, or calcic B horizons, produced by coalescing of calcareous nodules and tubules (Machette, 1985; Mack and James, 1992). We envisage autocyclic processes of lateral channel migration, abandonment, pedogenesis, and development of overbank swamps and marshes to explain the interbedding of channel conglomerate, sandstone, paleosols, and algal limestone. The laterally discontinuous nature of limestone interbeds supports this interpretation.

Color variations. It is possible to interpret the systematic color variation in facies associations 1 and 2 in terms of duration and frequency of sediment saturation by ground water, which was likely related to the time-averaged position of the water table relative to the fan surface. The lateral transition from red to green and tan sediments is seen at the lateral transition from medial to distal alluvial-fan facies (Fig. 3). Red coloration of facies association 1 and 2 is interpreted to be the product of early postdepositional oxidation of mafic minerals, forming clays and immature iron oxides and hydroxides. This commonly occurs when large volumes of oxidizing meteoric waters are flushed down through unconsolidated sands and gravels of alluvial fans, where the fan surface is maintained above the water table (Walker, 1967; Galloway and Hobday, 1983, p. 249). In contrast, green and tan sediments of facies associations 2 and 3 record deposition in topographically lower and flatter, distal parts of the ancient alluvial fan. In the distal fan, where the water table was shallow and periodically emergent, stagnant surface waters collected in marshes and swamps for time periods long enough to produce reducing conditions in shallow subsurface sediments. This interpretation is supported by the common occurrence of algal limestones interbedded in green variants of facies association 2. We thus attribute the lateral northward change from red to green and tan facies to a down-fan change from dominantly oxidizing to dominantly reducing shallow subsurface conditions, which resulted from down-fan shallowing and emergence of the water table.

Facies association 3: Mudstone, limestone, and sandstone

Description. Facies association 3 consists of thinly interbedded mudstone, claystone, limestone and sandstone (Fig. 5D). Limestone is similar to that of facies 2d. It contains fine algal laminae, typically has a fetid, organic odor when broken, and shows varying degrees of silica replacement. Bedding contacts are generally sharp and planar to wavy, and bed geometries are tabular to broadly lenticular. Planar lamination is the dominant internal sedimentary structure, with wave-ripple cross-lamination present in some beds. These deposits typically display regular alternation of mudstone and limestone in rhythmically bedded intervals (Fig. 5D). Variations of this facies association include thin, tabular-bedded, planar laminated, greenish fine- to coarse-grained sandstone and (rarely) pebbly sandstone that are interbedded with algal limestone and siltstone. Gradational vertical alternations between sand-rich and sand-poor facies are common.

Interpretation. We interpret facies association 3 to record relatively quiet-water deposition in a lacustrine and lacustrine-margin setting at the downstream termination of the low-gradient distal alluvial fan to the south (facies association 2). Planar-laminated mudstone, siltstone, and claystone were deposited by suspension sedimentation during slack water conditions, and limestone accumulated by organic precipitation of calcium carbonate during periods of little or no detrital input. This was a perennial lake that did not undergo seasonal drying, as evidenced by the lack of mud cracks and evaporite minerals. Interbedded fine- to coarse-grained sandstone and pebbly sandstone record periods of higher-energy (upper plane bed) traction sedimentation in a lacustrine-margin environment where sediment-laden floods debauched into shallow water. Sand-dominated deposits of facies association 2 are closely interbedded with facies association 3 in the northern part of the study area (section 3 in Fig. 3); the lateral transition between the two is quite gradual and indistinct, especially when compared with the abrupt down-fan fining seen at the lateral contact between facies association 1 and 2. This provides further evidence that the distal sandy part of the alluvial fan (facies association 2) was a low-gradient stream system that passed gradually northward into low-energy shoreline and lake environments (facies association 3).

Facies association 4: Megabreccia

Following the classification of Yarnold and Lombard (1989), we recognize three main facies types within the megabreccia facies association: (4a) matrix-rich breccia, (4b) jigsaw breccia, and (4c) crackle breccia.

Description. *Facies 4a* (matrix-rich breccia) consists of massive, thick to very thick bedded, disorganized sedimentary breccia that contains angular to subangular clasts ranging from pebble- to large boulder-size in a matrix of mixed, poorly sorted muddy pebbly sandstone (Fig. 6A). Matrix-rich breccia is commonly interbedded with pebbly sandstone deposits of facies association 2 (Fig. 6A). Large-scale soft-sediment deformation and loading features such as clastic dikes, flames, and ball-and-pillow structures commonly occur along the base of matrix-rich breccia in a basal zone of mixing, as much as 10–15 m thick, between substrate sediment and transported breccia material. Clasts and blocks in all breccia facies consist of variably mylonitized quartz-monzonite (adamellite), gabbro, and diabase. These lithologies match those of the War Eagle pluton, a 19 Ma syn-faulting plutonic complex that is currently exposed in the lower plate of the Whipple detachment fault in the area of the War Eagle detachment (Fig. 2; Yin and Dunn, 1992).

Facies 4b is jigsaw breccia of Yarnold and Lombard (1989). It contains angular, small pebble-sized clasts that are closely packed in a clast-supported "jigsaw puzzle" fabric (Fig. 6B) indicative of a small amount of relative motion between clasts. The matrix consists of finely comminuted, crushed crystalline rock having the same composition as the clasts. Jigsaw breccia is found in gradational contact with both matrix-rich breccia and crackle breccia.

Facies 4c (crackle breccia) consists of large blocks and slabs of fractured crystalline rock ranging to more than 100 m in size, in which lithologically intact crystalline rock is cut by multiple, closely spaced brittle fractures that have little or no offset (Fig. 6C). Original lithologic contacts such as intrusive contacts and compositional banding can be easily traced through large bodies of crackle breccia, in spite of a pervasive system of fractures that cause the rock to fall apart into small pieces when disturbed.

Contractional structures in conglomeratic sandstone facies (facies association 2) are found up to ~20 m below the base of the breccia (Fig. 6D). These structures consist of small drag folds, thrusts, and fold-thrust duplexes. Based on limited kinematic data obtained from basal grooves and lineations, stereonet plots of folded bedding, and reconstructed fold axes, the overall direction of shearing and transport of the megabreccia is generally toward the north-northeast. In the northern part of the study area, where the megabreccia unit is more than 100 m thick, crackle breccia is the dominant facies, with matrix-rich and jigsaw breccia making up only a minor portion of the unit at and near the base. In the south, by contrast, where the megabreccia unit is only about 35 m thick and is overlain by sandstone, it consists almost entirely of jigsaw and matrix-rich breccia facies, with little or no crackle breccia.

Interpretation. We interpret megabreccia facies in the Aubrey Hills to represent different parts of a large surficial rock avalanche, or "sturzstrom" (Hsü, 1975; Yarnold and Lombard, 1989). Megabreccia in the Aubrey hills correlates with an extensive breccia sheet (Copper Canyon landslide of Yin and Dunn, 1992; War Eagle landslide of Forshee and Yin, 1995), which is exposed to the southwest across Lake Havasu in the northeastern Whipple Mountains (Fig. 2). This landslide was transported a minimum of 10 km to the north-northeast into the north Whipple basin; it covers a present-day area of ~70–100 km^2 and comprises a volume of about 11 km^3 (Yin and Dunn, 1992; Forshee and Yin, 1993). The megabreccia consists of lower-plate rock that was derived from the north-facing War Eagle detachment fault in the north Whipple Mountains (Fig. 2; Yin and Dunn, 1992). Large rock avalanches are produced by extremely rapid slope movements and transport of large bodies of bedrock material (commonly greater than 10^6 m^3) into sedimentary basins that flank steep mountain ranges; they commonly travel at speeds greater than 50 km/hr and can travel horizontal distances that exceed their net vertical fall by as much as 30 times (Yarnold and Lombard, 1989; Melosh, 1979, 1987, 1990). Basal portions and margins of rock-avalanche deposits tend to be disrupted and mixed with substrate material (matrix-rich breccia), whereas internal parts of flows comprise large, coherent blocks of highly fractured bedrock (jigsaw breccia grading into crackle breccia) that experience little or no internal disruption (Yarnold and Lombard, 1989). The decrease in total thickness of megabreccia toward the south-southeast, and increase in the same direction in relative abundance of matrix-rich facies, is interpreted to represent an oblique across-transport transition from the center toward the lateral margin of the rock avalanche. We infer that rock debris experienced enhanced mixing with underlying substrate sediment and/or spalling of smaller flows at the lateral margin of the avalanche. This interpretation is consistent with the north-northeasterly transport direction deduced from basal shear lineations and deformation structures in underlying sedimentary rocks.

BASALT UNIT

Description. A prominent basalt unit, 80–100 m thick, is present within sequence III in the study area. The age of this basalt was determined from K-Ar whole-rock analysis to be 14.1 ± 0.3 Ma (J. Nakata, unpublished data, reported in Nielson and Beratan, 1990). Basalt occurs as a succession of sharp-based, laterally continuous layers that average 2–5 m thick. Layers of basalt maintain a relatively constant thickness over large distances, and there is no evidence for internal truncations or cross-cutting relationships. The basalt has a fine-grained groundmass

Figure 6. Outcrop photographs of facies association 4 (megabreccia) in the Aubrey Hills. (A) Disorganized matrix-rich breccia (bottom and top units) showing sharp, irregular bedding contacts with pebbly sandstone of probable debris-flow origin (middle unit). Note large load structure in upper left. Hammer for scale. (B) Jigsaw breccia (Facies 4b), showing uniform clast lithology, tight packing fabric, and granular matrix consisting of pulverized crystalline protolith. Hammer for scale. (C) Crackle breccia (Facies 4c) consisting of highly fractured but internally intact crystalline rock (quartz monzonite and diorite). Hammer for scale. (D) Small thrust fault and related fold in deformed conglomeratic sandstone about 20 m below the base of the megabreccia, northern part of study area. Hammer for scale. View looking southeast.

of pyroxene, plagioclase, and minor olivine, with rare to no phenocrysts and variable amounts of vesicle-filling secondary minerals (calcite and zeolite). Amygdules display systematic vertical trends in size and abundance within each bed, increasing upward from small (≤1 mm) and rare in the lower part, to large (~1 cm), abundant and closely spaced in the upper part of each bed. Although rare, a key aspect of layered basalts in the Aubrey Hills is the presence of thin (0–20 cm), discontinuous lenses of baked sandstone and siltstone at the top of basalt layers, which commonly are truncated by the sharp undulating base of the overlying basalt. More common are discontinuous thin veins and stringers of baked sandstone and siltstone, randomly oriented and concentrated in the lower half of basalt beds. The upper contact of the basalt unit is covered.

Interpretation. The Aubrey Hills basalt unit is interpreted to be a thick sequence of stacked flows that inundated the basin during an eruptive episode. The upward increase in vesicularity within single beds was probably produced by gas bubbles expanding and migrating upward during cooling of individual flows. The rare occurrence of discontinuous sandstone and siltstone lenses between basalt beds provides evidence that these are surficial flows, rather than a sequence of layered shallow sills. The lateral continuity of layering and the lack of internal truncations support this interpretation. Randomly oriented thin veins of baked siltstone are interpreted to be clastic dikes injected from thin interbedded sediments.

PALEOCURRENTS

Paleocurrents in sedimentary rocks were measured using unidirectional indicators (conglomerate clast imbrications and three-dimensional exposures of sandstone trough cross-bedding) and linear, or bidirectional features (channel axis orientations and basal erosional scours). Because of the overall poor sorting in conglomerate facies, subequant geometry of clasts, and limited development of cross-bedding in sandstone facies, paleocurrent measurements in the Aubrey Hills were difficult to obtain and few in number. Bidirectional data (Fig. 7A) show a preferred north-northeast-south-southwest alignment. Unidirectional data (Fig. 7B) show greater scatter, with all data falling into the northwest and northeast quadrants. Because of the clustering of unidirectional data into northeast and northwest quadrants (Fig. 7B), we infer that bidirectional data (Fig. 7A) also record flow toward the northeast and northwest quadrants, not southwest or southeast quadrants. By making this inference and assigning a unique current direction to each of the bidirectional data in Figure 7A, we combine all paleocurrent data in a single plot (Fig. 7C). This produces a data set with about 160° of angular variation and a paleocurrent vector mean of 017° (north-northeast). This is consistent with transport directions obtained from kinematic indicators along the base of the megabreccia and indicates derivation of sediment from a source area to the south in the ancient Whipple Mountains.

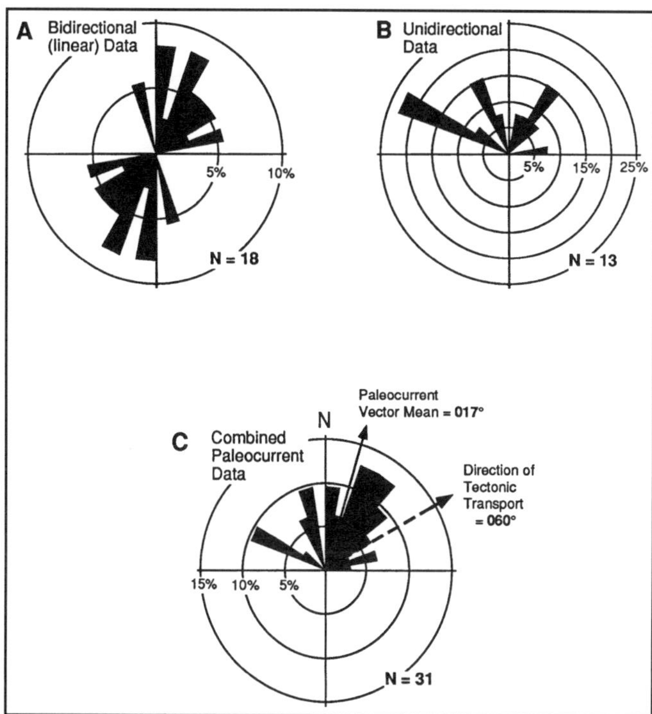

Figure 7. Rose diagrams of paleocurrent data, corrected for bedding dip, from sequence III sandstone and conglomerate facies in the study area. (A) Bidirectional (linear) data (channel axes, parting lineations, and erosional scours). (B) Unidirectional data (clast imbrications and three-dimensional exposures of cross bedding). (C) Combined paleocurrent data from A and B, in which unidirectional data from B are used to determine the direction of flow for linear data in A.

CONGLOMERATE CLAST COUNTS

Compositional analysis of conglomerate clasts was carried out to provide information about changes in source areas through time (Fig. 8). Clast counts were made at 20–40 m stratigraphic intervals in section 1 (Fig. 9), and lithologies of all clasts pebble-size or larger falling on a predetermined grid were recorded (Table 2). Clast compositions fall predominantly into two main categories: (1) volcanic rocks (basalt, tuff, rhyolite, and dacite); and (2) crystalline rocks (felsic to intermediate plutonic and metaplutonic rocks and minor quartzite). No mylonitic textures were observed in conglomerate clasts. Sedimentary rocks (limestone and sandstone) are intraformational rip-up clasts that make up less than 1% of the total clast population. Interpretation of conglomerate clast-count data, which follows, is aided by correlation with known source-area bedrock. The volcanic clast suite was most likely derived from a widespread, thick succession of Miocene volcanic rocks (sequence I and II) that underlie sequence III. Of note are clasts of the distinctive Peach Springs Tuff, which make up 1%–15 % of most sampled conglomerates (Table 2). Crystalline clasts are derived from pre-Cambrian and Cretaceous plutonic and metaplutonic rocks that make up the upper-plate crystalline basement for Miocene vol-

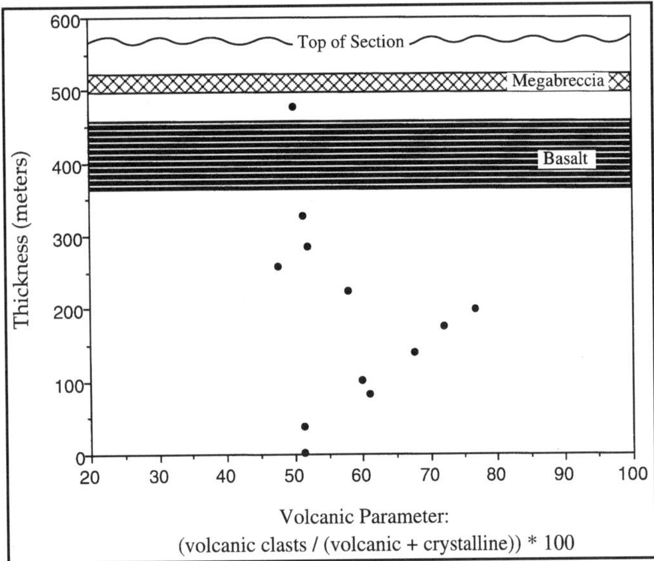

Figure 8. Plot of volcanic parameter versus stratigraphic thickness above the base of sequence 3 in measured section 1 (Figs. 3 and 9). See text for discussion.

canic rocks of sequence I. Nonmylonitic crystalline clasts may also be derived in part from nonmylonitic rocks in the lower plate of the Whipple detachment fault, which are currently exposed above the mylonitic front in the northwestern Whipple Mountains (Fig. 2).

We characterize the relative abundance of volcanic lithologies in the total clast suite using a volcanic parameter: (volcanic clasts/volcanic + crystalline clasts) * 100. The abundance of volcanic clasts increases gradually from about 50% at the base to 80% at 200 m (Fig. 8). Above 200 m, the volcanic parameter drops abruptly to about 50% and remains roughly constant for the rest of the section. The volcanic parameter is interpreted here as an approximate measure of the amount of Miocene volcanic "cover" rocks versus older crystalline rock that were exposed in the source at any given time. Vertical trends in the volcanic parameter (Fig. 8) are mimicked by sandstone petrographic data from the same stratigraphic interval (Dorsey and Roberts, 1992).

The gradual upsection increase in the volcanic parameter (lower 200 m, Fig. 8) is interpreted to reflect a progressive increase in the areal exposure of upper-plate volcanic rocks as

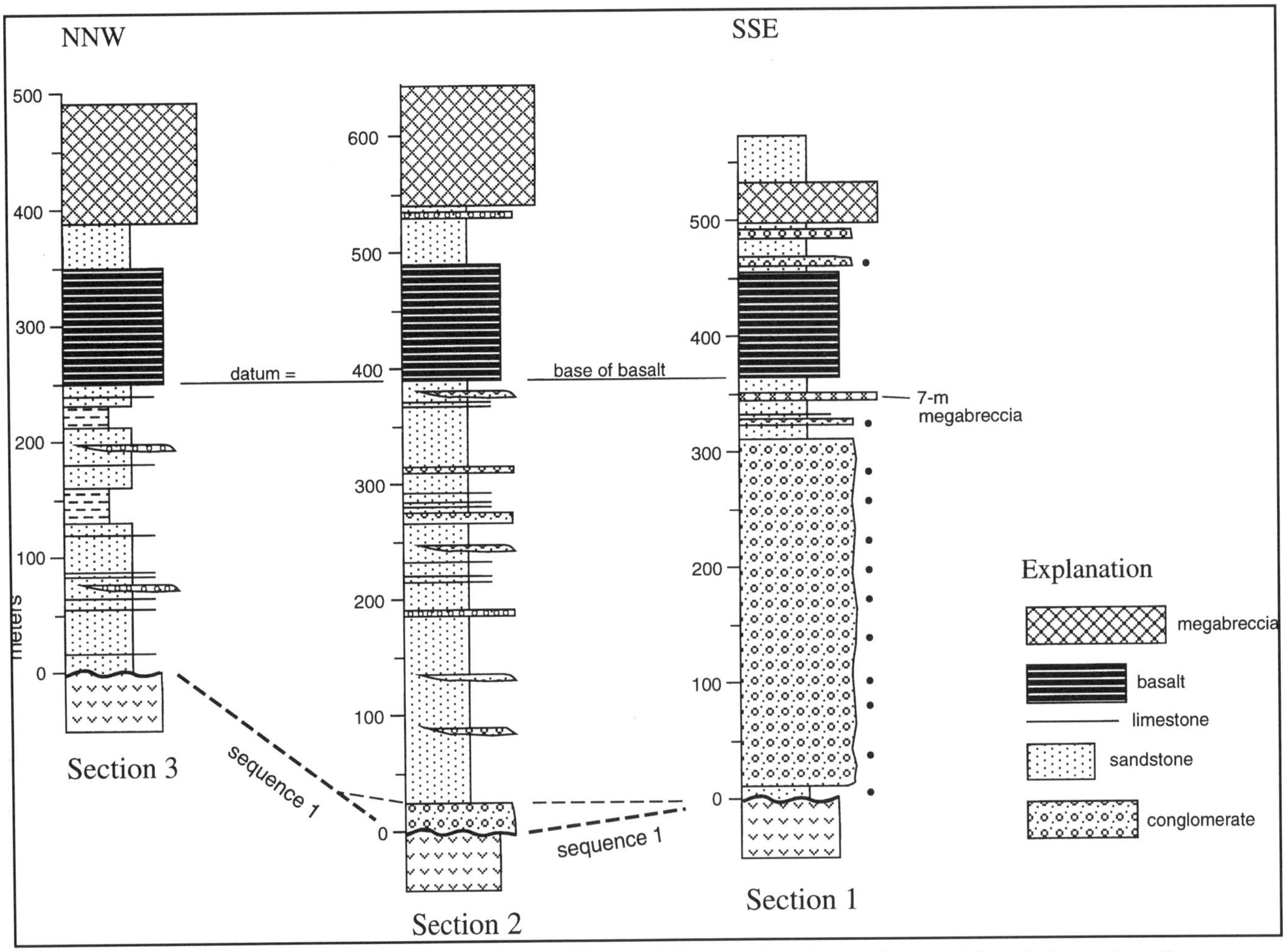

Figure 9. Compilation of measured sections in the northern Aubrey Hills, showing north-to-south lateral changes in sedimentary facies and total thickness. Black dots in section 1 indicate position of conglomerate clast counts in Figure 8.

TABLE 2. CLAST-COUNT DATA FOR CONGLOMERATES IN SECTION 1

	Stratigraphic Position in Meters											
	3	38	82	102	140	176	198	225	258	286	328	477
Peach SpringsTuff	0	1	6	3	16	15	8	8	9	5	2	4
Basalt	19	18	34	25	17	31	38	22	14	21	17	25
Tuff	8	6	13	4	3	1	0	5	7	8	6	13
Rhyolite	11	18	15	17	18	16	14	21	14	12	17	5
Rhyo-dacite	5	5	5	7	8	5	13	0	1	0	2	0
Dacite	12	6	8	6	8	5	12	6	1	0	4	1
Turkey-track porphry	0	2	3	4	3	4	4	3	5	8	8	5
Volcanic total	55	56	84	66	73	77	89	65	51	54	56	53
Granite	7	1	1	1	1	0	0	0	0	0	0	2
Graphic granite	11	2	13	11	2	3	3	4	3	1	3	0
Granodiorite	4	1	1	0	0	0	0	0	0	0	0	0
Diorite	12	9	11	8	8	3	3	4	7	4	4	5
Quartzite	4	0	12	1	0	0	0	0	0	2	2	0
Meta-plutonic	14	40	27	23	24	24	21	39	46	43	44	46
Plutonic total	52	53	54	44	35	30	27	47	56	50	53	53
Total	107	109	138	110	108	107	116	112	107	104	109	106
Volcanic Parameter	51.4	51.4	60.9	60.0	67.6	72.0	76.7	58.0	47.7	51.9	51.4	50.0

the proto–Whipple Mountains grew in overall size and topographic relief. During this time (post-18.5 Ma, pre-14.1 Ma), topography was apparently developing for the first time in the proto–Whipple Mountains, upper-plate volcanic rocks were being eroded, and lower-plate rocks were not yet exposed at the surface. A similar conclusion was reached by Beratan (1991) based on clast composition of age-equivalent conglomerates on the southern flank of the Whipple Mountains, which showed that the earliest uplift and erosion of upper-plate volcanic rocks in the proto–Whipple Mountains took place early in sequence III time (Beratan, 1991). The abrupt decrease in volcanic parameter at about 200 m above the base of sequence III (Fig. 8) was caused either by a shift in source area, to one having relatively more exposed crystalline rocks, or by removal of volcanic cover during rapid unroofing of one source (proto–Whipple Mountains). Because there is no change in paleocurrent directions and the compositional shift occurs in the middle of a continuous succession of mid-fan conglomerate (Figs. 8 and 9), we prefer the latter interpretation. Evidence for later surface exposure of the lower plate is provided by mylonitic rocks of the War Eagle pluton in the main megabreccia sheet (Fig. 9). The conglomerate clast-count data, combined presence of the mylonite-bearing megabreccia unit, therefore provide a step-like record of unroofing of upper-plate and then lower-plate crystalline rocks in the growing proto-Whipple Mountains, probably by a combination of erosional and tectonic denudation. This interpretation is similar to an unroofing sequence documented by Miller and John (1988) in upper-plate conglomerates of similar age in the Chemehuevi Mountains (Fig. 1).

STRATIGRAPHIC ARCHITECTURE

Three stratigraphic sections were measured in sequence III along well exposed transects that follow major arroyos in the Aubrey Hills (Figs. 3 and 9). A north-northwest–south-southeast stratigraphic panel (Fig. 10) was then constructed by combining data from the measured sections with geometries determined from detailed facies mapping. The panel in Figure 10 reveals the stratigraphic architecture of an alluvial fan and adjacent shallow lake that existed early in sequence III time, accumulating above the basal unconformity and prior to deposition of the widespread basalt unit. In the southern part of the study area, strata below the basalt are dominated by about 300 m of medial-fan sandy conglomerate of facies association 1 (Figs. 9 and 10). At 310 m in section 1 (Fig. 9), red conglomerate is abruptly overlain by a thin interval of sandstone and minor conglomerate (facies association 2). A 7 m thick breccia at about 350 m in section 1 contains War Eagle pluton lithologies and internal facies (matrix-rich breccia grading up into crackle breccia) identical to those seen in the large megabreccia above the basalt. The upper sandstone interval (310–365 m in section 1, Fig. 9) caps mid-fan conglomerate facies throughout the southern part of the study area (Fig. 10). To the north in section 2, in contrast to equivalent strata in section 1, deposits below the basalt consist almost entirely of distal-fan sandstone and minor conglomerate (facies association 2). Farther north, in section 3 (Fig. 9), the section beneath the basalt is considerably thinner and consists of distal-fan deposits of facies association 2 interbedded with lacustrine and lacustrine-margin mudstone, limestone, and

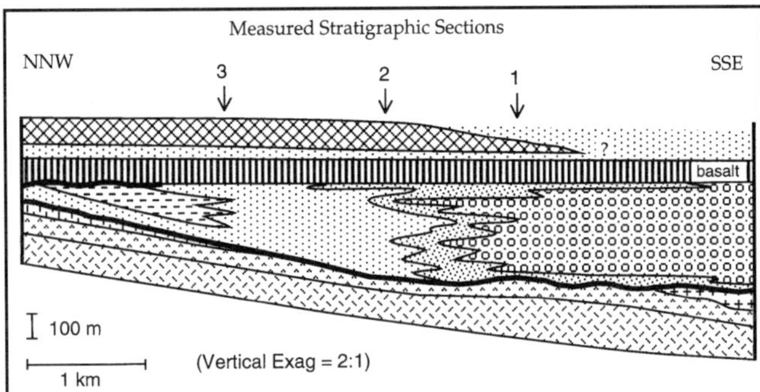

Figure 10. Stratigraphic panel for the northwestern Aubrey Hills, constructed from measured sections and detailed facies mapping. Note transition from medial alluvial-fan facies in the south to distal-fan and lacustrine facies in the north, pronounced northward decrease in thickness of sequence III deposits beneath the basalt, and truncation of lacustrine deposits along a local erosional unconformity beneath the basalt in the north. Facies patterns are same as in Figure 3, except as noted. Heavy lines denote unconformities.

minor sandstone (facies association 3). Sedimentary rocks north of section 3 thin substantially beneath a local erosional unconformity at the base of the basalt, and facies are dominated by fine-grained lacustrine and lacustrine-margin deposits (Figs. 3 and 10). Above the basalt, sedimentary facies show slight southward coarsening, and the megabreccia thins appreciably in the same direction (Figs. 9 and 10).

DISCUSSION

In the following section, we develop an interpretation of basin-scale and regional structural processes that controlled stratigraphic relations in the Aubrey Hills. This discussion is based on principles of dynamic stratigraphy, subsidence, and sedimentation, as outlined in recent papers on this subject (Leeder and Gawthorpe, 1987; Heller at al., 1988; Paola, 1988; Flemings and Jordan, 1989, 1990; Angevine et al., 1990; Schlische 1991, 1993; Paola et al., 1992; Heller and Paola, 1992; Travis and Nunn, 1994).

Subsidence-flux equilibrium

The boundary between medial and distal alluvial-fan facies below the basalt (Fig. 10) is approximately vertical over a stratigraphic interval of about 300 m (representing perhaps ~ 0.5–1 m.y.). This indicates that, during this time interval, a dynamic balance was maintained between the rate of sediment input from the south and the rate of subsidence in the basin, assuming constant base level. Rate of sediment input and rate of basin subsidence are two fundamental parameters that control the distribution and geometry of coarse- and fine-grained facies in tectonically active basins (Heller et al., 1988; Flemings and Jordan, 1989; Paola et al., 1992). Although gravelly detritus was continually supplied to the basin, as evidenced by the thick, conformable accumulation of red sandy conglomerate in the south, the lateral boundary between gravel-dominated mid-fan and sand-dominated distal-fan environments remained stationary because of the compensating effect of asymmetric basin subsidence. If there had been no subsidence during this time, gravels of the middle fan would have quickly prograded northward across the basin.

Disruption of equilibrium

The dynamic balance between subsidence and sediment input was disrupted shortly prior to deposition of the basalt. Evidence for this is seen in the abrupt upward fining from mid-fan conglomerate to distal-fan sandstone, conglomerate, and limestone in the southern part of the study area (Figs. 9 and 10). This abrupt vertical transition is interpreted to record a rapid retrogradation, or stepping back of the alluvial-fan system, which occurred when the ratio of subsidence rate to sediment-input rate increased suddenly. Retrogradation of this nature can occur in response to either an increase in the rate of subsidence or a decrease in the rate of sediment input. In the Aubrey Hills, the presence of a thin megabreccia unit below the basalt allows us to evaluate which of these two factors caused the change. Because steep topographic relief required to trigger rock avalanches commonly is produced by active seismicity, uplift, and slope oversteepening in the source (Krieger, 1977; Yarnold and Lombard, 1989), it is unlikely that breccia deposition coincided with a decrease in the rate of sediment delivery to the basin. We therefore conclude that the abrupt transition to sandstone and limestone records an increased rate of subsidence in the proximal part of the basin. The rate of sediment influx may have remained the same or increased slightly, but any increase was substantially less than that required to maintain the pre-existing balance between subsidence and sediment input.

Intrabasinal deformation

Sedimentary rocks in the Aubrey Hills are truncated beneath the basalt along an erosional unconformity that is restricted to the northernmost ~1 km of the study area; elsewhere, the basalt conformably overlies older sedimentary rocks (Figs. 3 and 10). Although some northward thinning of the section results from this erosional truncation, detailed mapping and tracing of key marker beds show that most of the stratigraphic thinning was produced by syndepositional asymmetric subsidence and tilting of the basin toward the south (above). Local erosion of the sequence in the north is interpreted to be the product of short-lived uplift of a small intrabasinal anticline. It is likely that this structure was active during disruption of basin equilibrium and emplacement of the 7-m-thick megabreccia deposit just below the basalt in section 1 (Fig. 9). All of these events took place shortly prior to emplacement of the basalt, and they all resulted from a pulse of structural deformation that affected the entire Aubrey Hills part of the basin and a nearby basin-flanking uplift (Whipple Mountains).

Basin models

In this section we consider different possible models for subsidence in the north Whipple basin during deposition of sequence III sediments below the basalt. As noted previously, the stratal geometries shown in Figure 10 are best explained by asymmetric subsidence and tilting of the basin toward the south or south-southeast. This geometry of tilting implies a horizontal axis of rotation that was oriented roughly east to west or east-northeast to west-southwest, which is approximately parallel to the direction of regional extension. In contrast, an extension-perpendicular axis of tilting (north to south, or north-northwest to south-southeast) would be expected if the basin evolved as a simple half graben that formed in response to east-northeast–directed extension. Thus any model for the Aubrey Hills part of the north Whipple basin must explain the observed southward thickening and coarsening of the sedimentary sequence and apparent tilting about an extension-parallel axis.

Figure 11 is a schematic diagram that illustrates the first hypothesis for basin subsidence. This model involves synbasinal tilting to the west-southwest because of progressive slip on a north-northwest–trending high-angle normal fault along the steep western margin of a classic sedimentary half-graben basin (e.g., Leeder and Gawthorpe, 1987). Recent models for extensional basins have shown that high-angle normal faults commonly grow along strike, increasing in map length at their ends through time (Schlische, 1991, 1993). Following a period of fault growth, nonuniform fault slip produces basin-fill sequences that thicken along strike, toward the locus of maximum fault displacement. It is possible that nonuniform fault slip in this type of setting produced the observed thickening and coarsening of strata toward the south-southeast in the Aubrey Hills (Figs. 10 and 11). However, this model predicts that there should be a large upthrown fault block in the upper plate located southwest of the Aubrey Hills that experienced uplift and erosion coincident with deposition in the Aubrey Hills. No such upthrown block exists to the southwest in the northeastern Whipple Mountains. Sub-basalt stratigraphy, which can be traced about 5 km to the southwest (Dunn, 1986; Dorsey et al., 1993), precludes application of the basin model in Figure 11 to development of stratigraphic relations in the Aubrey Hills.

Another possibility is that no differential subsidence took place during sedimentation, and that the southerly thickening of facies was produced solely by high-standing mountainous topography to the south which controlled deposition on a dipping fan surface (Fig. 12A). This model predicts that the most pronounced thickness change should occur in the coarse, proximal part of the alluvial fan, because this is where the steepest topographic gradients are found in alluvial-fan systems. In contrast, however, the observed stratigraphy reveals a relatively uniform thickness in mid-fan conglomerate, with most of the northward thinning seen in fine-grained distal-fan to lacustrine facies (Fig.10). Furthermore, as discussed above, facies architecture in the southern part of the study area records a dynamic balance between asymmetric basin subsidence and sediment input. We therefore eliminate the model in Figure 12A.

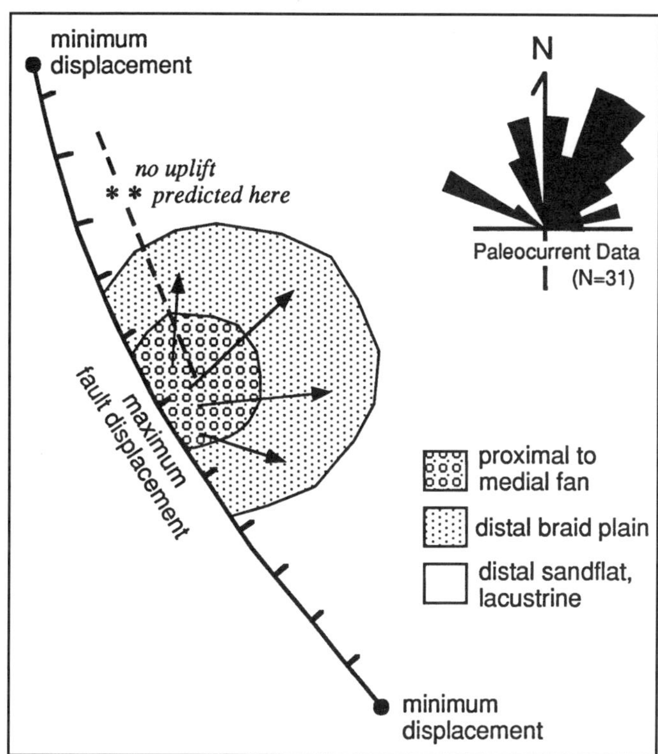

Figure 11. Schematic diagram illustrating plan view of facies and stratigraphic relationships produced by nonuniform slip in an evolving half-graben basin (Schlische, 1991, 1993). Dashed line shows approximate possible position of study area; progressive nonuniform slip could produce southward thickening and coarsening of strata along this line of section. We eliminate this model in this study for reasons discussed in the text.

The next possible model (Fig. 12B) involves asymmetric subsidence toward the south along a major oblique-slip transfer fault in the upper plate, which may have been controlled by an inflection in a corrugation of the underlying Whipple detachment fault (e.g., Beratan, 1991). In this hypothesis, southward tilting and subsidence is produced by the dip-slip component of oblique-slip displacement (Fig. 12B). A transfer fault oriented exactly parallel to the direction of extension (east-northeast) should experience primarily strike-slip, horizontal displacement as it accommodates different tilting histories in adjacent fault blocks. However, east-northeast–directed extension combined with slip on an east-west-striking (nonextension-parallel) transfer fault could produce a component of down-on-the-north displacement and southward tilting of a basin on the north side of such a fault. This model appears flawed because transfer faults in zones of continental extension play a role similar to that of oceanic transforms: they usually form parallel to the extension direction because they accommodate differential movement between structural blocks whose relative displacements are governed by regional extension (Lister et al., 1986). We therefore

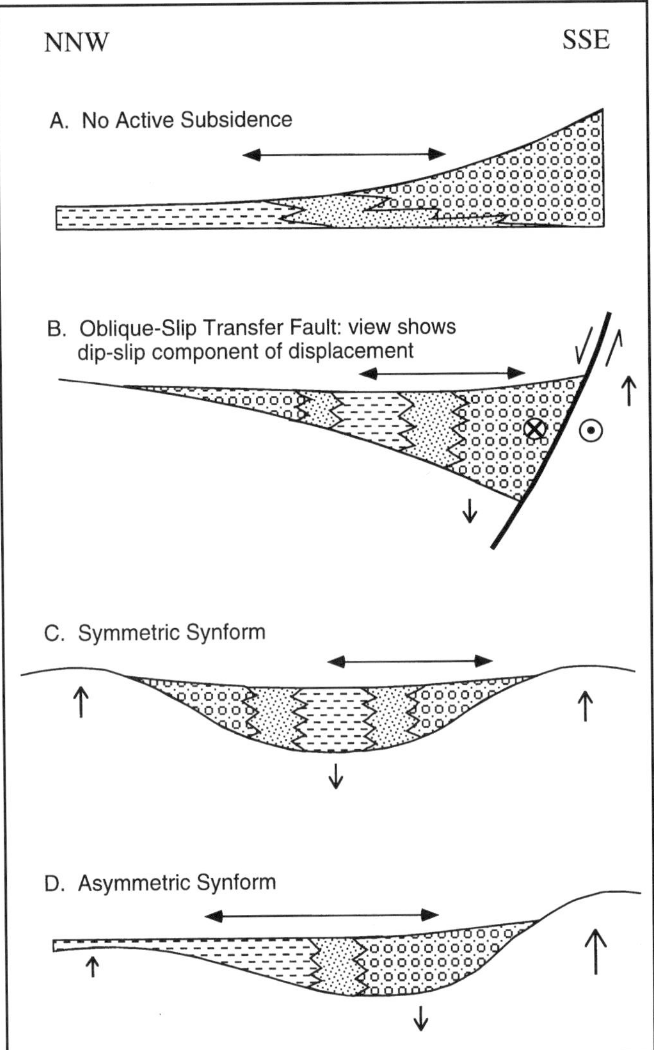

Figure 12. Four hypothetical models for asymmetric basin subsidence and southward tilting during deposition of sequence III. The model shown in D is our preferred interpretation. Horizontal double-barbed arrows indicate inferred position of study area in each model.

do not favor an oblique-slip transfer-fault origin for asymmetric subsidence in the Aubrey Hills, although it cannot be completely ruled out.

Figure 12C illustrates a basin in which subsidence is controlled by growth of a symmetrical syncline. In this model the synclinal axis is oriented east-northeast to west-southwest, perpendicular to the line of cross section and parallel to the direction of tectonic transport. We can eliminate this model because it produces the thickest sediment accumulation in the central, distal portion of the basin, which is opposite of the observed stratal geometries in the Aubrey Hills (Fig. 10). This problem is solved in Figure 12D by creating an asymmetric syncline in which the deep southern portion of the syncline is situated north of a large asymmetric anticline. We favor this basin model (Fig. 12D), because it can readily produce the stratal geometry and distribution of lithofacies seen in the Aubrey Hills (compare Figs. 10 and 12D). In addition, the small northern anticline in this model can explain short-lived uplift and erosion that took place in the northern part of the basin just prior to deposition of the basalt.

Structural model

Figure 13 presents a schematic, three-dimensional structural model for the Aubrey Hills portion of the north Whipple basin and surrounding areas, as they probably looked shortly after emplacement of the regionally extensive megabreccia (War Eagle landslide). By this time, upper-plate rocks had been mostly stripped off from the core of the Whipple Mountains by a combination of tectonic and erosional denudation, and lower-plate mylonitic rocks plus the War Eagle intrusive complex were exposed at the surface. This model builds upon previous studies that have established a history of lower-plate uplift and exhumation, juxtaposition of upper-plate rocks against the mylonitic lower plate, development of half-graben sedimentary basins, and development of extension-parallel folds in the detachment (Davis and Lister, 1988; Howard and John, 1987; Nielson and Beratan, 1990; Beratan, 1991; Yin and Dunn, 1992). Upper-plate rocks now located in the Aubrey Hills formed on the north flank of the rising Whipple Mountains (Fig. 13), and sometime later (after emplacement of the widespread War Eagle megabreccia) they were transported tectonically to the east-northeast into their present position (Fig. 2). The block diagram in Figure 13 illustrates two orthogonal, horizontal axes of tilting in the upper plate: (1) tilting about extension-perpendicular axes caused by slip on north-northwest–striking normal faults and asymmetric subsidence in classic half-graben sedimentary basins; and (2) extension-parallel tilting resulting from growth of extension-parallel folds in the detachment fault. Based on evidence from recent and ongoing studies, it appears that extension-parallel tilting probably was minor compared to the large displacements that were produced by extension-perpendicular tilting in half-graben basins (Nielson and Beratan, 1990; Beratan, 1990, 1991; Fedo and Miller, 1992; Dorsey et al., 1993).

The geologic map in Figure 2 shows two major extension-parallel antiforms and an intervening synform in the lower plate of the Whipple detachment fault. We interpret the northwestern antiform to be the large fold that formed the southern margin of the north Whipple basin (Figs. 12D and 13). The hinge and northwestern limb of this antiform project northeastward into the Aubrey Hills (Fig. 2). This presents an apparent discrepancy between the model in Figure 13 and present-day geologic relationships, because the broad northern limb of this fold impinges on the southern part of the study area, suggesting that this fold should have produced uplift and thinning (not thickening) toward the south in the northwestern Aubrey Hills. We believe the map relationships were produced by continued growth of the antiform *after* deposition of the basalt, as the zone of antiformal uplift grew and expanded laterally northward into the basin. We

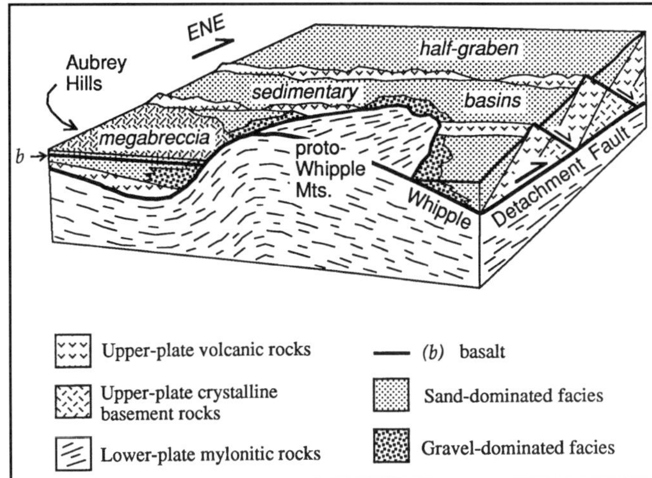

Figure 13. Schematic block diagram illustrating three-dimensional geometry of basin subsidence in the upper plate of the Whipple detachment fault, shortly after emplacement of the regionally extensive Copper Canyon landslide (megabreccia). See text for discussion.

postulate that this caused a change from prebasalt subsidence and southward tilting, to postbasalt uplift and tilting of upper-plate rocks back toward the north in this part of the basin. In the northeastern Whipple Mountains, pronounced thinning of sedimentary rocks above the same basalt provide evidence for postbasalt northward tilting in the basin (U. Becker, 1994, unpublished field data). Our interpretation of evolving fold geometries is consistent with a recent structural study of the northeastern Whipple Mountains (Yin and Dunn, 1992): anomalous west-northwest strikes of high-angle normal faults that cut the War Eagle landslide resulted from rotation of these structures about a large extension-parallel anticline in the north Whipple Mountains (Yin and Dunn, 1992). Postlandslide reorientation of faults in the northeastern Whipple Mountains therefore coincides well in style and timing with postbasalt folding and northward tilting that we infer for the north Whipple basin.

Relevance to previous work

Previous studies in the Whipple Mountains region have recognized the influence of extension-parallel structural features on the evolution of upper-plate basins. Nielson and Beratan (1990) and Beratan (1991) emphasized the importance of transfer faults in early segmentation of the upper plate into domains and basins of differential tilting. Transfer-fault-bounded blocks are interpreted to have experienced independent tilting histories that are recorded in sequence I based on contrasting ages and stratigraphic relationships in neighboring domains (Nielson and Beratan, 1990; Beratan, 1991). In this chapter, we have considered oblique-slip displacement on a transfer fault as a possible control on younger (sequence III) subsidence and sedimentation in the Aubrey Hills (Fig. 12B), and we conclude that this is a possible although unlikely basin model. It is possible that transfer faults were active mainly during sequence I time, and relatively unimportant during deposition of sequence III. Extension-parallel transfer faults and extension-parallel folds are, however, likely to produce similar stratigraphic and sedimentologic features such as lateral variations in grain size and lithofacies. These similarities can be expected to make distinction between the two mechanisms difficult.

Extension-parallel antiforms and synforms of detachment faults are widespread phenomena that have been alternately interpreted as (1) primary corrugations (mullions) that form at substantial depth early in the history of detachment faulting and involve no warping or folding of the upper plate (Spencer, 1985; John, 1987; Davis and Lister, 1988; Spencer and Reynolds, 1991); or (2) fold trains that grow in the detachment due to weak constriction and warping of upper-plate rocks and the faults that cut them (Bartley et al., 1990a; Yin, 1991; Yin and Dunn, 1992; Mancktelow and Pavlis, 1994). It is possible that both models are correct: if primary corrugations form early in the history of detachment faulting, and later a weak extension-normal compressional stress is applied, then we can expect folding to amplify pre-existing structural weaknesses such as fault corrugations in the proper orientation (e.g., Yin, 1991). The results of the above analysis, combined with compelling evidence for a deep-crustal origin of extension-parallel corrugations (e.g. John, 1987; Spencer and Reynolds, 1991), appear to support this idea.

The regular spacing of extension-parallel folds, combined with new evidence for their near-surface growth, suggests that a significant component of extension-normal constriction affected this and other detachment faults during mid-Tertiary extension in the southwestern United States. Other examples of extension-parallel folding of detachment faults include the Death Valley turtlebacks (Holm and Lux, 1991; Holm, 1992), the central Mojave desert (Bartley et al., 1990a, 1990b), the Virgin and Beaver Dam Mountains (Anderson and Bohannon, 1993; Anderson and Barnhard, 1993), and the Las Vegas region (Wernicke et al., 1988). In at least one example (Death Valley turtlebacks), it is clear that upper-plate sedimentary and volcanic rocks were warped and folded at shallow crustal levels (Holm and Lux, 1991; Holm, 1992). Extension-normal constriction is thought to result from differential crustal thinning and resultant pressure gradients in the crust (e.g., Wernicke et al., 1988; Wernicke, 1990; Block and Royden, 1990; Kruse et al., 1991). In a region undergoing northeast-southwest extension, northwest-southeast variations in the amount of extension and crustal thinning cause areas of relatively little thinning (i.e., greater crustal overburden) to feel greater pressure at depth than areas of strong extension and crustal thinning. These variations in gravitational potential may be sufficient to produce weak northwest-southeast compression and horizontal flow of crustal material along an extension-perpendicular vector, which results in extension-parallel folding. Although disagreement exists over the total amount and relative importance of extension-normal contraction in some areas (e.g., Wernicke et al., 1988; Anderson and Barnhard, 1993), many workers now agree that extension-normal shortening was an important result of middle Tertiary extension in the southern Basin and Range province.

CONCLUSIONS

1. Sedimentary rocks (sequence III) in the northwestern Aubrey Hills overlie the 18.5 Ma Peach Springs Tuff and older volcanic rocks along an angular unconformity. These rocks record deposition in a synextensional basin in the upper plate of the Whipple detachment fault.

2. Lithofacies consist of red sandy conglomerate, sandstone with minor conglomerate and limestone, and mudstone-limestone-sandstone. These deposits accumulated in middle alluvial-fan, distal alluvial-fan, and lacustrine to lacustrine-margin depositional environments, respectively. Paleocurrent data and paleogeographic relations indicate transport of sediment to the north-northeast during this time.

3. A megabreccia unit that caps the section can be correlated with a large breccia sheet located to the southwest in the northeastern Whipple Mountains. This breccia was derived from the War Eagle pluton in the lower plate of the Whipple detachment fault. Facies within the megabreccia include matrix-rich breccia, jigsaw breccia, and crackle breccia; these facies types and basal deformation features are typical of large rock avalanches, or sturzstroms.

4. A 100 m thick basalt, which has previously been dated at 14.1 ± 0.3 Ma, lies just below the base of the megabreccia. It contains common stringers of baked siltstone and sandstone and rare discontinuous interbeds of clastic sediments. This unit is a package of multiple, stacked basalt flows that were emplaced in rapid succession across a large area of the basin.

5. Conglomerate clasts are composed primarily of lower Miocene volcanic rocks and nonmylonitic plutonic and metaplutonic rocks. Vertical stratigraphic trends in clast composition, combined with information from the capping megabreccia, provide a step-like record of uplift and unroofing of the lower plate of the Whipple detachment fault.

6. Southward thickening and coarsening of alluvial-fan facies in the Aubrey Hills provides evidence for syndepositional growth of an extension-parallel syncline that grew on the north flank of a large extension-parallel antiform in the growing Whipple Mountains.

ACKNOWLEDGMENTS

This study has benefited from discussions with Kathi Beratan, Julia Miller, Gary Axen, Greg Davis, An Yin, Jane Nielson, Keith Howard, Ernie Duebendorfer, and Paul Umhoefer. We thank Jane Nielson and Kathi Beratan for introducing us to the study area. Thorough and critical reviews by Julia Miller, Kathi Beratan, and an anonymous reviewer of an earlier draft of the manuscript are greatly appreciated. Financial support of this research was provided by Northern Arizona University and by the Petroleum Research Fund, administered by the American Chemical Society. We thank Bart Wagoner and other workers at Lake Havasu State Park for permission to work on park land and for their unfailing generosity and helpfulness.

REFERENCES CITED

Anderson, R. E., and Barnhard, T. P., 1993, Aspects of three-dimensional strain at the margin of the extensional orogen, Virgin River depression area, Nevada, Utah, and Arizona: Geological Society of America Bulletin, v. 105, p. 1019–1052.

Anderson, R. E., and Bohannon, R. G., 1993, Three-dimensional aspects of the Neogene strain field, Nevada-Utah-Arizona tricorner area, *in* Lahren, M. M., Trexler, J. H., Jr., and Spinosa, C., eds., Crustal evolution of the Great Basin and Sierra Nevada: Geological Society of America, Cordilleran/Rocky Mountain Section Meeting, Guidebook: Reno, Nevada, University of Nevada, Department of Geological Sciences, p. 167–196.

Angevine, C. L., Heller, P. L., and Paola, C., 1990, Quantitative sedimentary basin modeling: Tulsa, Oklahoma, American Association of Petroleum Geologists Education Course Note Series No. 32, 133 p.

Bartley, J. M., Fletcher, J. M., and Glazner, A. F., 1990a, Tertiary extension and contraction of lower-plate rocks in the central Mohave metamorphic core complex, southern California: Tectonics, v. 9, p. 521–534.

Bartley, J. M., Glazner, A. F., and Schermer, E. R., 1990b, North-south contraction of the Mojave Block and strike-slip tectonics in southern California: Science, v. 248, p. 1398–1401.

Beratan, K. K., 1990, Basin development during Miocene detachment faulting, Whipple Mountains, southeastern California [Ph.D. thesis]: Los Angeles, University of Southern California, 267 p.

Beratan, K. K., 1991, Miocene synextension sedimentation patterns, Whipple Mountains, southeastern California: Implications for the geometry of the Whipple detachment system: Journal of Geophysical Research, v. 96, p. 12,425–12,442.

Block, L., and Royden, L. H., 1990, Core complex geometries and regional scale flow in the lower crust: Tectonics, v. 9, p. 557–567.

Buck, R., 1988, Flexural rotation of normal faults: Tectonics, v. 3, p. 647–657.

Bull, W. B., 1972, Recognition of alluvial-fan deposits in the stratigraphic record, *in* Rigby, J. K., and Hamblin, W. K., eds., Recognition of ancient sedimentary environments: Tulsa, Oklahoma, Society of Economic Paleontologists and Mineralogist Special Publication 16, p. 63–83.

Collinson, J. D., 1986, Alluvial sediments, in Reading, H. G., ed., Sedimentary environments and facies: Oxford, Blackwell Scientific, p. 20–62.

Davis, G. A., 1980, Problems of intraplate extensional tectonics, western United States, *in* Continental tectonics: Washington, D.C., National Academy of Sciences, National Research Council, p. 9–95.

Davis, G. A., 1983, Shear-zone model for the origin of metamorphic core complexes: Geology, v. 11, p. 342–347.

Davis, G. A., 1988, Rapid upward transport of mid-crustal mylonitic gneisses in the footwall of a Miocene detachment fault, Whipple Mountains, southeastern California: Geologisches Rundschau, v. 77, p. 191–209.

Davis, G. A., and Anderson, J. L., 1991, Low angle normal faulting and rapid uplift of mid-crustal rocks in the Whipple Mountains metamorphic core complex, southeastern California: discussion and field guide, *in* Walawender, M. J., and Hanan, B. B., eds., Geological excursions in southern California and Mexico: Guidebook for Geological Society of America 1991 Annual Meeting, Department of Geological Sciences, San Diego, California, San Diego State University, p. 417–446.

Davis, G. A., and Coney, P. J., 1979, Geologic development of the Cordilleran metamorphic core complexes: Geology, v. 7, p. 120–124.

Davis, G. A., and Lister, G. S., 1988, Detachment faulting in continental extension; Perspectives from the Southwestern U.S. Cordillera, *in* Clark, S. P., Burchfiel, B. C., and Suppé, J., eds., Processes in continental lithospheric deformation: Boulder, Colorado, Geological Society of America Special Paper 218, p. 133–159.

Davis, G. A., Anderson, J. L., Frost, E. G., and Shackelford, T. J., 1980, Mylonitization and detachment faulting in the Whipple-Buckskin-Rawhide Mountains terrane, southeastern California and western Arizona, *in* Crittenden, M. D., Coney, P. J., and Davis, G. H., eds., Cordilleran metamorphic core complexes: Boulder, Colorado, Geological Society of America Memoir 153, p. 79–130.

Davis, G. A., Lister, G. S., and Reynolds, S. J., 1986, Structural evolution of the Whipple and South Mountains shear zones, southwestern United State: Geology, v. 14, p. 7–10.

Dean, W. E., and Fouch, T. D., 1984, Lacustrine environments, *in* Scholle, P. A., Bebout, D. G., and Moore, C. H., eds., Carbonate depositional environments: Tulsa, Oklahoma, American Association of Petroleum Geologists Memoir 33, p. 97–130.

Dorsey, R. J., and Becker, U., 1995, Evolution of a large Miocene growth structure in the upper plate of the Whipple detachment fault, northeastern Whipple Mountains, California: Basin Research, v. 7, in press.

Dorsey, R. J., and Roberts, P., 1992, Petrographic record of Whipple Mountain uplift and unroofing, north Whipple basin, SW Arizona and SE California: Geological Society of America Abstracts with Programs, v. 24, no. 5, p. 20.

Dorsey, R. J., and Roberts, P., 1993, Structural controls on syn-extensional subsidence, Miocene north Whipple basin, Aubrey Hills, W Arizona: Geological Society of America Abstracts with Programs, v. 25, no. 5, p. 31.

Dorsey, R. J., Dunlap, J., and Becker, U., 1993, A large growth structure in the Miocene north Whipple basin: Implications for upper-plate evolution: Geological Society of America Abstracts with Programs, v. 25, no. 6, p. 352.

Dunn, J. F., 1986, The structural geology of the northeastern Whipple Mountains detachment fault terrane, San Bernadino County, California [M.S. thesis]: Los Angeles, California, University of Southern California, 172 p.

Fedo, C. M., 1990, Sedimentology and evolution of a Miocene half-graben basin, Colorado River extensional corridor, southeastern California [M.S. thesis]: Nashville, Tennessee, Vanderbilt University, 86 p.

Fedo, C. M., and Miller, J. M. G., 1992, Evolution of a Miocene half-graben basin, Colorado River extensional corridor, southeastern California: Geological Society of America Bulletin, v. 104, p. 481–493.

Flemings, P. B., and Jordan, T. E., 1989, A synthetic stratigraphic model of foreland basin development: Journal of Geophysical Research, v. 94, p. 3851–3866.

Flemings, P. B., and Jordan, T. E., 1990, Stratigraphic modeling of foreland basins: Interpreting thrust deformation and lithosphere rheology: Geology, v. 18, p. 430–434.

Forshee, A. J., and Yin, A., 1993, A model for initiation of a mid-Tertiary rock avalanche in the Whipple Mtns. area, SE California: Implications for seismicity along low-angle detachment faults: Geological Society of America Abstracts with Programs, v. 25, no. 6, p. 351.

Forshee, A. J., and Yin, A., 1995, Evolution of monolithologic-breccia deposits in supradetachment basins, Whipple Mountains, California: Basin Research, V. 7, in press.

Galloway, W. E., and Hobday, D. K., 1983, Terrigenous clastic deopositional systems: New York, Springer-Verlag, 423 p.

Heller, P. L., and Paola, C., 1992, The large-scale dynamics of grain-size variation in alluvial basins, 2: Application to syntectonic conglomerate: Basin Research, v. 4, p. 91–102.

Heller, P. L., Angevine, C. L., Winslow, N. S., and Paola, C., 1988, Two-phase stratigraphic model of foreland-basin sequence: Geology, v. 16, p. 501–504.

Holm, D. K., 1992, Structural, thermal and paleomagnetic constraints on the tectonic evolution of the Black Mountains crystalline terrain, Death Valley region, California, and implications for extensional tectonism [Ph.D. thesis]: Cambridge, Massachussetts, Harvard University, 237 p.

Holm, D. K., and Lux, D. R., 1991, The Copper Canyon Formation: A record of unroofing and Tertiary folding of the Death Valley turtleback surfaces: Geological Society of America Abstracts with Programs, v. 23, no. 2, p. 35.

Howard, K. A., and John, B. E., 1987, Crustal extension along a rooted system of imbricate low-angle faults: Colorado extensional corridor, California and Arizona, *in* Coward, M. P., Dewey, J. F., and Hancock, P. L., eds., Continental Extensional Tectonics: London Geological Society of London Special Publication 28, p. 299–311.

Howard, K. A., Nielson, J. E., Wilshire, H. G., Nakata, J. K., Goodge, J. W., Reneau, S. L., John, B. E., and Hansen, V. L., 1990, Preliminary geologic map of the Mohave Mountains area, Mohave County, western Arizona: U.S. Geological Survey Open-File Report 90-684, 56 p.

Hsü, K. J., 1975, Catastrophic debris streams (sturzstroms) generated by rockfalls: Geological Society of America Bulletin, v. 86, p. 129–140.

John, B. E., 1987, Geometry and evolution of a mid-crustal extensional fault system: Chemehuevi Mountains, southeastern California, *in* Coward, M. P., Dewey, J. F., and Hancock, P. L., eds., Continental extensional tectonics: London, Geological Society of London Special Publication 28, p. 313–335.

Krieger, M. H., 1977, Large landslides, composed of megabreccia, interbedded in Miocene basin deposits, southeastern Arizona: U.S. Geological Survey Professional Paper 1008, 25 p.

Kruse, S., McNutt, M., Phipps-Morgan, J., Royden, L., and Wernicke, B., 1991, Lithospheric extension near Lake Mead, Nevada: A model for ductile flow in the lower crust: Journal of Geophysical Research, v. 96, p. 4435–4456.

Leeder, M. R., and Gawthorpe, R. L., 1987, Sedimentary models for extensional tilt-block/half-graben basins, *in* Coward, M. P., Dewey, J. F., and Hancock, P. L., eds., Continental Extensional Tectonics: London, Geological Society of London Special Publication 28, p. 139–152.

Lucchitta, I., and Suneson, N., 1988a, Geologic map of the Artillery Peak NW Quadrangle, Mohave County, Arizona: U.S. Geological Survey Open-File Map, OF 85-277, scale 1 : 24,000.

Lucchitta, I., and Suneson, N., 1988b, Geologic map of the Planet 2 SW Quadrangle, Mohave County, Arizona: U.S. Geological Survey Open-File Map, OF 88-547, , scale 1 : 24,000.

Lucchitta, I., and Suneson, N., 1993, Tectonic stratigraphy near a metamorphic core complex: Lessons from the Casteneda-Signal area of west-central Arizona: Geological Society of America Abstracts with Programs, v. 25, no. 5, p. 111.

Machette, M. N., 1985, Calcic soils of the southwestern United States, *in* Weide, D. L., ed., Soils and Quaternary geology of the southwestern United States: Boulder, Colorado, Geological Society of America Special Paper 203, p. 1–21.

Mack, G. H., and James, W. C., 1992, Paleosols for sedimentologists: Cincinnati, Ohio, Geological Society of America Short Course Notes, 127 p.

Mancktelow, N. S., and Pavlis, T. L., 1994, Fold-fault relationships in low-angle detachment systems: Tectonics, v. 13, p. 668–685.

Melosh, H. J., 1979, Acoustic fluidization: A new geologic process?: Journal of Geophysical Research, v. 84, no. B13, p. 7513–7520.

Melosh, H. J., 1987, The mechanics of large rock avalanches: Boulder, Colorado, Geological Society of America Reviews in Engineering Geology, v. 7, p. 41–49.

Melosh, H. J., 1990, Giant rock avalanches: Nature, v. 384, p. 483–484.

Miall, A. D., 1978, Lithofacies types and vertical profile models in braided-river deposits; A summary, *in* Miall, A. D., ed., Fluvial sedimentology: Calgary, Alberta, Canadian Society of Petroleum Geologists Memoir 5, p. 597–604.

Miller, J. M. G., and John, B. E., 1988, Detached strata in a Tertiary low-angle normal fault terrane, southeastern California: A sedimentary record of unroofing, breaching, and continued slip: Geology, v. 16, p. 645–648.

Miller, J. M. G., and John, B. E., 1993, Evolution of a synextensional Miocene basin, W. Arizona: Geological Society of America Abstracts with Programs, v. 25, no. 5, p. 122.

Nemec, W., and Steel, R. J., 1984, Alluvial and coastal conglomerates: Their significant features and some comments on gravelly mass-flow deposits, *in* Koster, E. H., and Steel, R. J., eds., Sedimentology of gravels and conglomerates: Calgary, Alberta, Canadian Society of Petroleum Geologists Memoir 10, p. 1–31.

Nielson, J. E., 1986, Miocene stratigraphy of the Mojave Mountains, Arizona, and correlation with adjacent ranges, *in* Nielson, J. E., and Glazner, A. F., eds., Cenozoic stratigraphy, structure, and mineralization in the Mojave Desert, Field Trip 5: Fresno, California, Geological Society of America Cordilleran Section Guidebook, p. 15–24.

Nielson, J. E., and Beratan, K. K., 1990, Tertiary basin development and tectonic implications, Whipple detachment system, Colorado extensional corridor, California and Arizona: Journal of Geophysical Research, v. 95, p. 599–614.

Nielson, J. E., Lux, D. R., Dalrymple, G. B., and Glazner, A. F., 1990, Age of the Peach Springs Tuff, southeastern California and western Arizona: Journal of Geophysical Research, v. 95, p. 571–580.

Paola, C., 1988, Subsidence and gravel transport in alluvial basins, *in* Kleinspehn, K. L., and Paola, C., eds., New perspectives in basin analysis: New York, Springer-Verlag, p. 231–243.

Paola, C., Heller, P. L., and Angevine, C. L., 1992, The large-scale dynamics of grain-size variation in alluvial basins, 1: Theory: Basin Research v. 4, p. 73–90.

Roberts, P., 1992, Miocene basin evolution in the upper plate of the Whipple detachment fault, western Arizona [M.S. thesis]: Flagstaff, Arizona, Northern Arizona University, 136 p.

Rust, B. R., 1978, Depositional models for braided alluvium, *in* Miall, A. D., ed., Fluvial sedimentology: Calgary, Alberta, Canadian Society of Petroleum Geologists Memoir 5, p. 605–625.

Rust, B. R., and Koster, E. H., 1984, Coarse alluvial deposits, *in* Walker, R. G., ed., Facies models, (second edition): Toronto, Geological Society of Canada, p. 53–69.

Schlische, R. W., 1991, Half-graben basin filling models: New constraints on continental extensional basin development: Basin Research, v. 3, p. 123–141.

Schlische, R. W., 1993, Anatomy and evolution of the Triassic-Jurassic continental rift system, eastern North America: Tectonics, v. 12, p. 1026–1042.

Shackelford, T. J., 1976, Structural geology of the Rawhide Mountains, Mohave County, Arizona [Ph.D. thesis]: Los Angeles, California, University of Southern California, 175 p.

Spencer, J. E., 1984, Role of tectonic denudation in warping and uplift of low-angle normal faults: Geology, v. 12, p. 95–98.

Spencer, J. E., 1985, Miocene low-angle normal faulting and dike emplacement, Homer Mountain and surrounding areas, southeastern California and southernmost Nevada: Geological Society of America Bulletin, v. 96, p. 1140–1155.

Spencer, J. E., and Reynolds, S. J., 1989, Tertiary structure, stratigraphy, and tectonics of the Buckskin Mountains, *in* Spencer, J. E., and Reynolds, S. J., eds., Geology and mineral resources of the Buckskin and Rawhide Mountains, west-central Arizona, Arizona Geological Survey Bulletin 198, p. 103–167.

Spencer, J. E., and Reynolds, S. J., 1991, Tectonics of mid-Tertiary extension along a transect through west-central Arizona: Tectonics, v. 10, p. 1204–1221.

Spencer, J. E., Grubensky, M. J., Duncan, J. T., Shenk, J. D., Yarnold, J. C., and Lombard, J. P., 1989, Geology and mineral deposits of the central Artillery Mountains, *in* Spencer, J. E., and Reynolds, S. J., eds., Geology and mineral resources of the Buckskin and Rawhide Mountains, west-central Arizona: Arizona Geological Survey Bulletin 198, p. 168–183.

Travis, C. J., and Nunn, J. A., 1994, Stratigraphic architecture of extensional basins: Insights from a numerical model of sedimentation in evolving half grabens: Journal of Geophysical Research, v. 99, p. 15,653–15,666.

Walker, T. R., 1967, Formation of red beds in ancient and modern deserts: Geological Society America Bulletin, v. 78, p. 353–368.

Wernicke, B., 1981, Low-angle normal faults in the Basin and Range province: Nappe tectonics in an extending orogen: Nature, v. 291, p. 645–648.

Wernicke, B., 1985, Uniform-sense normal simple shear of the continental lithosphere: Canadian Journal of Earth Science, v. 22, p. 108–125.

Wernicke, B., 1990, The fluid crustal layer and its implications for continental dynamics, *in* Salisbury, M. H., and Fountain, D. H., eds., Exposed cross sections of the continental crust: Boston, Massachuttes, Kluwer Academic Publishers, v. 317, p. 509–544.

Wernicke, B., 1992, Cenozoic extensional tectonics of the U.S. Cordillera, *in* Burchfiel, B. C., Lipman, P. W., and Zoback, M. L., eds., The Cordilleran Orogen: Conterminous U.S.: Boulder, Colorado, Geological Society of America, The Geology of North America, v. G-3, p. 553–581.

Wernicke, B., and Axen, G., 1988, On the role of isostacy in the evolution of normal fault systems: Geology, v. 16, p. 848–851.

Wernicke, B., Axen, G., and Snow, J. K., 1988, Basin and Range extensional tectonics at the latitude of Las Vegas, Nevada: Geological Society of America Bulletin, v. 100, p. 1738–1757.

Yarnold, J. C., 1994, Tertiary sedimentary rocks associated with the Harcuvar core complex in Arizona (U.S.A.): Insights into paleogeographic evolution during displacement along a major detachment fault system: Sedimentary Geology, v. 89, p. 43–63.

Yarnold, J. C., and Lombard, J. P., 1989, A facies model for large rock-avalanche deposits formed in dry climates, *in* Colburn, I. P., Abbott, P. L., and Minch, J., eds., Conglomerates in basin analysis: A symposium dedicated to A. O. Woodford: Society of Economic Paleontologist and Mineralogists, Pacific Section, v. 62, p. 9–31.

Yin, A., 1991, Mechanisms for the formation of domal and basinal detachment faults: A three-dimensional analysis: Journal of Geophysical Research, v. 96, p. 14,577–14,594.

Yin, A., and Dunn, J. F., 1992, Structural and stratigraphic development of the Whipple-Chemehuevi detachment fault system, southeastern California: Implications for the geometrical evolution of domal and basinal low-angle normal faults: Geological Society of America Bulletin, v. 104, p. 659–674.

Manuscript Accepted by the Society April 21, 1995

Printed in U.S.A.

Geological Society of America
Special Paper 303
1996

Timing and character of deformation along the margin of a metamorphic core complex, west-central Arizona

Ivo Lucchitta
U.S. Geological Survey, 2255 North Gemini Drive, Flagstaff, Arizona 86001
Neil H. Suneson
Oklahoma Geological Survey, 100 East Boyd, Energy Center, Room 131, Norman, Oklahoma 73019

ABSTRACT

Stratigraphic and structural information from the Castaneda Hills–Signal area of west-central Arizona, between the Buckskin-Rawhide metamorphic core complex and the Colorado Plateau, shows that the first event associated with formation of the highly extended terranes (HET) in this region was upwarping in the area of the metamorphic core complex ca. 27 Ma. Basalt of lithospheric-mantle derivation was erupted about 19 Ma, signaling the beginning of volcanic activity. After this time, and probably before 13 Ma, extension characteristic of the HET occurred and was manifested by mylonitization of lower-plate rocks, detachment faulting, (probably) listric faulting and strong tilting of blocks in the upper plate, mineralization, and silicic volcanism of lower-crust derivation. HET tectonism evolved into basin-range (BR) tectonism, accompanied by the eruption of mantle-derived basalts during the period 13–12 Ma; however, by about 8 Ma tectonism ceased altogether, and volcanism was restricted to emplacement of minor volumes of megacryst-bearing basalt derived from the asthenospheric mantle.

The data do not support domino or rolling-hinge models for upper-plate deformation, or simple-shear models for the crustal-scale evolution of the HET. Instead, they suggest a thermally driven, dominantly pure-shear mechanism. We propose a thermotectonic model driven by an asthenospheric upwelling that heats up the crust and applies divergent stresses to it, causing HET-type tectonism. With time, the upwelling migrates northeast, allowing the crust to gradually cool and become atectonic, while the crust of the nearby Colorado Plateau heats up and begins to rift.

INTRODUCTION

The origin and mechanism of continental extension are of great current interest, largely because continental extension is poorly understood in comparison to ocean-basin and continent-margin tectonism. Much work on this problem is being done in the Basin and Range Province of the North American Cordillera, which has attracted attention for more than 100 yr because of its spectacular extensional features. Gilbert (1875), Powell (1877), and King (1878) were the first to suggest that the ranges were uplifted relative to the basins along high-angle normal faults, whereas Longwell (1945) recognized the importance of low-angle normal faults in this province. The province has come under renewed scrutiny in the past two decades because of the recognition of additional kinds of extensional features—metamorphic core complexes (MCCs) and highly extended terranes (HETs). Anderson's (1971) pioneering work on listric normal faults, Proffett's (1977) on rotated normal faults, Armstrong's (1972) on extensional denudation (detachment) faults, and Davis's (1975) on "gneiss domes" focused geologists' attention on some recently discovered basin-range structures. A major consequence of this work has been the realization that the archi-

Lucchitta, I., and Suneson, N. H., 1996, Timing and character of deformation along the margin of a metamorphic core complex, west-central Arizona, *in* Beratan, K. K., ed., Reconstructing the History of Basin and Range Extension Using Sedimentology and Stratigraphy: Boulder, Colorado, Geological Society of America Special Paper 303.

tecture of both MCCs and HETs consists of a lower plate, commonly composed of lineated mylonitic gneiss, that has deformed in a ductile fashion; an upper plate that has deformed in a brittle fashion along subparallel normal faults; and a detachment fault separating the plates. The detachment fault is subhorizontal to gently dipping and separates domains that differ markedly in structural style and degree of metamorphism.

More recently, several models have been advanced to explain features displayed in the upper structural levels of MCCs and HETs. Lucchitta and Suneson (1981) showed that the upper plate of the Buckskin-Rawhide metamorphic core complex was not a gravity-glide sheet. They showed that this upper plate was structurally continuous with the nearby Colorado Plateau and that it was more appropriate to view the lower plate as the active tectonic element. Wernicke (1981) proposed that detachment faults are regional in extent and originated at low angles. Anderson (1971) showed that rocks above the detachment fault in the El Dorado Mountains of Nevada extend along listric faults that merge downward with the detachment, whereas Chamberlin (1983) and Gans and Miller (1983) suggested that upper-plate faults are planar and parallel, and intersect the detachment at moderate to high angles causing domino-style tilting. Alternatively, a detachment fault may initiate with steep dip and subsequently rotate to a lower dip angle (Buck, 1988; Hamilton, 1988; Wernicke and Axen, 1988). Parsons and Thompson (1993) suggested that magmatic inflation of the middle crust resulted in low-angle faulting and MCC-style deformation.

Initially, studies of extension associated with MCCs and HETs focused primarily on structural features within the MCCs themselves (e.g., Rehrig and Reynolds, 1980). More recently, the importance of the syntectonic strata overlying the detachment faults has been emphasized (Lucchitta, 1993). The reason for this interest is that upper-plate rocks contain detailed information on the timing, history, and character of deformation. In this chapter, we use the stratigraphy of upper-plate rocks to interpret their depositional environment, age relations, extent, source terranes, and presence or absence of unconformities. By so doing, we are able to derive a history of middle and late Cenozoic deformation for the Castaneda Hills–Signal area of west-central Arizona, to evaluate models proposed by others for the origin of highly extended terranes, and to advance our own.

LOCATION

The Castaneda Hills–Signal area of west-central Arizona (Fig. 1a) is in the eastern part of the Colorado River extensional corridor of Howard and John (1987), a domain of the Basin and Range Province characterized by HET-type extension. The area also extends to within about 20 km of the Colorado Plateau to the northeast (Fig. 1b). This is one of the few places where a nearly continuous belt of outcrops extends from the HET to the little-faulted Colorado Plateau, a characteristic that influenced the siting of the Consortium for Continental Refraction Profiling (COCORP) (Hauser et al., 1987) and Pacific-Arizona Crustal Experiment (PACE) (McCarthy et al., 1991) seismic experiments. The proximity of the Colorado Plateau represents a constraint that is critical for understanding the processes of HET extension.

This report is based on stratigraphic and stuctural data obtained while mapping in detail four 7.5' quadrangles and adjacent areas (Fig. 1).

GEOLOGIC FRAMEWORK

The Castaneda Hills–Signal area includes the northern edge of the Buckskin-Rawhide metamorphic core complex (Fig. 1b). This complex, similar to others of the North American Cordillera (Coney, 1980), shows the threefold architecture characteristic of MCCs. In the Castaneda Hills–Signal area, upper-plate faults strike northwest, dip northeast, and terminate at the Rawhide detachment fault. Upper-plate strata face southwest, suggesting a southwest relative-transport direction of the lower plate. The Rawhide detachment fault is undulating, with a regional dip of about 15° northeast toward the Colorado Plateau. The upper plate is lithologically and structurally continuous with the Plateau in the sense that no major structural break separates the two.

Cenozoic deformation in the Castaneda Hills–Signal area is recorded in the lower-plate mylonitic gneiss and in the mostly Neogene sedimentary and volcanic strata of the upper plate. In the Buckskin Mountains, Bryant and Wooden (1989) have shown that the pervasively mylonitized Swansea Plutonic Suite is 21.6 ± 1.5 Ma (U-Pb, zircon), indicating that at least some of the mylonitization is early Miocene or younger. This relatively young age confirmed Rehrig and Reynolds's (1980) suggestion, based on a 25.3 ± 0.6 Ma biotite age (K-Ar) on lower-plate mylonitic gneiss from the nearby Harcuvar Mountains, that mylonitization associated with the MCC in this area is middle Tertiary. Shackelford (1976) recognized that the mylonitic fabric in the Rawhide Mountains formed partly by northeast-southwest extension, but he associated it with thrusting. Subsequently, Davis et al. (1980) and Rehrig and Reynolds (1980) established the association of the mylonitic gneisses with lower-plate extension.

We divide the upper-plate sedimentary and volcanic strata into sequences bounded by major unconformities, in a manner similar to that employed by Eberly and Stanley (1978) for southwest Arizona. Each sequence contains several map units (basal arkose, upper basin beds, etc.); (Fig. 2) that can be recognized throughout our map area. We have avoided using formal names such as the Artillery and Chapin Wash Formations (Lasky and Webber, 1949) because of the difficulty in objectively identifying these deposits that are characterized by internal unconformities and rapid changes in thickness and lithology. The units forming each sequence are an objective entity. However, the significance of each sequence is best understood when the sequence is related with a distinctive tectonic regimen that obtained while the sequence was being deposited. Consequently,

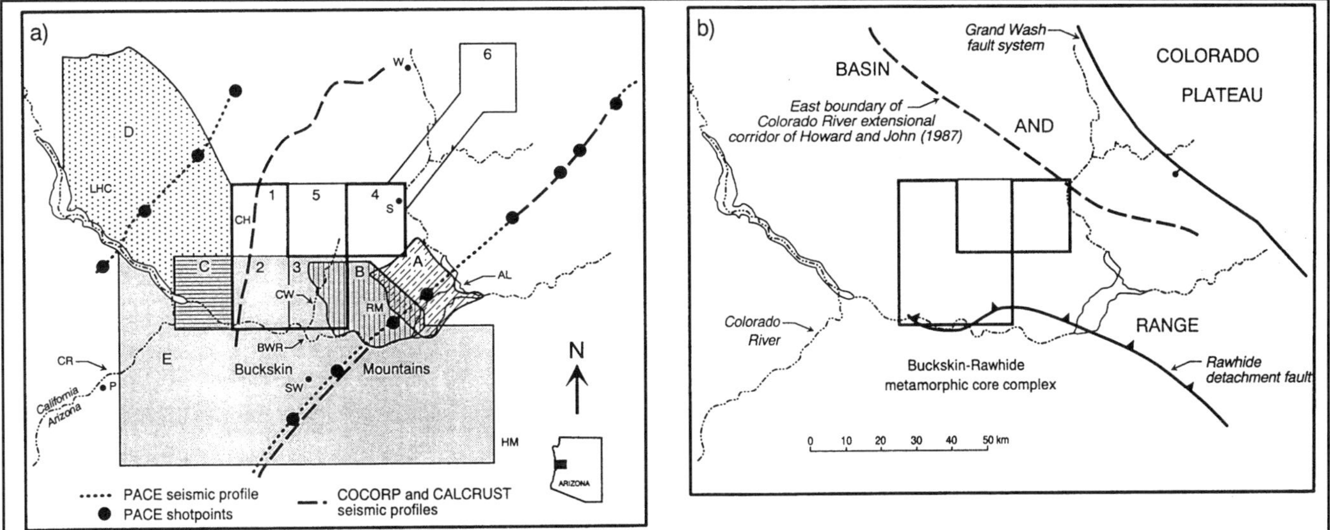

Figure 1. Location maps. a. Location map and sources of geologic data. Numbers 1 through 4 within heavy outlines refer to quadrangles published by authors; numbers 5 and 6 within light outline refer to unpublished mapping by authors; letters A through E refer to mapping published by other workers. 1. Castaneda Hills 7.5' quadrangle (Lucchitta and Suneson, 1994a); 2. Castaneda Hills SW 7.5' quadrangle (Lucchitta and Suneson, 1994b); 3. Centennial Wash 7.5' quadrangle (Lucchitta and Suneson, 1994c); 4. Signal 7.5' quadrangle (Lucchitta and Suneson, 1994d); 5. McCracken Peak 7.5' quadrangle (I. Lucchitta and N. H. Suneson, unpublished mapping, 1976); 6. Big Sandy strip and Elephant Mountain 7.5' quadrangle (I. Lucchitta and L. S. Beard, unpublished mapping, 1980). A. Artillery Mountains region (Lasky and Webber, 1949); B. Rawhide Mountains (Shackelford, 1989); C. Monkey's Head 7.5' quadrangle (Sherrod, 1988); D. Mohave Mountains area (Howard et al., 1990); E. Buckskin and Rawhide Mountains (Spencer and Reynolds, 1989). Geographic features: AL—Alamo Lake; BWR—Bill Williams River; CH—Castaneda Hills; CR—Colorado River; CW—Centennial Wash; HM—Harcuvar Mountains; LHC—Lake Havasu City; P—Parker; RM—Rawhide Mountains; S—Signal; SW—Swansea; W—Wickiup. b. Tectonic setting of the Castaneda–Signal area.

we have attached an interpretive tectonic label to each sequence and subsequence (Fig. 2).

From oldest to youngest, the sequences and subsequences are as follows:

Sequence I (HET)
- Subsequence IA (pre-tectonic)
- Subsequence IB (tumescence/arching/doming)
- Subsequence IC (syn-extension)
- Sequence II (Basin-range, BR)
- Sequence III (post-tectonic)

Strata of sequence I typically dip steeply and have been tilted along faults that characterize HETs. Sequence II was deposited in basins bordered by high-angle normal faults with significant structural relief. These basins and the associated ranges are those still visible today as geomorphic features. The youngest sequence (III) reflects the cessation of tectonism and the establishment of Colorado River through-flowing drainage.

The oldest Tertiary sedimentary rocks in the Castaneda Hills–Signal area are late Oligocene and perhaps somewhat younger to the south (Swansea area, 21.8 ± 0.5 Ma, Spencer and Reynolds, 1989) and west (Aubrey Hills, 23 ± 2.4 Ma, Nielson and Beratan, 1990). Tilting associated with extension of upper-plate strata began after ca. 19 Ma and probably before ca. 15 Ma; by ca. 8 Ma, deformation in the area had largely ceased (Lucchitta and Suneson, 1993).

STRATIGRAPHY OF THE UPPER PLATE

Basement rocks

Upper-plate basement rocks in the Castaneda Hills–Signal area include (1) Proterozoic granitic and metamorphic rocks intruded by thin aplite, pegmatite, and diabase dikes; (2) mylonitic gneiss and associated coarse-grained mafic plutonic rocks; and (3) Mesozoic metavolcanic and metasedimentary rocks. The most widespread Proterozoic units are the 1410 Ma (Gray et al., 1989) Signal Granite (Lucchitta and Suneson, 1982) and quartz-feldspar-biotite gneiss and amphibolite that are probably Early Proterozoic in age (Gray et al., 1989). The mylonitic gneiss (gneiss of Centennial Wash of Lucchitta and Suneson, 1994c) crops out only in a small area along the east side of the Centennial Wash 7.5' quadrangle (Fig. 1a) and is juxtaposed against the Signal Granite along a shear zone meters to a few hundreds of meters thick that shows increasing-downward cataclasis and

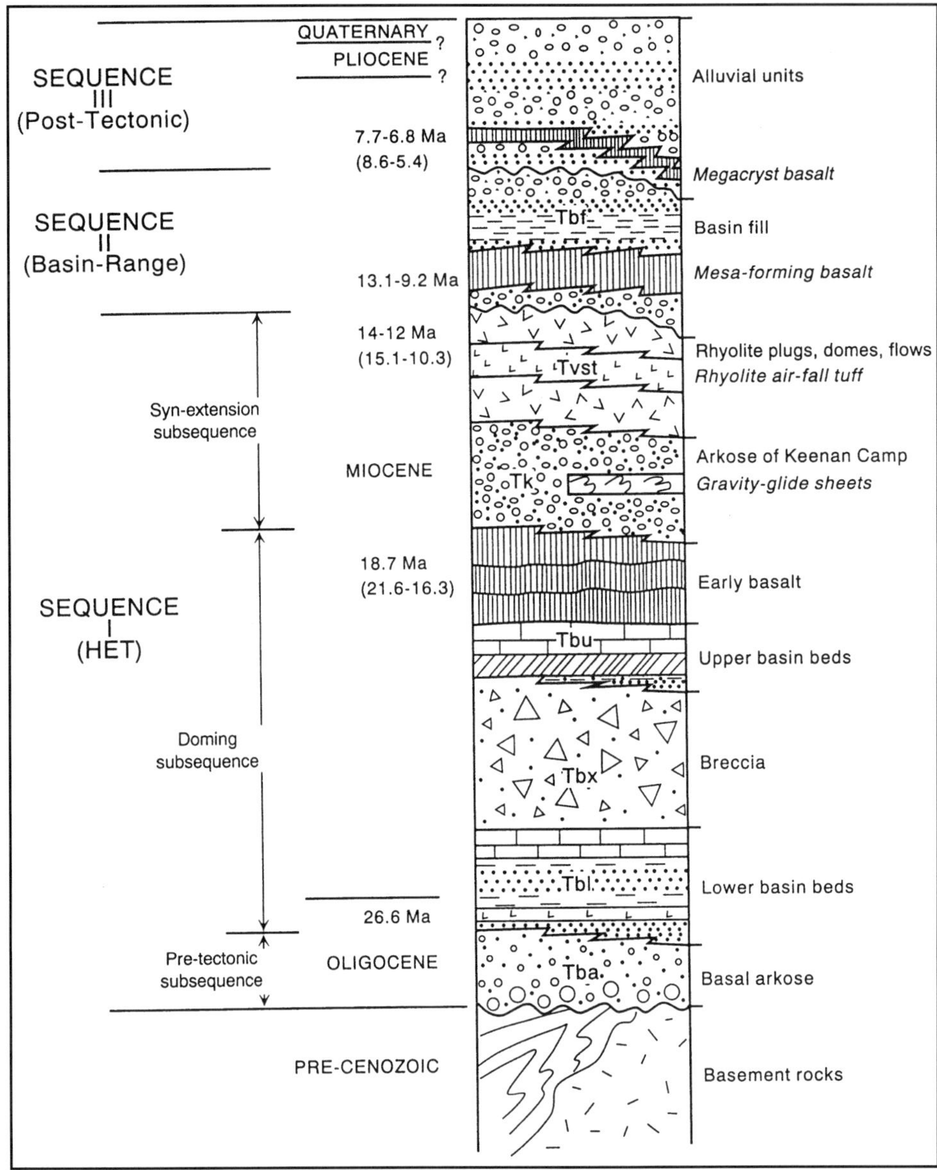

Figure 2. Stratigraphy of upper-plate. Lithologic symbols used are conventional. Isotopic ages in parentheses give range of ages for units. Isotopic ages not in parentheses give most commonly reported ages and can be taken to approximate age of unit. Abbreviations (Tba, etc.) are the same as those used in Figure 5.

mylonitization of granite into gneiss. The mylonitic gneiss is intruded by latest-kinematic to post-kinematic 90 Ma gabbro and diorite (Lucchitta and Suneson, 1994c). Dikes of diorite similar in composition to those of the Centennial Wash unit also intrude gneissic rocks near Castaneda Hills (Lucchitta and Suneson, 1994a). The Proterozoic rocks and mylonitic gneiss are unconformably overlain by the Tertiary sedimentary and volcanic strata described below. Greenschist-facies metavolcanic and metasedimentary rocks crop out widely in the southern part of the Castaneda Hills–Signal area; they probably correlate with the Triassic Buckskin Formation, Triassic(?) and Jurassic Vampire Formation, and Jurassic Planet Volcanics of Reynolds and Spencer (1989). In the Castaneda Hills–Signal area, the schistose rocks are nowhere juxtaposed against the Proterozoic units and are only in fault contact with the Tertiary sedimentary and volcanic strata; the rocks are identical in lithology to the widespread breccias and a few of the gravity-glide sheets that are part of the Tertiary sequence I.

Rocks similar in lithology and age to the Signal Granite and associated Proterozoic granitic and metamorphic rocks extend from the Castaneda Hills–Signal area northeast to the Colorado Plateau, where they are buried beneath Paleozoic strata. These rocks are not present to the south. In contrast, the Mesozoic metavolcanic and metasedimentary rocks are widespread to the southwest, but are absent north of the area. The observation that the separate basement terranes consist of distinct and contrasting rock types is critical for determining basin-sedimentation patterns and uplift histories.

Sequence I (HET)

Subsequence IA (pre-tectonic). The oldest Tertiary unit in the Castaneda Hills–Signal area is a red arkosic coarse-grained sandstone and pebble-, cobble-, and locally, boulder conglomerate (basal arkose, Fig. 2). The basal arkose is sheet-like, of regional extent, and composed exclusively of northern-terrane basement rock types. Locally, the basal arkose coarsens to the northeast. In places, the basal arkose overlies a regolith meters to tens of meters thick developed on the Proterozoic Signal Granite.

Subsequence IB (doming). The lower basin beds (Fig. 2) consist of sandstone, mudstone, limestone, and airfall tuff. A tuff layer about 10 m above the base of the lower basin beds is about 26.5 Ma ($^{40}Ar/^{39}Ar$, laser fusion, 26.28 ± 0.01 Ma on sanidine, 26.57 ± 0.15 Ma on biotite, Lucchitta and Suneson, 1993). This fine-grained tuff is the oldest Cenozoic volcanic deposit in the area.

A breccia unit (Fig. 2) is present throughout the Castaneda Hills–Signal area, where it has a strike length of at least tens of kilometers and extends beyond the borders of our map area. The clasts of the breccia were derived exclusively from the southern basement terrane. Most are clast-supported and tens of centimeters in size; some are meters and a few are tens of meters in size. Bedding is absent, except in a few places where the breccia passes laterally into weakly bedded reworked facies. The breccia is monolithologic at most outcrops and can easily be mistaken for bedrock in poor exposures. The breccia locally shows an inverted stratigraphy, with Mesozoic metamorphic rocks near the base of the unit, Paleozoic metamorphic rocks in the middle, and Proterozoic(?) granitic rocks in the uppermost part. Gravity-glide sheets described below show the same rock units in the same inverted order. Locally, the breccia incorporates masses tens to hundreds of meters in size identical to the gravity-glide sheets. The breccia is exposed as much as 20 km northeast of the trace of the Rawhide detachment fault. At its northeast terminus, it is deposited directly on Proterozoic basement (Lucchitta and Suneson, 1994d).

The upper basin beds (Fig. 2) consist of sandstone, mudstone, limestone, and gypsum. The early basalt (Fig. 2; basalt member of Artillery Formation of Lasky and Webber, 1949) underlies and locally is interbedded with the lower part of the arkose of Keenan Camp. This basalt is of regional extent; in the area, it has a calculated volume of 8 km^3 (Suneson and Lucchitta, 1983) and is the first evidence of local volcanism (Suneson and Lucchitta, 1983). The early basalt is deeply weathered, thus difficult to date radiometrically. The most reliable published isotopic dates range from 16.3–21.6 Ma (Reynolds et al., 1986). Our data indicate an age of ca. 19 Ma (Suneson and Lucchitta, 1983) for the basalt.

Subsequence IC (Syn-extension). The arkose of Keenan Camp (Fig. 2) is the most widespread unit in the Castaneda Hills–Signal area. This unit, partly equivalent to the Chapin Wash Formation of Lasky and Webber (1949), is a red arkosic sandstone and conglomerate with clasts as much as 1 m in diameter. The clasts in the conglomerate were derived from northern and southern source terranes and are of many different rock types, including the underlying early basalt. Hundreds-of-meters- to kilometer-sized gravity-glide sheets of Mesozoic metamorphic rocks, Paleozoic metacarbonate rocks and metasandstone, and Proterozoic(?) granitic rocks are complexly interlayered with the arkose along its southern exposure near the trace of the Rawhide detachment fault. The rock types of the gravity-glide sheets are identical to those that form the breccia and, like the breccia, are locally stacked in inverse stratigraphic sequence. A conspicuous difference between gravity-glide sheets and breccia is that only a few of the sheets are composed of the Mesozoic metamorphic rocks, which are the predominant component of the breccia. In most places younger units overlie the arkose of Keenan Camp in angular unconformity. Locally, however, the basin fill of sequence II overlies the uppermost part of the arkose of Keenan Camp with a disconformable contact that is very inconspicuous in the field because the basin fill and upper part of the arkose of Keenan Camp are lithologically similar.

The upper part of the arkose of Keenan Camp is interbedded with mostly 14–12 Ma rhyolite flows and tuffs that have a calculated volume in the area of 34 km^3 (Suneson and Lucchitta, 1983; Fig. 2). The flows and tuffs are associated with rhyolite plugs. Together, these silicic rocks form numerous volcanic complexes within the area.

Many units in sequence I are chemically altered, especially the early basalt and the arkose of Keenan Camp; and some of the rhyolitic intrusive rocks are unusually high in potassium as a result of potassium metasomatism. Areas of disseminated manganese, uranium, copper, hematite, silver, and gold mineralization are widespread but mostly of low grade. Concentrations of some of these elements are present locally along upper-plate faults, where they have been prospected extensively. Gangue minerals include calcite, quartz, barite, and fluorite (Spencer et al., 1989a).

Sequence II (Basin-range, BR).

Most of the strata of sequence II (BR) (Fig. 2) are the fill of present-day basins, composed of locally derived fanglomerate and lesser amounts of mudstone and sandstone, some tuffaceous. Clasts are similar to rocks in nearby present-day highlands and grain size generally increases toward those sources.

Basalt lava flows cap prominent flat to gently tilted mesas and are interbedded with the basin fill. These basalts are 13.1–9.2 Ma in age, and have a calculated volume in the area of 3 km^3 (Suneson and Lucchitta, 1983). In contrast to the early basalt of subsequence IB and the associated basin beds, the basalt in sequence II is unaltered. The basin fill is dull-colored, poorly indurated, and unmineralized.

Sequence III (Post-tectonic)

Sequence III (Fig. 2) consists chiefly of piedmont-slope deposits and, to a lesser extent, gravel and sand of more distant derivation along major drainages such as the Bill Williams River. Surfaces and deposits are graded to present-day or ancestral through-flowing drainages that are part of the Colorado River drainage network. Unaltered 8.6–5.4 Ma (mostly 7.7–6.8 Ma) megacryst-bearing basalt with a calculated volume in the area of 0.4 km^3 (Suneson and Lucchitta, 1983) is locally interbedded with sequence III strata. Like the underlying sequence II (BR) strata, deposits of sequence III are unmineralized.

STRUCTURAL DATA

Strata of sequence I are tilted southwest along kilometer-spaced normal faults that strike northwest, dip northeast, and are down to the northeast. Some of these faults dip less than 45°. Older strata generally dip more steeply than younger strata (Lucchitta and Suneson, 1993). Disconformities and unconformities range from local to regional, and facies change abruptly from proximal to distal. Because of these complexities, coupled with incomplete exposures, unconformities that are of regional, as opposed to local, significance are difficult to recognize unambiguously in the field.

In contrast to sequence I faults, those associated with sequence II (BR) have northerly strikes, spacing measured in tens of kilometers, displacements measured in hundreds to thousands of meters, and dips typically greater than 60°. Tilt of intervening blocks typically is less than 15°. Sense of displacement is not constant, resulting in a horst-and-graben structure. Most sequence II faults are bounding faults for present-day basins and ranges. Where exposed, these faults commonly show growth-fault relations to the sequence II basin fill. Within the area mapped, several sequence II faults swing to southeast strikes as they are traced southward, paralleling and adopting the strikes of sequence I structures.

Faults that locally offset sequence III strata are normal faults that are few in number, small in displacement, and strike north. The youngest offset the 8.6–5.4 Ma megacryst-bearing basalt a few meters at most. Younger deposits are not faulted.

For sequence I faults, the spacing between faults increases northeast, toward the Colorado Plateau, whereas the degree of tilt decreases in the same direction. No such systematic changes can be seen for sequence II faults. Indeed, large normal faults form the western edge of the Plateau in this area.

INTERPRETATIONS

Environments of deposition and paleogeography

Basal arkose (Fig. 3a). The composition and northward coarsening of clasts indicate that the basal arkose was derived from the northern source terrane. Its sheet-like geometry suggests deposition on a surface of low relief by systems of streams flowing generally southwest, during or at the end of a considerable period of stability and low relief indicated by the thick regolith underlying the basal arkose. We interpret the basal arkose to be a piedmont-slope deposit laid down southwest of the Mogollon Highlands. The Mogollon Highlands were defined by Cooley and Davidson (1963) and Young (1987) to be a northwest-trending uplift of Laramide age that marked the southern margin of the Colorado Plateau and was the source of the rim gravels on the Plateau. The regolith may signal a long period of tectonic stability following the Laramide disturbance.

Lower basin beds (Fig. 3b). The composition and fine grain size of the lower basin beds indicate a low-energy lacustrine depositional environment, brought about by ponding of the southwest-flowing streams that deposited the basal arkose. The ponding resulted from the rise of a topographic barrier to the south of the outcrop area of the lower basin beds. One way of forming such a barrier and associated depositional basin is by tilting blocks. However, we infer that tilting of blocks had not yet started at this time because we can document no difference between the dips of the lower basin beds and those of the basal arkose (see discussion in a following section). Because tilting of blocks occurred along faults that bound the blocks, we further infer that faulting was not yet taking place, so drainage-blocking scarps did not exist. The absence of scarps is further supported by the lack of coarse clastic material within the lower basin beds, contrary to what one would expect had a fault scarp existed. Consequently, we believe that the topographic barrier responsible for formation of the basin was the result of upwarping rather than faulting, and we further believe the upwarp was gentle during deposition of the lower basin beds.

The upwarping started at ca. 26.5 Ma, as shown by the age of the tuff near the base of the lower basin beds. This tuff is an air-fall deposit and does not represent the inception of volcanism within the Castaneda Hills–Signal area.

Breccia (Fig. 3b). The very coarse grain size and widespread occurrence of the breccia suggest close proximity to its source, substantial relief of the source area, an unusual mechanism of deposition, or all three.

Breccia units such as that in the Castaneda Hills–Signal area appear to be characteristic of HETs in general. Topping (1993, and references cited therein) discusses the Amargosa Chaos (megabreccia) in Death Valley, California; megabreccias are present in many of the areas described in Sherrod and Nielson (1993). Such breccias commonly are interpreted as the product of long-runout slides or rock avalanches (e.g., Yarnold, 1993), following Shreve's (1968) analysis of the Blackhawk landslide.

Our work in the Castaneda Hills–Signal area does not support the rock-avalanche interpretation:

(1) The breccia is of regional extent, which is not consonant with its origin as a rock avalanche.

(2) There is no evidence for tilting during or before breccia deposition (as described below), so no evidence for faulting to

produce the steep and high scarp necessary to generate a run-out slide.

(3) There is no evidence for a steep scarp to be found in the facies of strata above and below the breccia. These strata are fine-grained basin beds lacking the coarse facies that would be present at the foot of a high and steep scarp.

(4) Similar rock types and stratigraphic succession in breccia and gravity-glide sheets, and inclusion of gravity-glide sheets within the breccia, all imply closely related mechanisms of emplacement.

We prefer the interpretation that sheets of rock peeled off the developing high to the south in what is now the Buckskin-Rawhide area and slid down a gentle slope, lubricated by pore overpressure (Lucchitta, 1990), and breaking up possibly because of hydrofracturing mechanisms.

Upper basin beds. The upper basin beds represent a return a to low-energy, lacustrine environment of deposition. The beds show that the southern source terrane that previously supplied the breccia no longer contributed significant debris to the basin flanking it to the north.

Early basalt (Fig. 3c). The early basalt is considered to be mantle-derived in origin on the basis of its strontium-isotope ratios (Suneson and Lucchitta, 1983). Thickening of the basalt to the west suggests its derivation from that direction. In the rest of the area, the early basalt is relatively thin but widespread, showing that the topography over which it flowed was subdued,

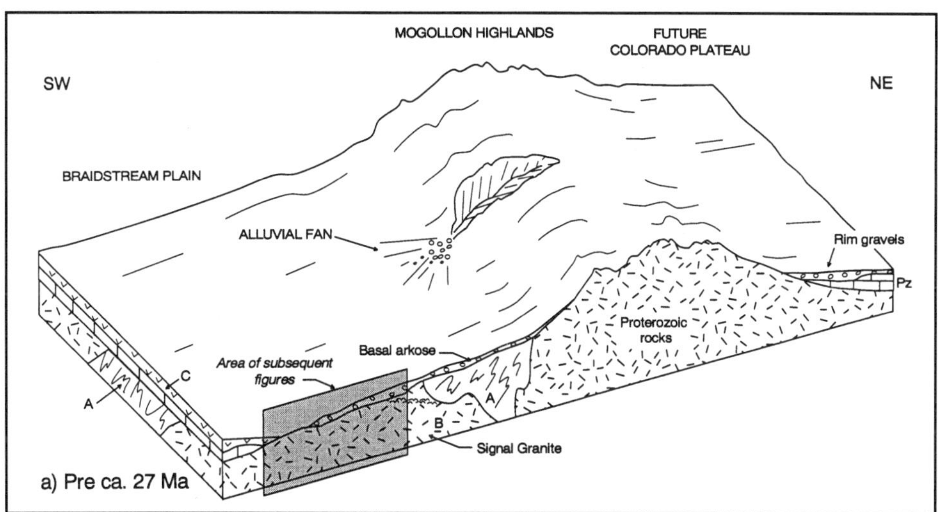

Figure 3. Paleogeography and environment of deposition of sequence I (HET) strata. (a) Rectangle shows area covered by Figures 3b to 3f (not to scale). Basal arkose (Tba) time. A—Proterozoic gneiss and amphibolite. B—90-Ma mylonitic shear zone; C—wedge of Paleozoic metacarbonate and metaquartzite, and Mesozoic metavolcanic and metasedimentary rocks. A and B are not shown in subsequent figures.

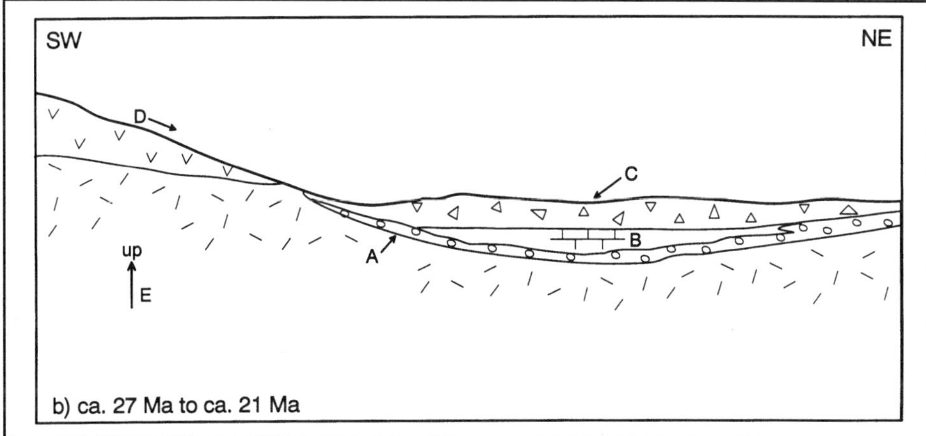

Figure 3b. Lower basin beds (Tbl) and breccia (Tbx) time. A—Basal arkose (subsequence IA); B—lower basin beds (subsequence IB); C—breccia (subsequence IB); D—derivation of breccia from southwest; E—uplift of core-complex area. A, B, and C are combined in subsequent figures.

with no significant barriers. This, in turn, is not consonant with early initiation of faulting.

Arkose of Keenan Camp (Fig. 3, d, e and f). The arkose of Keenan Camp represents a return to a high-energy environment. For the first time in this area, the strata record a multiplicity of source areas, in contrast to the lower parts of the section, where individual units were derived chiefly from either the northern or southern source terranes. Also making their first appearance as clasts are highly altered granitic rocks not present in the northern source terrane. We interpret this to indicate that by now the basement of the southern source area had been stripped at least in places of its Paleozoic and Mesozoic cover. Among the lithologies represented in the clasts is the early basalt, which underlies the arkose of Keenan Camp. This is the first time that rocks from units lower in the Tertiary basin-filling section are incorporated within upper units. We believe that this signals the beginning of tilting and, therefore, movement along faults responsible for the tilting.

Silicic volcanic rocks (Fig. 3e). Strontium-isotope data suggest that the silicic volcanic rocks were derived from the lower crust, indicating that this part of the crust had attained sufficiently high temperatures to melt in places. High crustal temperatures, combined with extension, are also indicated by the ability of relatively small bodies of viscous silicic magma to reach the surface without solidifying.

The alteration, metasomatism, mineralization, and resulting bright colors that are common in the rocks of sequence I probably resulted from hydrothermal-circulation processes operating

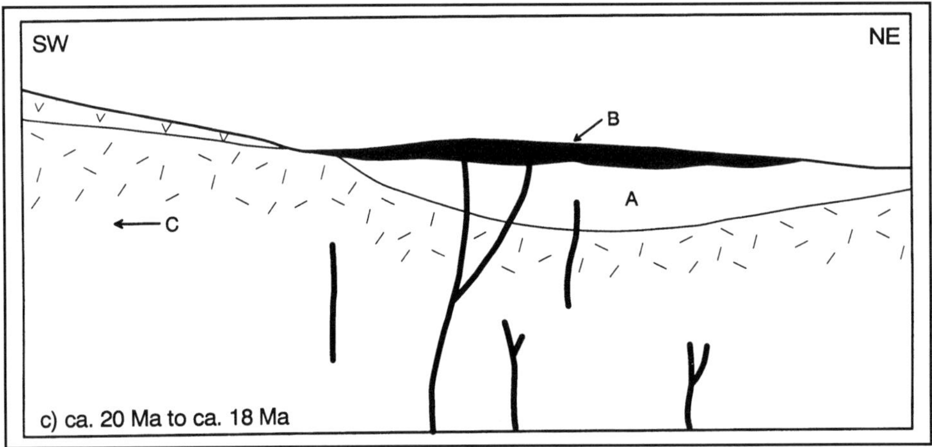

Figure 3c. Early basalt (Teb) time. A—Strata of subsequence IA and early subsequence IB; B—early basalt (Artillery Basalt); C—intrusion of early basalt dikes possibly associated with slight extension. In subsequent diagrams, right edge is fixed (to Colorado Plateau), left edge moves southwest. A and B are combined in subsequent figures.

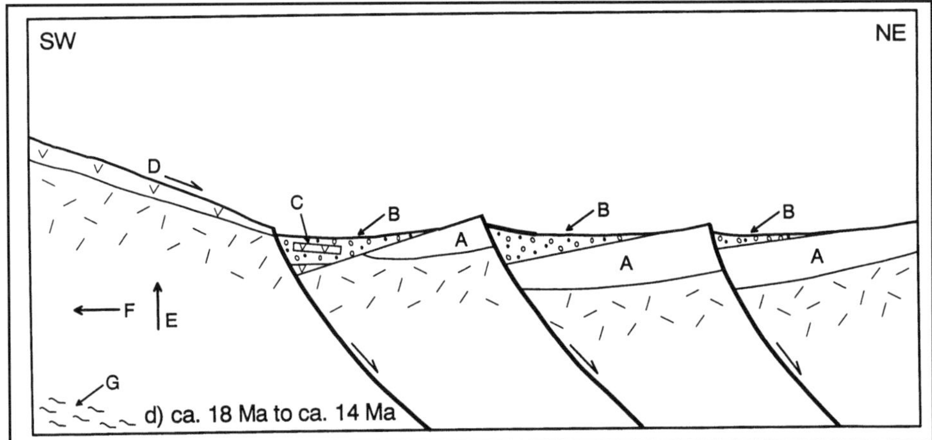

Figure 3d. Early arkose of Keenan Camp (Tk) time. A—Strata of subsequences IA and IB; B—early arkose of Keenan Camp strata (early subsequence IC); C—gravity-glide blocks; D—reactivated slope; E—uplift; F—extension; G—beginning of development of mylonitic gneiss. A and B are combined in subsequent figures.

during HET-type faulting. This is consonant with mechanisms we propose for emplacement of breccia and gravity-glide sheets, and for the deep alteration of the early basalt, which contrasts markedly with the freshness of sequence II basalts that are only a few million years younger.

Basin fill (Fig. 4). Deposition of the basin fill under conditions of interior drainage is indicated by the local derivation of clastic material, the absence of far-traveled material, and the very fine grain size of material present along the axes of basins. Interior drainage is also indicated by the pancake shape of the mesa-forming basalts interbedded with the basin fill. These basalts show no evidence of flowing down drainages. Only at the transition between sequences II and III do basalts begin to have elongate shapes suggestive of drainage systems. These basalts mark the beginning of integration of drainage in the area.

Strontium-isotope ratios are consistent with derivation from the lithospheric mantle (Suneson and Lucchitta, 1983), which we interpret as evidence for the waning of the thermal pulse that had peaked during emplacement of the silicic volcanic rocks.

The absence of alteration and mineralization in sequence II rocks shows that the circulation of hydrothermal fluids had ceased by this time.

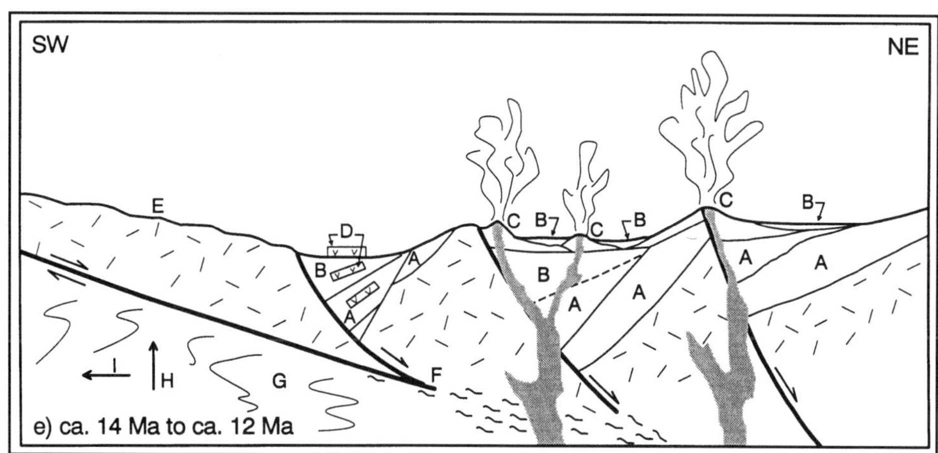

Figure 3e. Middle arkose of Keenan Camp (Tk) and silicic volcanic rocks (Tvs) time. A—Subsequences IA, IB, and early IC (fan dips). B—Middle subsequence IC (fan dips); C—silicic volcanic rocks (subsequence IC); D—gravity-glide sheets in inverted stratigraphic order; E—active slope denuded to Proterozoic(?) basement (pre-Proterozoic rocks still present upslope to southwest); F—basal detachment fault grading into zone of ductile shear; G—mylonitic gneiss; H—uplift; I—strong extension. A, B, and C are combined in subsequent figures.

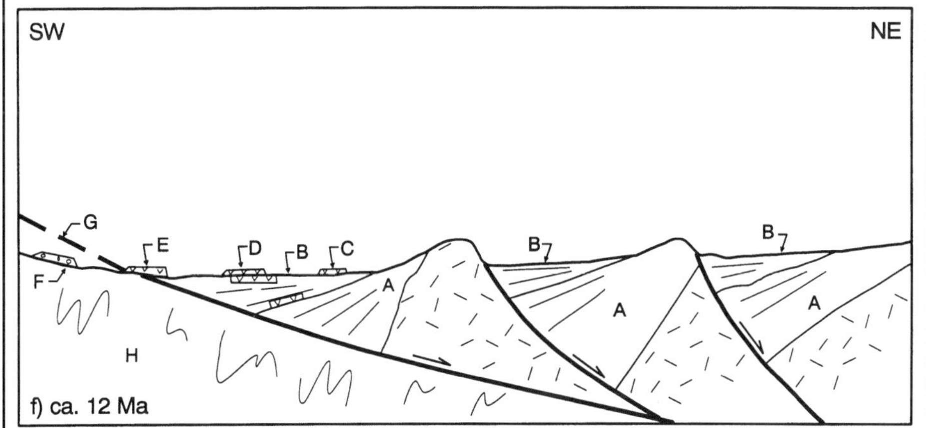

Figure 3f. Late arkose of Keenan Camp (Tk) time. A—Subsequences IA, IB, and early and middle IC; B—late subsequence IC, locally grading upward into earliest sequence II; C—gravity-glide block of subsequence IC rocks; D—gravity-glide block of Proterozoic(?) plutonic rocks overlyng glide block of Paleozoic metasedimentary rocks; E—glide block of Paleozoic metasedimentary rocks emplaced across eroded trace of detachment fault; F—subsequence IC rocks emplaced on lower-plate mylonitic gneiss; G—eroded part of detachment fault; H—lower-plate mylonitic gneiss.

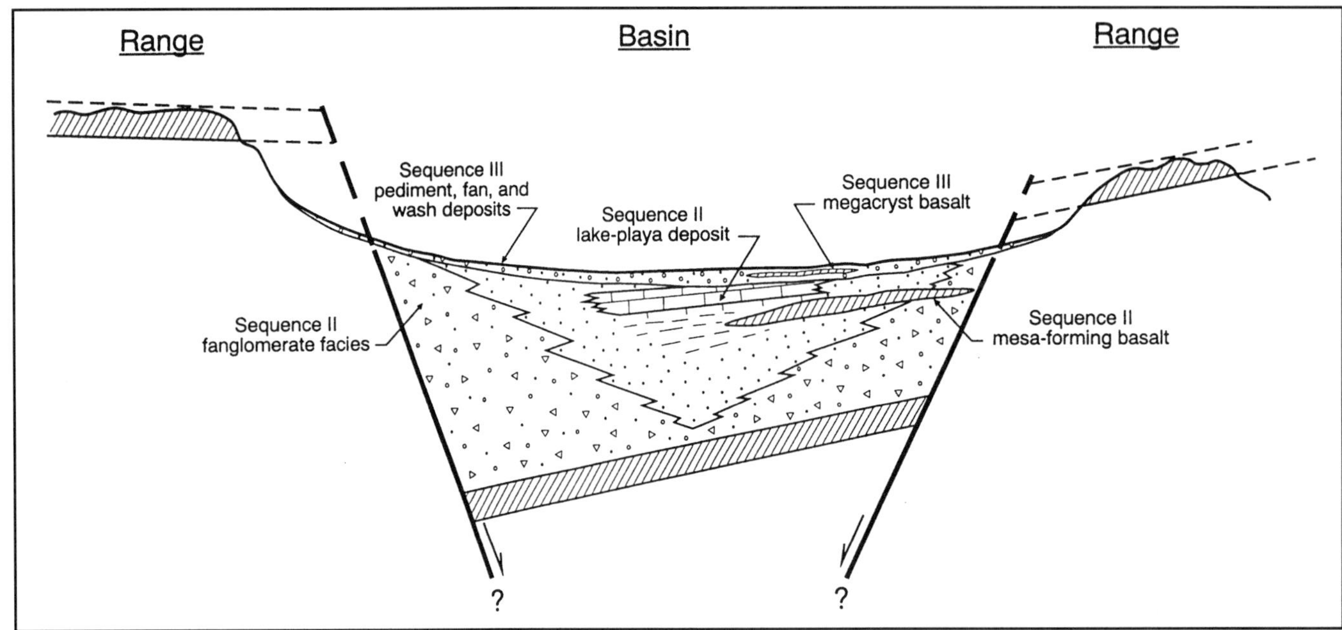

Figure 4. Geologic and topographic relations in a basin-range basin during sequence II and III time. Lithologic symbols are conventional (not to scale).

Deposits of sequence III (Fig. 4). These deposits postdate integration of drainage in the area into the Colorado River drainage network.

The megacrystic basalt has strontium-isotope ratios consistent with its derivation from the asthenospheric mantle (Suneson and Lucchitta, 1983), in contrast to the lithospheric mantle and lower-crust derivation of the mesa-forming basalt and silicic volcanic rocks, respectively. We take this to indicate the progressive cooling of the lithosphere.

Structural development

In some of our early work (Lucchitta and Suneson, 1983), we interpreted the entire sequence I (HET) section as characterized by fan dips, i.e., by a gradual stratigraphically upward decrease in dip from steep to moderate (e.g., Fig. 5a). The structural implication of fan dips is that tilting of strata, movement on upper-plate faults, and movement on the basal detachment into which the upper-plate faults merge or terminate, are all synchronous with deposition.

Because a time-related change in dip is critical for understanding the structural history of strata in the Castaneda Hills–Signal area, and because the existence of such a change cannot be verified well in the field, we attempted to determine the structural characteristics of sequence I (HET) by developing a statistical technique for analyzing dips (Lucchitta and Suneson, 1993). The technique consists of grouping dips of map units by quadrangle (Fig. 5, b–e), and by unit for all quadrangles (Fig. 5f), and plotting the means and standard deviations for each grouping.

Our analysis shows that the dips of the lower part of the section (basal arkose through upper basin beds, subsequences IA and IB) are similar (i.e., homoclinal) and average 45°, whereas dips of the arkose of Keenan Camp and interbedded rhyolite (subsequence IC) are considerably less (32°) (Fig. 5f). Mean dip of the basin fill of sequence II is much lower (8°, Fig. 5f). These observations suggest that units older than the arkose of Keenan Camp were deposited before faulting and stratal tilting began. Tilting probably started as the arkose of Keenan Camp began to be deposited and continued throughout Keenan Camp and rhyolite time. We interpret this phase and style of deformation as reflecting severe extension in the upper plate, i.e., the formation of the HET. The 19 Ma early basalt is interbedded with the lower part of the arkose of Keenan Camp. We conclude that tilting, i.e., HET-type faulting, started ca. 19 Ma and ended before eruption of the mesa-forming basalts of the BR sequence, which are about 13 Ma in age. Map relations and the generally low dips of strata of sequence II (BR) show that post–13 Ma extension occurred along faults with different strikes (north) and wider spacing than those of previous faults; the absence of significant stratal tilting suggests that the BR faults are planar and penetrate relatively deeply. Faulting and tilting had essentially ceased by the time sediments and basalts related to through-flowing drainage (sequence III) were being deposited (ca. 8 Ma).

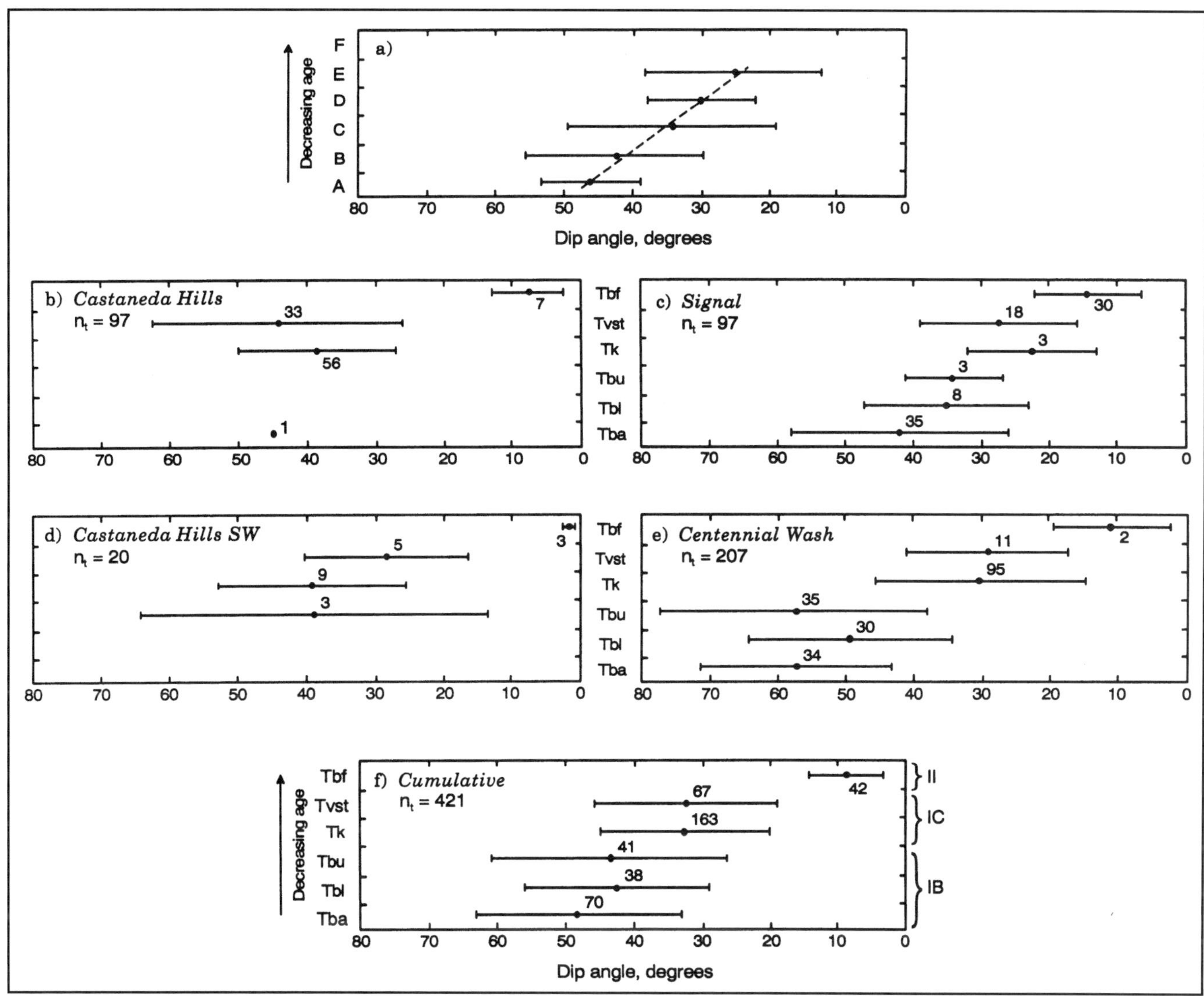

Figure 5. Dip diagrams. In assembling data sets, dips obviously affected by faulting or intrusion were not used. (a) Diagram showing hypothetical change in dip with time when rotation is synchronous with deposition (fan dips). (b–e) Dip diagrams for the Castaneda Hills–Signal area. Dips obtained from the geologic maps are plotted for each stratigraphic unit and are grouped by quadrangle: (b) Castaneda Hills 7.5' quadrangle; (c) Signal 7.5' quadrangle; (d) Castaneda Hills SW 7.5' quadrangle; (e) Centennial Wash 7.5' quadrangle. (f) Cumulative plot for all quadrangles. Average dip for subsequences IA (pretectonic) and IB (doming) is 45°. Average dip for subsequence IC (syn-extension) is 32°. Average dip for sequence II (BR) is 8°.

TECTONIC-STRATIGRAPHIC SYNTHESIS

The information and interpretations derived from stratigraphy, structure, and the geochemistry and isotopic ages of volcanic rocks in the Castaneda Hills–Signal area are summarized in a tectonostratigraphic diagram (Fig. 6). Stratigraphic and structural data provide information about the tectonic evolution of the upper crust, whereas data on the volcanic rocks allow correlating this evolution with events in the lower crust and upper mantle.

The oldest sediments (basal arkose) were deposited on a low-relief surface sloping southwest in the Castaneda Hills–Signal area. This surface, which was of regional extent, reflects a substantial period of tectonic quiescence—long enough to allow deep weathering and the development of low relief. The surface consisted of a braidstream plain at the foot of higher terrane to the northeast between the Castaneda Hills–Signal area and the Colorado Plateau. This terrane probably was the Mogollon Highlands, which formed during the Laramide orogeny and constituted the southwestern edge of the Colorado Plateau.

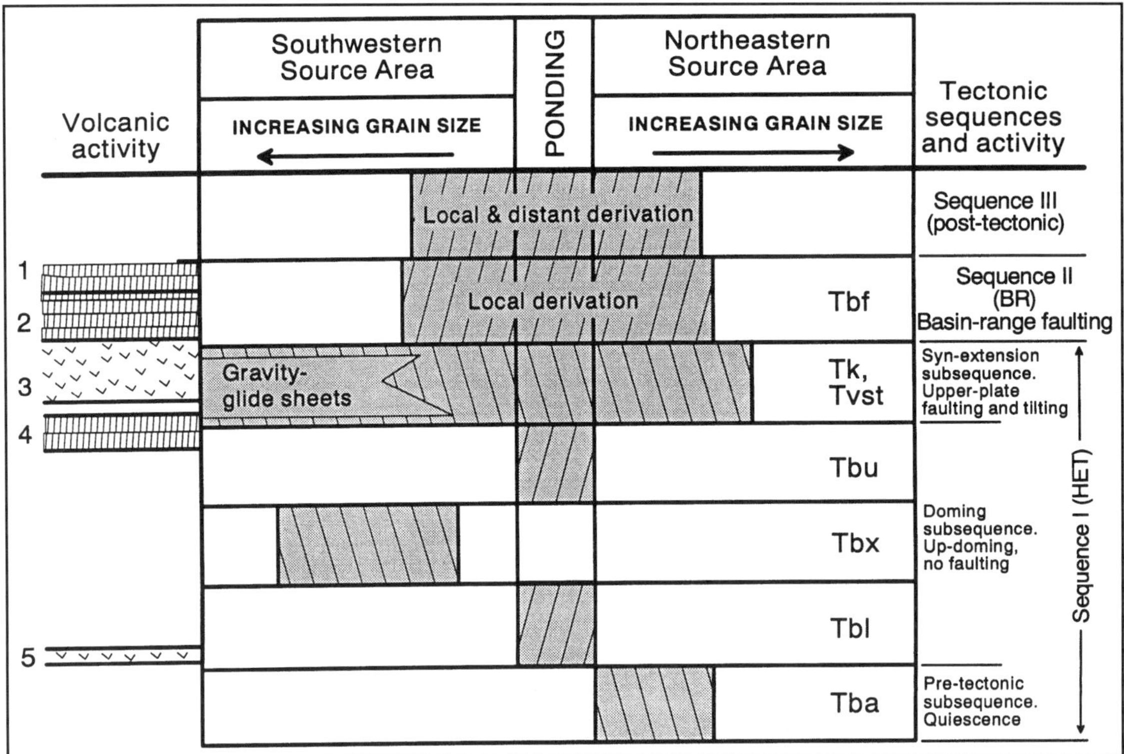

Figure 6. Tectonostratigraphic diagram. Main part of diagram specifies grain size and source area for sedimentary deposits. Grain size and energy of deposition increase to right and left of axis, which represents fine grain size and low energy of deposition (lacustrine conditions). Right column lists stratigraphic sequences and associated tectonic activity. Left column lists volcanic activity. (V-pattern represents silicic rocks; vertical pattern represents mafic rocks). 1—Megacryst basalt; 2—mesa-forming basalt; 3—rhyolite plugs, domes, flows, and tuffs; 4—early basalt; 5—air-fall tuff in lower basin beds. Left column shows composition and timing of volcanism (no scale). Map-unit symbols are same as used in Figure 5.

In the late Oligocene, the southwesterly drainage was ponded by a highland area developing to the south, probably as a result of upwarping; this resulted in the deposition of the lower basin beds. In the early Miocene, continued (and abrupt?) upwarping of the terrain to the south, combined with the onset of hydrothermal circulation, caused deposition of the breccia, possibly as a series of catastrophic slides initiated by hydrofracturing and pore overpressure. Lacustrine conditions were re-established in upper-basin–bed time as a result of lowering of the southern highlands through erosion (and/or subsidence?), possibly accompanied by a temporary cessation or decrease in hydrothermal circulation. There is no evidence within our area for upper-plate faulting, block rotation, movement on the Rawhide detachment fault, or local volcanism throughout the period of time represented by the basal arkose, lower basin beds, breccia, and upper basin beds (≥26.6 Ma to ca. 19 Ma).

At ca. 19 Ma, basalt of upper-mantle origin (early basalt) flowed into the Castaneda Hills–Signal area chiefly from a volcanic complex in the Mohave Mountains area to the west-northwest. At this time, relief within the basin was minor, allowing the early basalt to spread as thin sheets over a wide area. The beginning of volcanic activity probably occurred at approximately this time throughout the lower Colorado Extensional Corridor. Stratigraphic columns in Sherrod and Nielson (1993) typically show ages of the oldest volcanic rocks as being in the 19–22 Ma range. As is the case with the early basalt, these old volcanic rocks typically are highly altered and difficult to date reliably. We suspect the real ages are approximately the same, or differ at most by 1–2 m.y.

Shortly after emplacement of the early basalt, basins began to be formed in the Castaneda Hills–Signal area by faulting and tilting. The arkose of Keenan Camp was deposited in these basins; the sediments were derived partly from local source terranes, including the previously high terrane to the south, and partly from more distant terranes. Clasts of the early basalt within the arkose of Keenan Camp show that, for the first time in this Tertiary section, older stratigraphic units were exposed to erosion, presumably on ridges formed by the rotation of blocks. The fanning of dips in the arkose of Keenan Camp suggests that upper-plate faulting, block rotation, stratal tilting, and movement on the Rawhide detachment fault (i.e., HET-type deformation) were synchronous with basin formation and sedimentation. Parts of the arkose of Keenan Camp were being cut by the detachment fault, indicating that it was relatively close to the topographic surface. Large gravity-glide sheets slid off the resurgent terrane to the south (now exposed as the Buckskin-Rawhide metamorphic core complex), perhaps owing to renewed or intensified hydrothermal activity, and were deposited in inverted stratigraphic sequence along the north flank of the complex. The metamorphic core complex was stripped down to the lower-plate mylonite gneiss. Rhyolite derived from the lower crust was erupted mostly between ca. 14 Ma and 12 Ma, toward the end of deposition of the arkose of Keenan Camp.

The Rawhide detachment fault juxtaposes mylonitic gneiss of the lower plate against brittlely deformed upper-plate strata; therefore, at least some movement on the fault must postdate the mylonitization. However, the fault and mylonite are intimately related: they show the same northeast-directed kinematic indicators (first documented by Shackelford, 1976), which are different in orientation from the 90 Ma indicators in nearby mylonites of the upper plate (Lucchitta and Suneson, 1994c); furthermore, the fault and mylonites are contiguous over a large area. Upper-plate tilting (and presumed movement on the Rawhide detachment fault into which upper-plate faults sole) occurred between ca. 19 Ma and 12 Ma, and probably mostly between 15 Ma and 12 Ma, in the Castaneda Hills–Signal area. The age of mylonitization is more difficult to determine. At least some mylonitization occurred after emplacement of the 21.6 ± 1.5 Ma Swansea Plutonic Suite (Bryant and Wooden, 1989), but the overall time for the beginning and end of the mylonitic event are unknown. Regardless, at least some, if not most, of the lower-plate rocks below the Rawhide detachment fault were mylonitized while the fault was active. These interpretations and the evidence for voluminous rhyolitic magmatism and extensive circulation of hydrothermal fluids suggest a relation between (1) lower-plate mylonitization, (2) upper-plate stratal tilting and movement on the Rawhide detachment fault, and (3) a "hot" middle and upper crust.

After ca. 12 Ma, rhyolitic volcanism largely ceased, the crust cooled, HET-type deformation ended, and the least-principal-stress direction rotated clockwise about 45° to east-west. Strong tilting of small blocks was replaced by modest tilting of large blocks, resulting in the wide horsts and grabens that form present ranges and basins. Locally derived fanglomerate was deposited in the basins; the basin fill includes basalt flows derived from the lithospheric mantle but no volcanic material derived from the crust, suggesting that the crustal thermal event was waning. The widely separated, but large, faults resulted in interior drainage within the structural basins.

Between ca. 8.5 Ma and 7.5 Ma, tectonism ceased; this allowed the interior-drainage systems to become integrated. Volcanism was represented only by minor megacryst-bearing basalts of asthenospheric-mantle origin, suggesting that the crust and upper mantle had cooled to the point that they no longer harbored magma.

Figure 7 summarizes the middle and late Cenozoic tectonic history of the Castaneda Hills–Signal area, as reconstructed from stratigraphic, structural, and geochemical information. A long period of stability was followed in late Oligocene time by doming (but little or no faulting) in what is now the metamorphic core complex area. Severe extension (HET-type deformation) occurred several million years later and was synchronous with heating of the upper mantle and crust. The extension caused the Rawhide detachment fault to form and at least some mylonitization of lower-plate rocks. This thermotectonic event lasted at most ca. 6 m.y. and was characterized by (1) extreme northeast-southwest extension, which included upper-plate faulting, rotation of blocks, and basin formation; detachment faulting; and at least some ductile deformation in the lower plate; (2) continued rise of the metamorphic core complexes; (3) melting of the lower crust with associated high geothermal gradients; (4) rhyolitic volcanism; and (5) hydrothermal activity that altered and mineralized much upper-plate strata. After this peak of tectonic and volcanic activity, the crust began to cool, and the character of faulting changed to classic basin-range horst-and-graben structure. In the latest Miocene, 5–11 m.y. after HET-type heating, faulting, and volcanism in the Castaneda Hills–Signal area, the middle Miocene thermotectonic event was over.

MODELS OF BASIN-RANGE EXTENSION

Data from the Castaneda Hills–Signal area are useful for testing several models of basin-range extension.

The characteristic horst-and-graben structure of the Great Basin has long attracted attention (e.g., Nolan, 1943, and references cited therein). Basin-range faults were commonly seen as high-angle normal faults that extended with little change in dip to considerable depth, where they died out in a zone of plastic deformation (e.g., models A and B of Stewart, 1978, p. 16).

The pioneering work of Misch (1960) shifted attention to horizontal discontinuities by recognizing the existence of two distinct vertically stacked structural domains with different histories and mechanical behavior. The lower domain ("infrastructure" of Armstrong and Hansen, 1966) was characterized by ductile behavior; the upper one ("suprastructure") by brittle behavior. These domains were separated by a low-angle "denudation fault" (Armstrong, 1972). Lee et al. (1970) and Armstrong (1972) showed that such "denudation" faults could be middle Tertiary rather than Mesozoic in age, as had been suggested by previous workers. Compton et al. (1977) showed that the metamorphic fabric of the "infrastructure" was imprinted on a middle Tertiary pluton.

Geologists' attention shifted to the southern Basin and Range province, which differs from the northern Basin and Range in having few Quaternary faults and thermal springs and a lower average elevation (Eaton, 1982). Anderson (1971) was the first to recognize the "suprastructure" in this region. He described blocks rotated along closely spaced low-angle normal faults that dip in the opposite direction to the facing of strata in the blocks, and suggested that the blocks had rotated along concave-upward (listric) faults that merged downward into

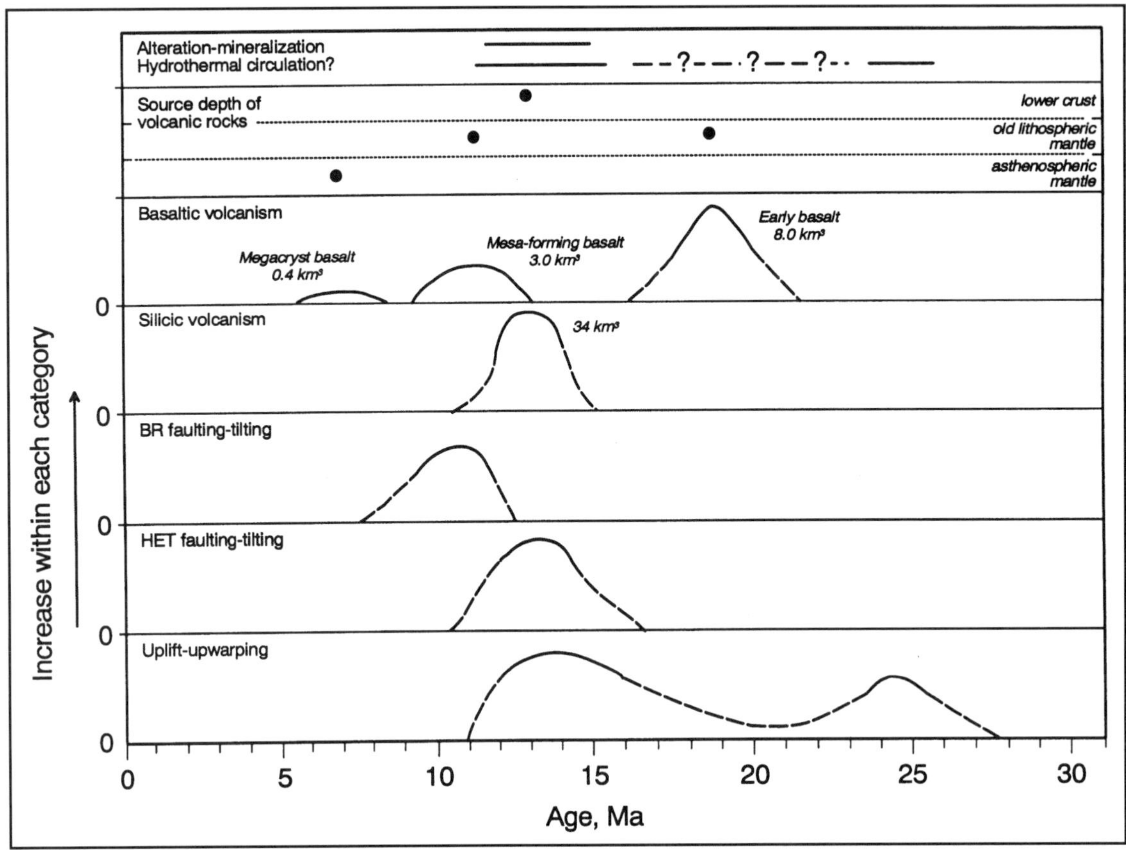

Figure 7. Summary diagram showing timing and relative intensity of various tectonic and igneous (or igneous-related) events, as well as source of volcanic rocks as determined from strontium isotope ratios.

"complex low-angle zones of detachment" (p. 43) or "a single detachment fault" (p. 49). Anderson (1971) also proposed that the detachment fault in the El Dorado Mountains of southern Nevada has at least 15,000 ft (4,500 m) of displacement and separates units that have different structural histories.

Recognition that Anderson's (1971) "thin-skin" style of extensional faulting in the suprastructure was related to a belt of metamorphic core complexes (infrastructure) came in a series of abstracts published by the Geological Society of America in 1977 (Davis, 1977; Rehrig and Reynolds, 1977; Shackelford, 1977), but the issue was controversial at the time. By 1980 (Crittenden et al., 1980), the structural relation between brittlely extended upper-plate rocks and ductilely extended lower-plate rocks was established and previously enigmatic geologic features such as the Death Valley 'turtlebacks' and Amargosa Chaos, younger-over-older "thrusts," large gravity-glide sheets, low-angle to subhorizontal normal faults, and "young" metamorphic rocks were placed into the broader context of southern Basin and Range extension. The resulting explosion of geologic mapping led to the recognition of Cenozoic detachment faults in many places. As mapping progressed, it was realized that many such detachments are part of the now well-known geologic trinity: ductile lower plate, subhorizontal detachment fault, and brittle upper plate cut by many closely spaced subparallel faults that tilt intervening blocks. These features commonly are associated with magmatism (Parsons and Thompson, 1993).

Much debate followed concerning the relative timing of lower-plate mylonitization and upper-plate faulting. One view was that the lower plate deformed in a ductile fashion in response to the extension that also produced the detachment and upper plate faults (e.g., Rehrig and Reynolds, 1980). An alternative was that the mylonitization was considerably older than, and unrelated to, the detachment fault (e.g., Davis et al., 1980).

The upper plate received its share of debate. Some of the early papers that described the closely spaced, low-angle, upper-plate normal faults recognized their extensional nature (e.g., Moores et al., 1968; Anderson, 1971); however, other studies viewed the upper plate as gravity-glide masses, that is to say, nontectonic in origin (e.g., Davis et al., 1980; Shackelford, 1980). Because the Castaneda Hills–Signal area is close to the Colorado Plateau, we recognized that the upper plate of the Buckskin-Rawhide metamorphic core complex was continuous with the Plateau so could not be a gravity-glide sheet as had been held previously. At a U.S. Geological Survey conference in November 1980, we proposed that the lower plate was best viewed as the active ("allochthonous") element and the upper plate the passive element, which failed in tension by shear-coupling to the lower plate across the detachment fault (Lucchitta and Suneson, 1981). Based on our observation that the Rawhide detachment fault dips toward the Colorado Plateau and does not shoal, as would be expected of gravity-glide sheets, we suggested that the Rawhide detachment fault dies out toward the craton through a zone of distributed shear (Fig. 8). According to this concept, lower-plate metamorphism and mylonitization would also decrease to the northeast.

In a 1980 personal communication to Terry Shackelford (cited in Shackelford, 1981), Brian Wernicke proposed that the Rawhide detachment fault extends beneath the Colorado Plateau, that the lower plate of the Buckskin-Rawhide metamorphic core complex is autochthonous, and that the upper plate (including the Colorado Plateau) moved to the northeast. In June 1981, Wernicke (1981) reversed plate movement and suggested that "zones of ductile shear" formed where the lower plate was drawn out from beneath the upper plate. As applied to the Castaneda Hills–Signal area, the 1981 Wernicke model requires the detachment fault to continue under the Colorado Plateau at a moderate dip, perhaps penetrating the entire crust into the upper mantle (Fig. 9). Movement on such a fault would bring originally hot, deep rocks to the surface by a process of simple shear. While appealing in its transcrustal nature, the model does not explain (1) the absence of known active, low-angle faults in modern extending continental orogens (e.g., Jackson, 1987) or (2) the domal character of metamorphic core complexes (e.g., Coney, 1980). In addition, the model suggested that rotational planar normal faults accommodated much, if not most, of the extensional strain in the upper plate (Wernicke and Burchfiel, 1982); this requires uniform upper-plate tilting and basin formation, a feature not observed in regional stratigraphic studies (Nielson and Beratan, this volume). Despite Coney's (1980, p. 20) observation that "the ignimbrite flare-up is very important because the metamorphic core complexes seem to have evolved just prior to and during the ignimbrite outburst," the Wernicke (1981) model ignores possible effects of Cenozoic magmatism on extension in the Basin and Range.

Several papers address problems inherent in the Wernicke (1981) model. Spencer (1984) recognized the role of isostatic uplift in response to tectonic denudation, and Buck (1988) suggested that detachment faults form as high-angle normal faults and are abandoned and subsequently rotated to low angles as extension continues ("rolling hinge" model, Fig. 10A). This model accommodates objections (1) and (2) above. As applied to the Castaneda Hills–Signal area, it requires diachronous movement on listric (as opposed to rotational planar) faults that become older away from the edge of the Colorado Plateau. It also requires that upwarping be late and follow faulting and largely ignores the effects of crustal heating.

Lucchitta and Suneson (1993) addressed the implications of the rotational planar (or "domino") and listric styles of upper-plate faulting. As a first approximation, the domino model requires that all faults be of the same age and all blocks rotate at the same time and by the same amount. Regional variations in the filling of structural basins and differences in the age of unconformities should be minor or absent (Fig. 10B). In contrast, the listric-fault model for upper-plate structure, as originally proposed by Anderson (1971), allows diachroneity of faulting and regional variations in basin filling (Fig. 10C). Lucchitta and Suneson (1993) noted variations in basin-filling histories in the Castaneda Hills–Signal area, and Beratan and Nielson (this volume) made the same observation across the entire central part of the Colorado River extensional corridor.

Our data from the Castaneda Hills–Signal area show that faulting is diachronous and becomes younger toward the Colorado Plateau and that basins have variable sedimentary histories (Lucchitta and Suneson, 1993). Our data also show that differential uplift preceded the onset of volcanism and HET-type faulting (Figs. 3, 6, and 7). These characteristics contradict the requirements of the rolling-hinge model.

Implications of the various models are listed in Table 1.

CONCLUSIONS AND PREFERRED MODEL

Our data and those of other workers in nearby areas show that (1) HET-type faulting was diachronous, (2) the ages of unconformities vary, and (3) upwarping preceded faulting and volcanism. These characteristics support the listric model for upper-plate structure, but not the rotational planar (domino)

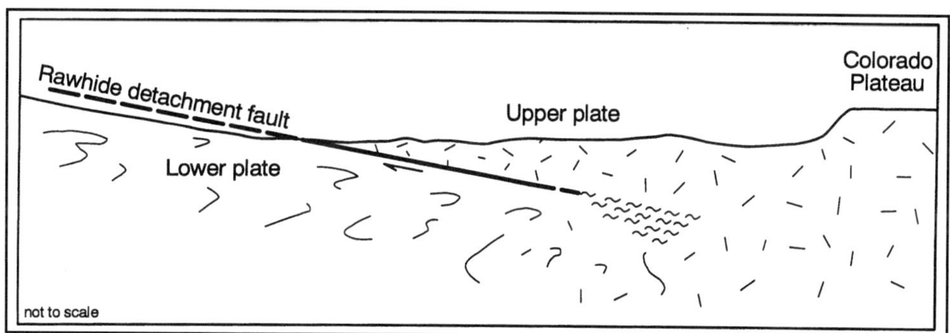

Figure 8. Inferred relation between Rawhide detachment fault and Colorado Plateau (not to scale).

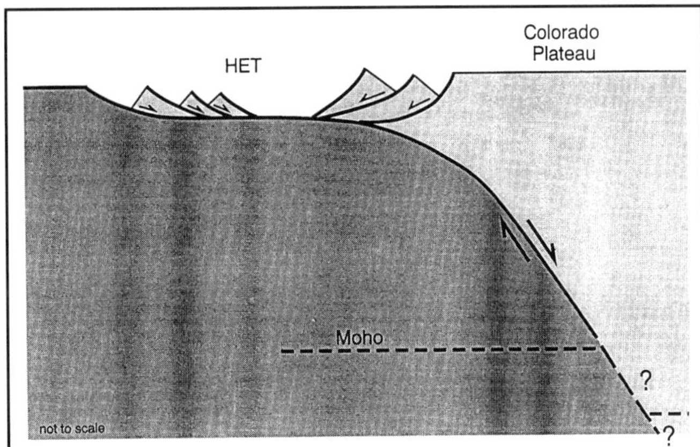

Figure 9. Simple-shear model for relation between detachment fault and Colorado Plateau (not to scale).

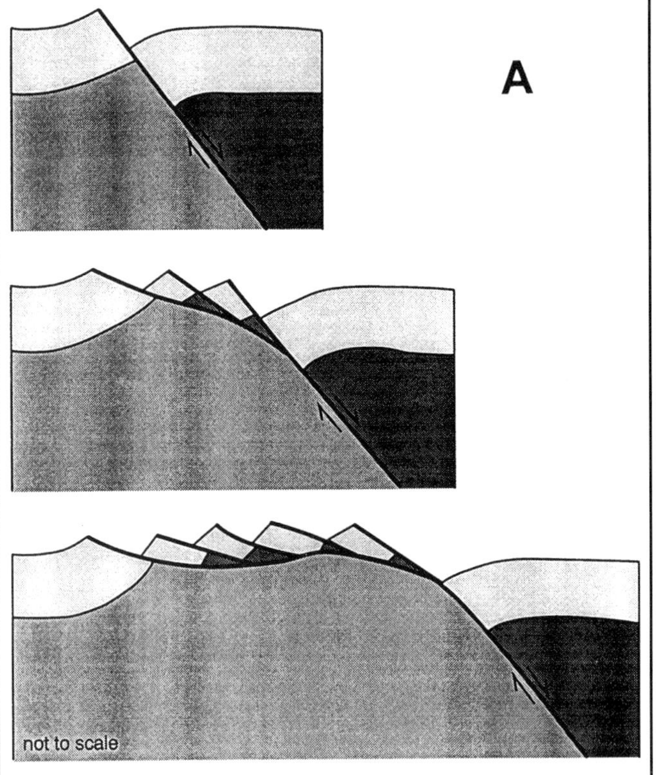

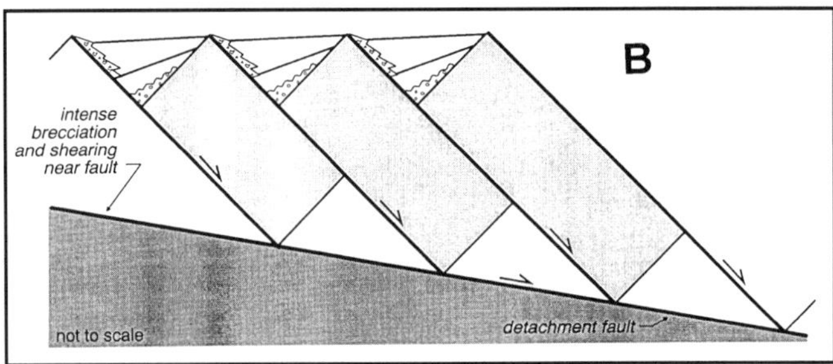

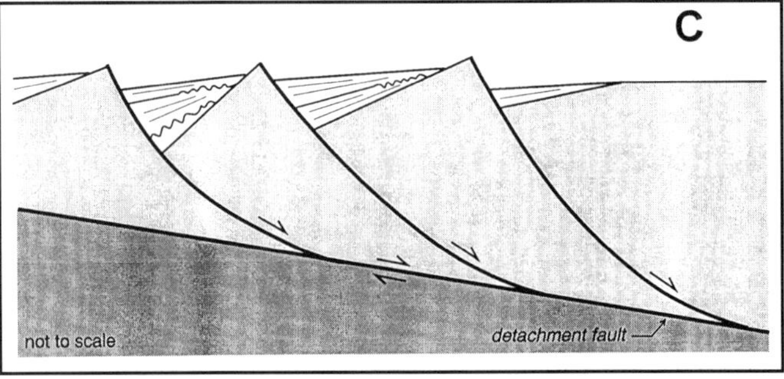

Figure 10. Models for upper-plate structure (not to scale). (A) Rolling-hinge, showing origin of low-angle detachment fault as high-angle normal fault. Thinned footwall rises isostatically. New prolongation of high-angle fault is created when older part of the fault has flattened to a position incapable of accommodating further strain. (B) Domino. Implications of model are that all faults must move at the same time and that sediments deposited in resulting troughs should be similar and of the same age. (C) Listric. Implications of model are that faults and deposits can be diachronous and that deposits in troughs can differ in character and age.

model; furthermore, they do not support the simple-shear model for crustal extension, nor its derivative, the rolling-hinge model, which requires faulting to precede uplift. In our opinion, thermally driven mechanisms involving predominantly pure shear are more likely (Fig. 11).

The role of magmatism is of critical importance to the origin of the HETs and metamorphic core complexes, but has received surprisingly little attention. Coney (1980, p. 20) recognized that a "massive thermal disturbance . . . caused much low-angle normal faulting." Eaton (1982, p. 412) was even clearer: "In places, the two features went hand-in-hand, listric faulting occurring in a regional extensional stress field in the shallow crust above passively emplaced plutons or in association with major caldera collapse. In other localities, maximum episodes of fault movement occurred during apparent peaks in igneous activity." More recently, Parsons and Thompson (1993, p. 247) observed that "low-angle, core-complex-style faulting almost never occurs without accompanying magmatism." Crustal heating as an essential component of HET-type deformation is evidenced in the Castaneda Hills–Signal area by the start of block tilting shortly after the 19 Ma mantle-derived basaltic volcanism; coincidence of severe (peak?) faulting and maximum crustal heating (rhyolites derived from the lower crust) at ca. 14–12 Ma; and simultaneous cessation of faulting and most volcanism at ca. 8 Ma (Fig. 7; Lucchitta, 1990). Additional evidence for heating during HET-type faulting includes the hydrothermal alteration and mineralization in rocks that were deformed by HET-type faults (sequence I), but not the younger faults; the presence of gravity-glide sheets that probably were emplaced along low slopes, suggesting that unusual mechanisms such as hydrothermally induced pore overpressure were required (Lucchitta, 1990); the mylonitic fabric recorded in some Tertiary plutons in the lower plate; and the common middle Tertiary isotopic and fission-track ages of the lower-plate rocks. Evidence that crustal cooling influences upper-plate faulting includes the change to basin-range-type tectonism with time and its eventual cessation, as well as the migration in time of volcanism and tectonic activity northeast toward and onto the Colorado Plateau. The model proposed in this chapter is constrained by the geochronologic data on lower-plate rocks observed by other workers in the area (Bryant et al., 1991; Foster et al., 1993; Richard et al., 1990), in particular, the middle Miocene (16–10 Ma), eastward-younging cooling ages (zircon and apatite fission tracks, feldspar and biotite ^{40}Ar-^{39}Ar, biotite K-Ar), and the history of upper-plate mineralization and potassium-metasomatism (Spencer et al., 1989a; Spencer et al., 1989b; Spencer and Welty, 1989).

TABLE 1. COMPARISON BETWEEN CHARACTERISTICS REQUIRED BY THE VARIOUS MODELS, AND FEATURES OBSERVED IN THE FIELD*

Models	Movement History of Various Basins		Younging Direction		Sedimentary Fill		Time of Upwarping	
	Required	Observed	Required	Observed	Required	Observed	Required	Observed
Listric	Diachronous ok	Diachronous		Younger NE	Variable	Variable		Early
Domino	Synchronous				Similar			
Simple shear	Diachronous ok						Late	
Rolling hinge	Diachronous		Younger NE				Late	
Thermal	Diachronous ok		Younger NE best				Early	

*Blank boxes = No requirement or does not apply.

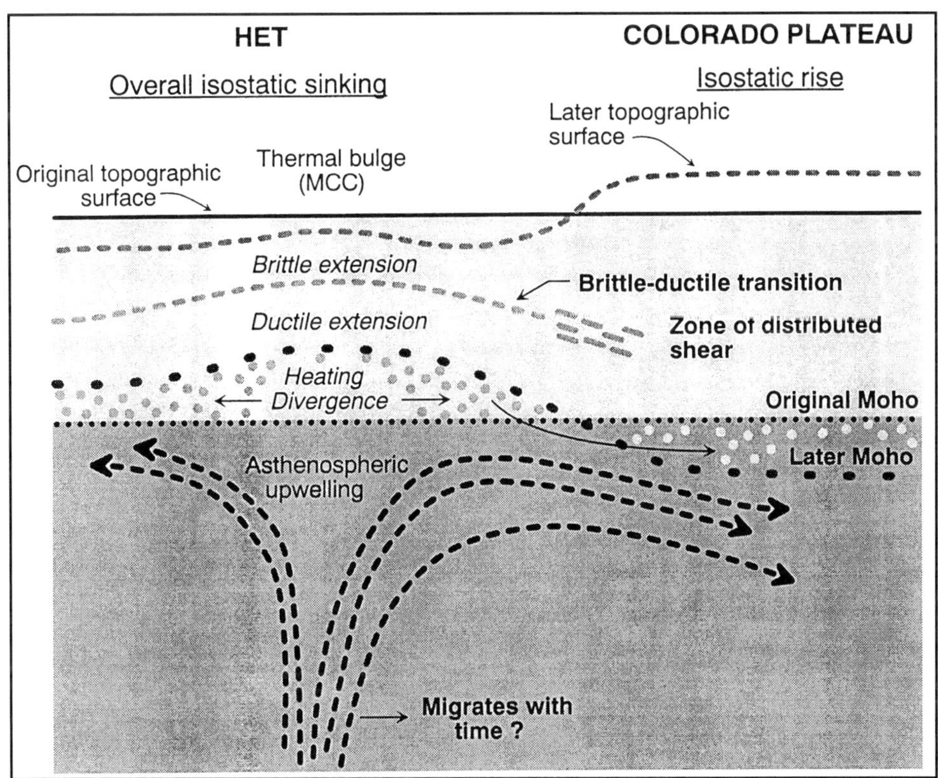

Figure 11. Thermal model showing relation of lithospheric processes to asthenospheric upwelling.

Thermotectonic model

In the following discussion, Figure 12 presents our interpretation of tectonic events in the upper crust, whereas Figure 13 provides a lithospheric overview. Both are modified from the thermotectonic model presented by Lucchitta (1990).

In the early Tertiary, the Castaneda Hills–Signal area was tectonically quiescent (Figs. 12a and 13a); a southwest-sloping deeply weathered surface that developed mostly on the Proterozoic Signal Granite was dissected by braided streams carrying pebbles, cobbles, and locally boulders derived from (the Mogollon) highlands to the northeast. Beginning in the late Oligocene, hot asthenosphere (mantle plume?) began to heat the overlying lithosphere in the general area of the Buckskin-Rawhide metamorphic core complex and HET of the Colorado River extensional corridor (Figs. 12b and 13b). The heating produced a decrease in density and a thermal expansion that caused parts of the crust to rise. Inhomogeneities in the heating or temperature anomalies residual from the Laramide event may have localized the rising areas (Fig. 12a). As the heating continued, the upper mantle partially melted; basalt rose through the crust that was being heated and was beginning to fail owing to tensile stresses being applied to its base by the divergence at the top of the plume (Figs. 12c and 13c). (Alternatively, extension within the crust may have been caused by the migration of the Mendocino triple junction [Glazner and Bartley, 1984] or some other poorly understood plate interaction. Regardless, the time-space association of heating and extension is clear in the Castaneda Hills–Signal area.) As heating and extension continued, rocks hotter than about 250–300 °C deformed ductilely at geologically significant rates (Lynch and Morgan, 1987), with strain possibly concentrated in anastomosing shear zones that collectively formed megaboudins (Fig. 12d). Overlying (and cooler) rocks deformed brittlely, and the zone between megaboudins and upper plate was severely deformed and mylonitized and represented a ductile-deformation zone. In effect, the mylonite zone marks the 250–300 °C isotherm, and the listric faults that accommodated extension in the brittle zone merged with the mylonites (e.g., Lister and Davis, 1989, fig. 20a). This mylonite zone is subhorizontal because the intrusion of plutons into the middle or upper crust changes the stress field orientation so that low-angle faults are more likely to accommodate the strain over the area of "thermal softening" (Parsons and Thompson, 1993, p. 248).

In the Castaneda Hills–Signal area, the paroxysmal phase of HET-type faulting occurred when the lower crust reached temperatures sufficiently high for partial melting; the 250–300 °C isotherm rose relatively close to the surface as silicic magma was intruded into the middle and upper crust (Figs. 12e and 13d). With time, the isotherm marking the transition (mylonite zone) came close enough to the topographic surface to interact with ground water and hydrothermal convective systems were established (Fig. 12d). Hydrothermal circulation resulted in a steep thermal front and allowed rocks with temperatures in the 250–300 °C range to be close (<2 km?) to the topo-

Figure 12. Proposed model for events in the middle to upper crust during evolution of the HET.

(a) Pretectonic time. 1. Well-developed regolith indicates low relief and tectonic stability. 2. Faults, if any are inactive. 3. Crust is cool. 250–300 °C isotherm is deep in the crust. 4. Isotherms may show subdued culminations reflecting relict heat from Laramide tectonism.

(b) Updoming time. 1. Beginning of heating, possibly accentuated in areas of relict heat, and possibly accompanied by beginning of intrusion of basaltic dikes in lower crust. 2. Isotherms rise in hotter areas, accompanied by thermal uplift of upper crust. 3. In areas of highest geothermal gradient, isotherms impinge on ground water, setting up a convective circulation system and resulting in a steep thermal front. 4. Pore overpressure associated with convective system, together with tensile stresses caused by updoming, causes failure of surface layers through hydrofracturing. Resulting breccias slide downslope.

(c) Early basalt time. 1. Widespread eruption of basalt. No evidence of faulting at surface. 2. Crust continues heating. 3. Continued interaction with ground water causes convective systems and steep thermal front to continue. 4. Former uplifted areas lowered through combination of erosion and sinking caused by thinning of crust.

(d) Rhyolite time (early phase). 1. Magma rises to middle crust. Overlying isotherms rise and are compressed against ground water. Isotherms on northeast side of northeast-migrating thermal disturbance (leading edge) are closer together and steeper than on southwest trailing edge. Intrusion of magma bodies causes middle crust to extend. 2. Extension below the 250-300 °C isotherm occurs by ductile shear along zones, forming megaboudins. 3. Rocks near 250–300 °C isotherm and within thermal front above rising magma fail and form major zone of ductile shear. 4. Rocks above the 250–300 °C isotherm fail in brittle fashion along faults that rotate intervening blocks and extend down to zone of ductile shear. 5. Vigorous hydrothermal circulation continues. 6. Hydrothermal circulation produces pore overpressure that causes surface rocks to "float" and slide down topographic slope as gravity-glide sheets.

(e) Rhyolite time (peak phase). 1. Rhyolite magma intrudes upper crust in quantity and vents to the surface. Intrusion causes (and/or reflects) extension. 2. Locus of maximum heating shifts northeast and is located above intruding magma. At thermal culmination, isotherms are compressed against ground water, forming steep thermal front. 250–300 °C isotherm may be 2 km or less below topographic surface. Isotherm marked by zone of ductile shear, which "consumes" base of rotated blocks formed earlier. 3. Vigorous hydrothermal convection above culmination. 4. At trailing edge of thermal disturbance, rocks cool. Rocks previously at or below the 250–300 °C isotherm are now above it and begin to fail in brittle fashion, forming a detachment fault, which preferentially, but not necessarily, follows thermally weakened rocks of ductile shear zone. 5. At leading edge, isotherms are closely spaced because of northeast migration of thermal disturbance. 6. Certain areas continue to rise owing to combination of higher heat and crustal underplating. Hydrothermal convection continues. Pore overpressure results in sliding of gravity-glide sheets down the topographic surface.

(f) Rhyolite time (waning phase, transitional to BR). 1. Intrusion of rhyolitic magma shifts northeast. 2. Culmination of 250–300 °C isotherm shifts northeast, and detachment fault extends itself northeast, with decreasing displacement. At dot, fault grades laterally into zone of shear. 3. Along the trailing edge, isotherms and zone of ductile shear sink. 4. "Fossil" zones of ductile shear combine into broad shear zone. 5. Continued but decreasing extension is accommodated by high-angle normal faults that cut detachment and "bottom" in deep zone of ductile shear, so are widely spaced, have major structural relief, and show minor rotation. 6. Erosion has cut through oldest part of detachment fault. 7. Late-phase syntectonic deposits blanket detachment fault locally.

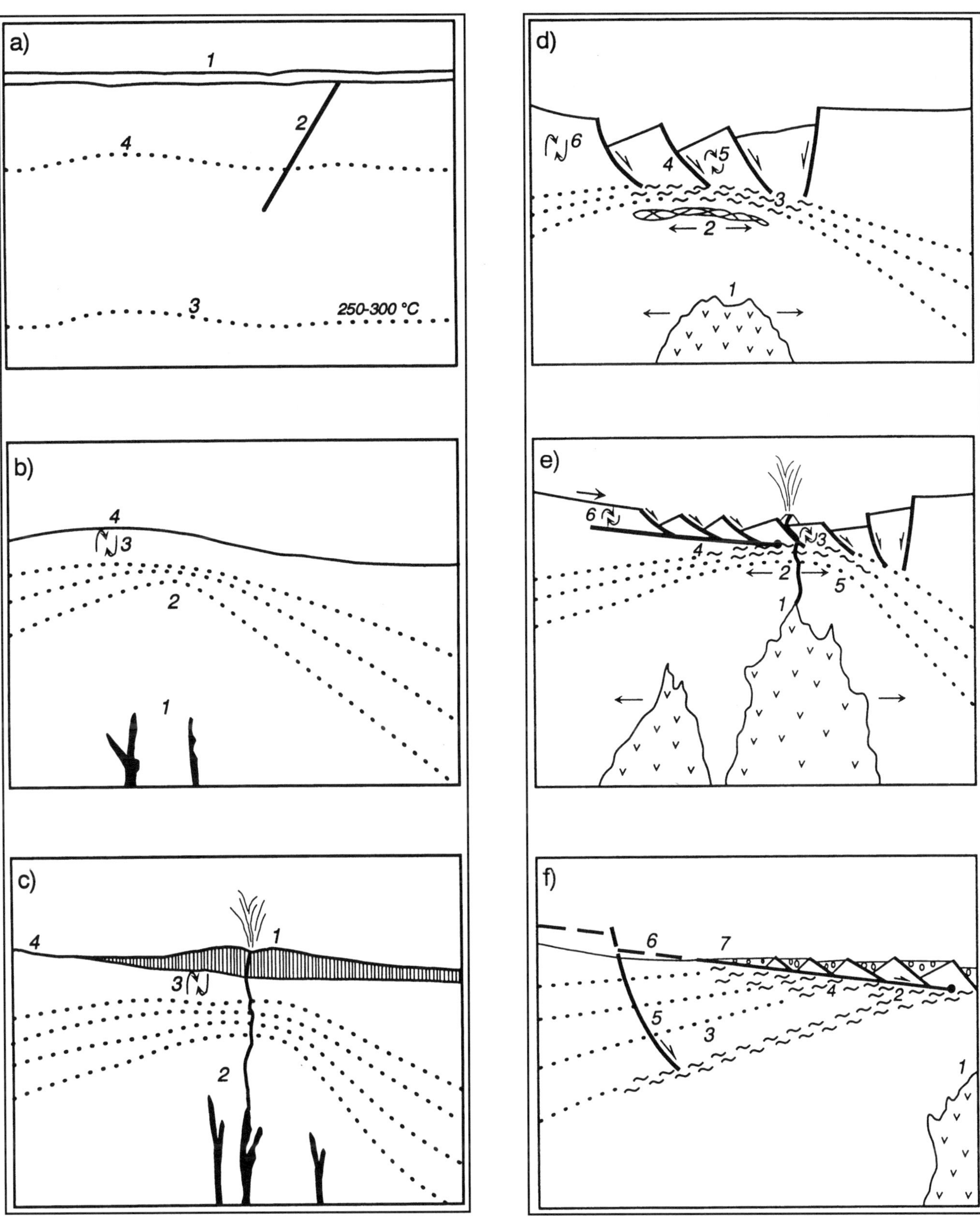
a)
1
2
4
3
250-300 °C
b)
4
3
2
1
c)
4
1
3
2
d)
6
4
5
3
2
1
e)
6
3
4
2
5
1
f)
6
7
4
2
5
3
1

graphic surface. Surface manifestations of the hydrothermal circulation were widespread alteration, mineralization, and potassium-metasomatism. The Rawhide detachment fault formed at the site of the slightly cooled, but thermally weakened, brittle-ductile transition behind the locus of maximum heating, which was migrating to the northeast (Fig. 12e). In effect, the tip of the detachment fault migrated as the dome in the 250–300 °C isotherm migrated. Because the fault was at or near the brittle-ductile transition, it probably was not seismogenic. As extension continued on both sides of the axis of maximum heating, blocks between HET-type faults were rotated to progressively steeper dips.

Eventually, the thermal pulse waned. As the crust cooled, continued extension was initially accommodated by movement on a subhorizontal detachment fault that formed at the top of the previous ductile-deformation zone and associated mylonites, where the HET-type faults terminated (Fig. 12f). At ca. 12 Ma, silicic volcanism largely ceased, mantle-derived basalt erupted, the least-principal stress direction rotated to east-west, and the crust became sufficiently cool that basin-range faults soled at a much deeper decoupling horizon, with the consequence that the faults were more widely spaced and stratal rotation was less than for HET faults (Figs. 12f and 13e). By ca. 8 Ma, the area had cooled completely and tectonism ceased (Fig. 13f).

As shown by Sass et al. (1994), the metamorphic core complex terrane is marked by lower heat flow than surrounding areas: metamorphic core complex terrane, 67 mW/m^2; southern Basin and Range province, 83 mW/m^2; Arizona transition zone, 86 mW/m^2. This is a puzzling situation considering that the area of the metamorphic core complexes seems to have been so hot at ca. 15 Ma. Conductive processes alone would not be enough to bring about such a degree of cooling in this time interval. We suggest that the low heat flow may result partly from the efficient removal of heat by convection in the form of silicic and basaltic volcanic rocks and hydrothermal systems, partly from enhanced heat flow resulting from stretching and thinning of the crust (which is a form of convection), and partly from depletion of radioactive elements from lower-plate crustal rocks through leaching by the hydrothermal solutions during HET time. Radioactive elements are now disseminated in the sequence I basin beds.

The cause of the cessation of heating in the Castaneda Hills–Signal area in particular, and the metamorphic core complexes in general, may be that the asthenospheric plume migrated away from the area of the complexes, in our case toward the Colorado Plateau, where it is manifested by heating, volcanism, and faulting that become younger to the northeast. The older features of this kind are now in the transition zone of Arizona; the younger ones are manifested by the young volcanic fields that rim the southern edge of the Plateau and by the seismic activity that is going on to this day. Migration of the subcrustal heat source away from the MCC would also help explain rapid cooling of those areas.

The thermotectonic model presented here takes into account the geologic features observed in the area between the Buckskin-Rawhide metamorphic core complex and the nearby Colorado Plateau, and attempts to explain the succession of tectonic and volcanic events that can be inferred for this area. The model has several implications that differ substantially from those in which crustal heating is not a factor:

- HET-type upper- and lower-plate extension occurred in areas of thermally weakened crust.
- Pure shear was the dominant deformational mechanism; the mylonitic gneisses of the Buckskin-Rawhide metamorphic core complex were not drawn out from beneath the Colorado Plateau.
- Extension of the lower plate was accompanied by intrusion of mostly silicic plutons with local eruption of associated rhyolitic lavas.
- The upper crust was sufficiently hot during silicic magmatism to allow rocks relatively near the surface to deform ductilely, which occurred at geologically significant rates at temperatures as low as 250–300 °C; the HET-type faults that

Figure 13. Proposed tectonic model for evolution of HET: a lithospheric perspective.

(a) Pretectonic time. 1. Ground surface. 2. Relatively thick crust inherited from Laramide event. 3. Lithospheric mantle. 4. Asthenospheric mantle.

(b) Updoming time. Upwelling of hot asthenospheric mantle heats lithospheric mantle, which in turn heats base of crust, causing decrease in density and thermally driven uplift.

(c) Early basalt time. 1. Continued upwelling of hot asthenospheric mantle. Upwelling axis moves northeast with time. 2. Lithospheric mantle hot enough to melt locally, producing early basalt. 3. Base of crust thinned either by phase changes, or subcrustal transfer, or both. Uplift temporarily ceases owing to thinning of crust. 4. Crust still cool enough to fail in tension, allowing basaltic magmas to rise to surface.

(d) Rhyolite time. 1. Locus of upwelling of hot asthenospheric mantle moves northeast. 2. Base of crust continues to thin owing to phase changes and/or subcrustal transfer. Displaced crustal material begins to accumulate to northeast (under Colorado Plateau). 3. Lower crust hot enough to melt locally. Rhyolitic and basaltic magmas reside in lower and middle crust. 4. Crust hot enough to deform ductilely even at high levels. Rhyolitic melts can rise through hot crust (even though it is deforming ductilely) because of density contrast. Basaltic melts are mostly prevented from rising. 5. Thinning of crust causes lowering of topographic surface, but especially hot areas form thermal blisters with positive relief.

(e) Basin-range time. 1. Asthenospheric upwelling weakens while migrating northeast. 2. Crustal thinning decreases and shifts northeast. 3. Crust cools, deforming in brittle fashion and allowing basaltic magmas (mesa-forming basalt) to rise from lithospheric mantle. Rhyolitic melts no longer reside in lower and middle crust. 4. Areas to northeast begin to collapse owing to thinning of crust in that area (Arizona transition zone).

(f) Post-tectonic time. 1. Hot asthenospheric upwelling no longer present. 2. Lithospheric mantle cooled. 3. Crust thin and cool. 4. Brittle crust (and lithospheric mantle?) allow melts to rise from the asthenospheric mantle (megacryst basalt).

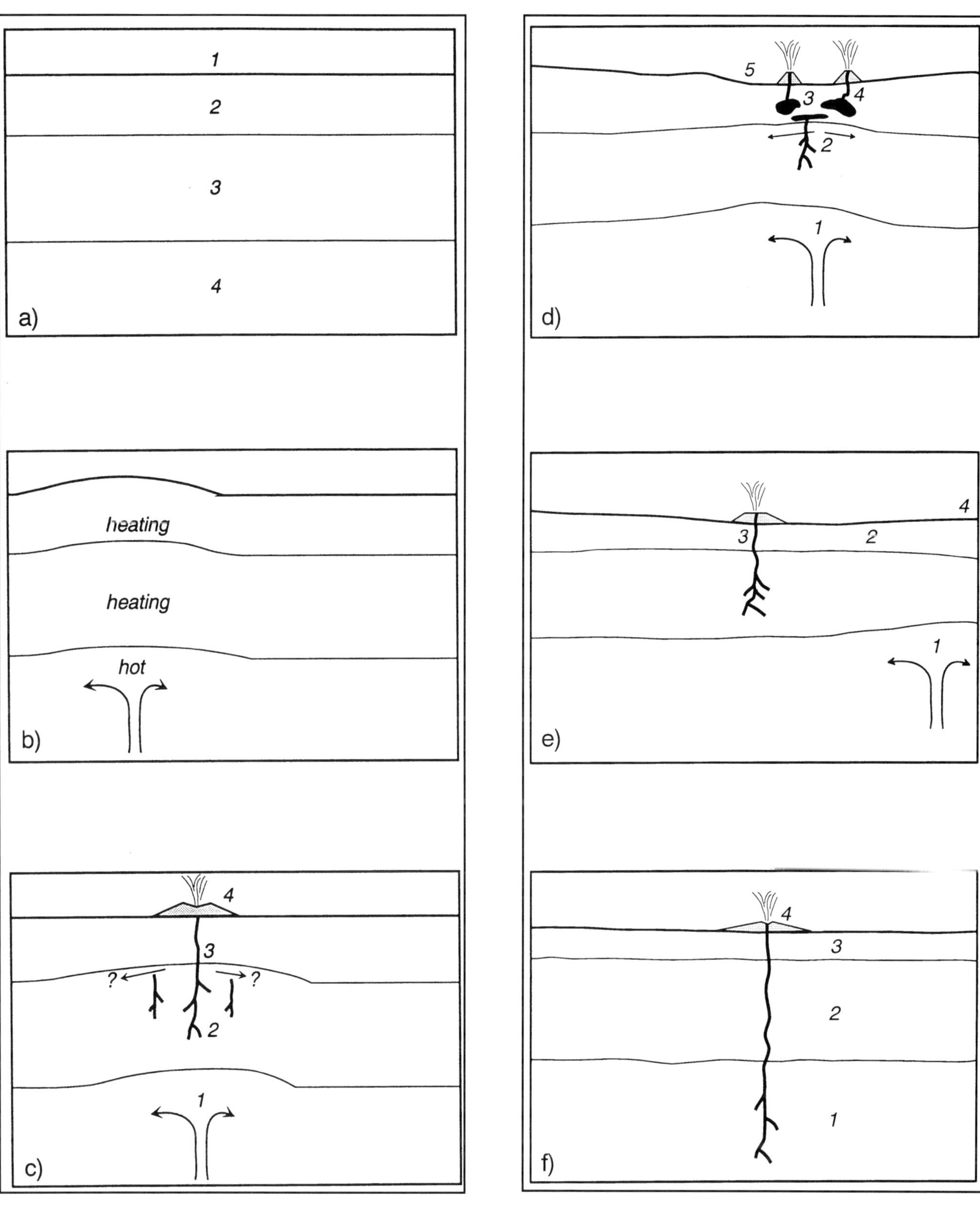
1
2
3
4
a)
heating
heating
hot
b)
4
3
?
?
2
1
c)
5
3
4
2
1
d)
4
3
2
1
e)
4
3
2
1
f)

accommodated extension in the brittle rocks above the ductile zone soled in a zone of distributive shear (mylonite zone).

• With cooling, the mylonite zone started to deform brittlely and became a detachment fault. The upper-plate faults, which previously merged into the mylonite zone, now soled into the detachment fault. Although regional in extent, this fault has variable and not necessarily large offset across it.

The model predicts that the timing and character of MCC deformation (including mylonitization, low-angle faulting, and detachment faulting) evolve in space and time as crustal heating begins, wanes, ceases, and/or migrates.

ACKNOWLEDGMENTS

Careful reviews by Wendell Duffield, U.S. Geological Survey, and Paul Morgan, Northern Arizona University, have greatly improved both the scientific lucidity and the clarity of expression is chapter, as have an edit by Kathi Beratan of Pittsburgh University and perceptive comments by Jane Nielson of the U.S. Geological Survey. To all these individuals we owe sincere thanks.

The work that led to the paper was supported by the Reactor Hazards, Pacific-Arizona Crustal Experiment (PACE) and National Geologic Mapping programs of the U.S. Geological Survey. Suneson's participation was supported in part by the Oklahoma Geological Survey.

REFERENCES CITED

Anderson, R. E., 1971, Thin skin distension in Tertiary rocks of southeastern Nevada: Geological Society of America Bulletin, v. 82, p. 43–58.

Armstrong, R. L., 1972, Low-angle (denudation) faults, hinterland of the Sevier orogenic belt, eastern Nevada and western Utah: Geological Society of America Bulletin, v. 83, p. 1729–1754.

Armstrong, R. L., and Hansen, E., 1966, Cordilleran infrastructure in the eastern Great Basin: American Journal of Science, v. 264, p. 112–127.

Bryant, B., and Wooden, J. L., 1989, Lower-plate rocks of the Buckskin Mountains, Arizona: A progress report: *in* Spencer, J. E., and Reynolds, S. J., eds., Geology and mineral resources of the Buckskin and Rawhide Mountains, west-central Arizona: Arizona Geological Survey Bulletin 198, p. 47–50.

Bryant, B., Naeser, C. W., and Fryxell, J. E., 1991, Implications of low-temperature cooling history on a transect across the Colorado Plateau—Basin and Range boundary, west central Arizona: Journal of Geophysical Research, v. 96, p. 12,375–12,388.

Buck, W. R., 1988, Flexural rotation of normal faults: Tectonics, v. 7, p. 959–975.

Chamberlin, R. M., 1983, Cenozoic domino-style faulting in the Lemitar Mountains, New Mexico: A summary: Socorro, New Mexico Geological Society, 34th Field Conference, Region II, Guidebook, p. 111–118.

Compton, R. R., Todd, V. R., Zartman, R. E., and Naeser, C. W., 1977, Oligocene and Miocene metamorphism, folding, and low-angle faulting in northwestern Utah: Geological Society of America Bulletin, v. 88, p. 1237–1250.

Coney, P. J., 1980, Cordilleran metamorphic core complexes: an overview: *in* Crittenden, M. D., Jr., Coney, P. J., and Davis, G. H., eds., Cordilleran metamorphic core complexes: Geological Society of America Memoir 157, p. 7–31.

Cooley, M. E., and Davidson, E. S., 1963, The Mogollon Highland—Their influence on Mesozoic and Cenozoic sedimentation: Arizona Geological Society Digest, v. 6, p. 7–35.

Crittenden, M. D., Jr., Coney, P. J., and Davis, G. H., 1980, Cordilleran metamorphic core complexes: Geological Society of America Memoir 153, 490 p.

Davis, G. A., Anderson, J. L., Frost, E. G., and Shackelford, T. J., 1980, Mylonitization and detachment faulting in the Whipple–Buckskin–Rawhide Mountains terrane, southeastern California and western Arizona: *in* Crittenden, M. D., Jr., Coney, P. J., and Davis, G. H., eds., Cordilleran metamorphic core complexes: Geological Society of America Memoir 153, p. 79–129.

Davis, G. H., 1975, Gravity-induced folding off a gneiss dome complex, Rincon Mountains, Arizona: Geological Society of America Bulletin, v. 86, p. 979–990.

Davis, G. H., 1977, Characteristics of metamorphic core complexes, southern Arizona: Geological Society of America Abstracts with Programs, v. 9, p. 944.

Eaton, G. P., 1982, The Basin and Range province: Origin and tectonic significance: Annual Review of Earth and Planetary Sciences, v. 10, p. 409–440.

Eberley, L. D., and Stanley, Y. B., Jr., 1978, Cenozoic stratigraphy and geologic history of southwestern Arizona: Geological Society of America Bulletin, v. 89, p. 921–940.

Foster, D. A., Gleadow, A. J. W., Reynolds, S. J., and Fitzgerald, P. G., 1993, Denudation of metamorphic core complexes and the reconstruction of the transition zone, west central Arizona: Constraints from apatite fission track thermochronology: Journal of Geophysical Research, v. 98, p. 2167–2185.

Gans, P. B., and Miller, E. L., 1983, Style of mid-Tertiary extension in east-central Nevada: Utah Geological and Mineralogical Survey Special Studies 59, p. 107–139.

Gilbert, G. K., 1875, Report on the geology of portions of Nevada, Utah, California, and Arizona examined in the years 1871 and 1872: Report, U.S. Geographic and Geological Surveys West of the One Hundredth Meridian (Wheeler), v. 3, p. 17–187.

Glazner, A. F., and Bartley, J. M., 1984, Timing and tectonic setting of Tertiary low-angle normal faulting and associated magmatism in the southwestern U.S.: Tectonics, v. 3, p. 385–396.

Gray, F., Miller, R. J., Pitkin, J. A., Bagby, W. C., Hassemer, J. R., McCarthy, J. H., Jr., Hanna, W. F., Callas, M. L. C., and Lane, M. E., 1989, Mineral resources of the Arrastra Mountain/Peoples Canyon wilderness study area, La Paz, Mohave, and Yavapai Counties, Arizona: U.S. Geological Survey Bulletin 1701, 24 p.

Hamilton, W., 1988, Detachment in the Death Valley region, California: U.S. Geological Survey Bulletin 1790, p. 51–85.

Hauser, E. C., Gephart, J., Latham, T., Oliver, J., Kaufman, S., Brown, L., and Lucchitta, I., 1987, COCORP Arizona transect: Strong crustal reflections and offset Moho beneath the Transition Zone: Geology, v. 15, p. 1103–1106.

Howard, K. A., and John, B. E., 1987, Crustal extension along a rooted system of imbricate low-angle faults: Colorado River extensional corridor, California and Arizona, *in* Coward, M. P., Dewey, J. F., and Hancock, P. L., eds., Continental extensional tectonics: London, Geological Society of London Special Publication 28, p. 299–311.

Howard, K. A., Nielson, J. E., Wilshire, H. G., Nakata, J. K., Goodge, J. W., Reneau, S. L., John, B. E., and Hansen, V. L., 1990, Preliminary map of the Mohave Mountain area, Mohave County, western Arizona: U.S. Geological Survey Open-File Report 90-684, scale 1:48,000.

Jackson, J. A., 1987, Active normal faulting and crustal extension, *in* Coward, M. P., Dewey, J. F., and Hancock, P. L., eds., Continental extensional tectonics: London, Geological Society of London Special Publication No. 28, p. 3–17.

King, C., 1878, Systematic geology: Report U.S. Geological Exploration 40th Parallel, Vol. 1: Washington, D.C., Government Printing Office, 803 p.

Lasky, S. G., and Webber, B. N., 1949, Manganese resources of the Artillery Mountains region, Mohave County, Arizona: U.S. Geological Survey Bulletin 961, 86 p.

Lee, D. E., Marvin, R. F., Stern, T. W., and Peterman, Z. E., 1970, Modification of potassium-argon ages by Tertiary thrusting in the Snake Range, White Pine County, Nevada: U.S. Geological Survey Professional Paper 700-D, p. D92–D102.

Lister, G. S., and Davis, G. A., 1989, The origin of metamorphic core complexes and detachment faults formed during Tertiary continental extension in the northern Colorado River region, U.S.A.: Journal of Structural Geology, v. 11, p. 65–93.

Longwell, C. R., 1945, Low-angle normal faults in the Basin and Range province: Transactions of the American Geophysical Union, v. 26, p. 107–118.

Lucchitta, I., 1990, Role of heat and detachment in continental extension as viewed from the eastern Basin and Range province in Arizona: Tectonophysics, v. 174, p. 77–114.

Lucchitta, I., 1993, Foreword, *in* Sherrod, D. R., and Nielson, J. E., eds., Tertiary stratigraphy of highly extended terranes, California, Arizona, and Nevada: U.S. Geological Survey Bulletin 2053, p. iii–iv.

Lucchitta, I., and Suneson, N. H., 1981, Observations and speculations regarding the relations and origins of mylonitic gneiss and associated detachment faults near the Colorado Plateau boundary in western Arizona: U.S. Geological Survey Open-File Report 81-503, p. 53–55.

Lucchitta, I., and Suneson, N. H., 1982, Signal granite (Precambrian), west-central Arizona: U.S. Geological Survey Bulletin 1529-H, p. H87–H90.

Lucchitta, I., and Suneson, N. H., 1983, Mid- and late Cenozoic extensional tectonism near the Colorado Plateau boundary in west-central Arizona: Geological Society of America Abstracts with Programs, v. 15, no. 5, p. 405.

Lucchitta, I., and Suneson, N. H., 1993, Dips and extension: Geological Society of America Bulletin, v. 105, p. 1346–1356.

Lucchitta, Ivo, and Suneson, N.H., 1994a, Geologic map of the Castaneda Hills quadrangle, Mohave County, Arizona: U.S. Geological Survey Geologic Quadrangle Map GQ 1720, scale 1:24,000.

Lucchitta, I., and Suneson, N. H., 1994b, Geologic map of the Castaneda Hills Southwest quadrangle, Mohave and La Paz Counties, Arizona: U.S. Geological Survey Geologic Quadrangle Map GQ 1719, scale 1:24,000.

Lucchitta, I., and Suneson, N. H., 1994c, Geologic map of the Centennial Wash quadrangle, Mohave and La Paz Counties, Arizona: U.S. Geological Survey Geologic Quadrangle Map GQ 1718, scale 1:24,000.

Lucchitta, I., and Suneson, N. H., 1994d, Geologic map of the Signal quadrangle, Mohave County, Arizona: U.S. Geological Survey Geologic Quadrangle Map GQ 1709, scale 1:24,000.

Lynch, H. D., and Morgan, P., 1987, The tensile strength of the lithosphere and the localization of extension, *in* Coward, M. P., Dewey, J. F., and Hancock, P. L. eds., 1987, Continental Extension Tectonics: London, Geological Society of London Special Publication 28, p. 53–65

McCarthy, J., Larkin, S. P., Fuis, G. S., Simpson, R. W., and Howard, K. A., 1991, Anatomy of a metamorphic core complex: seismic refraction/wide-angle reflection profiling in southeastern California and western Arizona: Journal of Geophysical Research, v. 96, p. 12,259–12,291.

Misch, P., 1960, Regional structural reconnaissance in central-northeast Nevada and some adjacent areas: observations and interpretations: Intermountain Association of Petroleum Geologists, 11th Annual Field Conference, Guidebook, p. 17–42.

Moores, E. M., Scott, R. B., and Lumsden, W. W., 1968, Tertiary tectonics of the White Pine–Grant Range region, east-central Nevada, and some regional implications: Geological Society of America Bulletin, v. 79, p. 1703–1726.

Nielson, J. E., and Beratan, K. K., 1990, Tertiary basin development and tectonic implications, Whipple detachment system, Colorado River extensional corridor, California and Arizona: Journal of Geophysical Research, v. 93, p. 599–614.

Nolan, T. B., 1943, The Basin and Range province in Utah, Nevada, and California: U.S. Geological Survey Professional Paper 197-D, p. 141–196.

Parsons, T., and Thompson, G. A., 1993, Does magmatism influence low-angle normal faulting?: Geology, v. 21, p. 247–250.

Powell, J. W., 1877, Report of the Geographical and Geological Survey of the Rocky Mountain region: Government Printing Office, Washington, D.C., 19 p.

Proffett, J. M., Jr., 1977, Cenozoic geology of the Yerington district, Nevada, and implication for the nature and origin of Basin and Range faulting: Geological Society of America Bulletin, v. 88, p. 247–266.

Rehrig, W.A., and Reynolds, S.J., 1977, A northwest zone of metamorphic core complexes in Arizona: Geological Society of America Abstracts with Programs, v. 9, p. 1139.

Rehrig, W. A., and Reynolds, S. J., 1980, Geologic and geochronological reconnaissance of a northwest-trending zone of metamorphic core complexes in southern and western Arizona: *in* Crittenden, M. D., Jr., Coney, P. J., and Davis, G. H., eds., Cordilleran metamorphic core complexes: Boulder, Colorado, Geological Society of America Memoir 153, p. 131–157.

Reynolds, S. J., and Spencer, J. E., 1989, Pre-Tertiary rocks and structures in the upper plate of the Buckskin detachment fault, west-central Arizona, *in* Spencer, J. E., and Reynolds, S. J., eds., Geology and mineral resources of the Buckskin and Rawhide Mountains, west-central Arizona: Arizona Geological Survey Bulletin 198, p. 67–102.

Reynolds, S. J., Florence, F. P., Welty, J. W., Roddy, M. S., Currier, D. A., Anderson, A. V., and Keith, S. B., 1986, Compilation of radiometric age determinations in Arizona: Arizona Bureau of Geology and Mineral Technology Bulletin 197, 258 p.

Richard, S. J., Fryxell, J. E., and Sutter, J. F., 1990, Tertiary structure and thermal history of the Harquahala and Buckskin Mountains, west central Arizona: Implications for denudation by a major detachment fault system: Journal of Geophysical Research, v. 95, p. 19,973–19,987.

Sass, J. H., Lachenbruch, A. H., Galanis, S. P., Jr., Morgan, P., Priest, S. S., Moses, T. H., and Munroe, R. J., 1994, Thermal regime of the southern Basin and Range province: 1. Heat flow from Arizona and the Mojave Desert of California and Nevada: Journal of Geophysical Research, v. 99, p. 22,093–22,119.

Shackelford, T. J., 1976, Structural geology of the Rawhide Mountains, Mohave County, Arizona [Ph.D. dissert.]: Los Angeles, University of Southern California, 176p.

Shackelford, T. J., 1977, Late Tertiary tectonic denudation of a Mesozoic(?) gneiss complex, Rawhide Mountains, Arizona: Geological Society of America Abstracts with Programs, v. 9, p. 1169.

Shackelford, T. J., 1980, Tertiary tectonic denudation of a Mesozoic–early Tertiary(?) gneiss complex, Rawhide Mountains, western Arizona: Geology, v. 8, p. 190–194.

Shackelford, T. J., 1981, Tertiary tectonic denudation of a Mesozic–early Tertiary(?) gneiss complex, Rawhide Mountains, western Arizona: Reply: Geology, v. 9, p. 51–52.

Shackelford, T. J., 1989, Structural geology of the Rawhide Mountains, Mohave County, Arizona, *in* Spencer, J. E., and Reynolds, S. J., eds., Geology and mineral resources of the Buckskin and Rawhide Mountains, west-central Arizona: Arizona Geological Survey Bulletin 198, p. 15–46.

Sherrod, D. R., 1988, Preliminary geologic map of the Monkey's Head 7.5-minute quadrangle, Mohave County, Arizona: U.S. Geological Survey Open-File Report 88-899, scale 1:24,000.

Sherrod, D. R., and Nielson, J. E., eds., 1993, Tertiary stratigraphy of highly extended terranes, California, Arizona, and Nevada: U.S. Geological Survey Bulletin 2053, 250 p.

Shreve, R. L., 1968, Boulder, Colorado, The Blackhawk Landslide: Geological Society of America Special Paper 108, 47 p.

Spencer, J. E., 1984, Role of tectonic denudation in warping and uplift of low-angle normal faults: Geology, v. 12, p. 95–98.

Spencer, J. E., and Reynolds, S. J., 1989, Tertiary structure, stratigraphy, and tectonics of the Buckskin Mountains, *in* Spencer, J. E., and Reynolds, S. J., eds., Geology and mineral resources of the Buckskin and Rawhide

Mountains, west-central Arizona: Arizona Geological Survey Bulletin 198, p. 103–167.

Spencer, J. E., and Welty, J. W., 1989, Geology of mineral deposits in the Buckskin and Rawhide Mountains, *in* Spencer, J. E., and Reynolds, S. J., eds., Geology and mineral resources of the Buckskin and Rawhide Mountains, west-central Arizona: Arizona Geological Survey Bulletin 198, p. 223–254.

Spencer, J. E., Grubensky, M. J., Duncan, J. T., Shenk, J. D., Yarnold, J. C., and Lombard, J. P., 1989a, Geology and mineral deposits of the central Artillery Mountains, *in* Spencer, J. E., and Reynolds, S. J., eds., Geology and mineral resources of the Buckskin and Rawhide Mountains, west-central Arizona: Arizona Geological Survey Bulletin 198, p. 168–183.

Spencer, J. E., Shafiqullah, M., Miller, R. J., and Pickthorn, L. G., 1989b, K-Ar geochronology of Miocene extension, volcanism, and potassium metasomatism in the Buckskin and Rawhide Mountains, *in* Spencer, J. E., and Reynolds, S. J., eds., Geology and mineral resources of the Buckskin and Rawhide Mountains, west-central Arizona: Arizona Geological Survey Bulletin 198, p. 184–189.

Stewart, J. H., 1978, Basin-Range structure in western North America: A review, *in* Smith, R. B., and Eaton, G. P. eds., Cenozoic tectonics and regional geophysics of the western Cordillera: Boulder, Colorado, Geological Society of America Memoir 152, p. 1–31.

Suneson, N. H., and Lucchitta, I., 1983, Origin of bimodal volcanism, southern Basin and Range province, west-central Arizona: Geological Society of America Bulletin, v. 94, p. 1005–1019.

Topping, D. J., 1993, Paleogeographic reconstruction of the Death Valley extended region: Evidence from Miocene large rock-avalanche deposits in the Amargosa Chaos basin, California: Geological Society of America Bulletin, v. 105, p. 1190–1213.

Wernicke, B., 1981, Low-angle normal faults in the Basin and Range province—Nappe tectonics in an extending orogen: Nature, v. 291, p. 645–648.

Wernicke, B., and Axen, G., 1988, On the role of isostasy in the evolution of normal fault systems: Geology, v. 16, p. 848–851.

Wernicke, B., and Burchfiel, B. C., 1982, Modes of extensional tectonics: Journal of Structural Geology, v. 4, p. 105–115.

Yarnold, J. C., 1993, Rock-avalanche characteristics in dry climates and the effect of flow into lakes: Insights from mid-Tertiary sedimentary breccias near Artillery Peak, Arizona: Geological Society of America Bulletin, v. 105, p. 345–360.

Young, R. A., 1987, Landscape development during the Tertiary, in Colorado Plateau, *in* Graf, W. L., ed., Geomorphic systems of North America: Boulder, Colorado, Geological Society of America, Centennial Special Volume 2, p. 265–276.

Manuscript Accepted by the Society April 21, 1995

Printed in U.S.A.

Geological Society of America
Special Paper 303
1996

Tests of detachment fault models using Miocene syntectonic strata, Colorado River extensional corridor, southeastern California and west-central Arizona

Kathi K. Beratan
Dept. of Geology and Planetary Science, 321 Old Engineering Hall, University of Pittsburgh, Pittsburgh, Pennyslvania 15260
Jane E. Nielson
U.S. Geological Survey, MS 975, 345 Middlefield Road, Menlo Park, California 94025-0628

ABSTRACT

Several genetic models have been proposed to explain the observed geometric relations of regional-scale low-angle normal faults (detachment faults) that developed in areas of extreme (>100%) crustal extension. In this chapter, we test three general models by comparing the time of upper-plate deformation, structural evolution, and stratigraphic relations predicted by the models, with those found in synextensional Miocene strata in the Colorado River extensional corridor (CRec) of southeastern California and western Arizona. The *evolving shear-zone* model, which incorporates field evidence from areas such as the CRec, suggests that the faults initiated at low angles (<15°). The *rolling hinge* and *lower-crustal flow* models attempt to reconcile field evidence with studies of modern seismogenic normal faults and theoretical mechanical arguments which suggest that earthquakes are not generated on gently dipping (<15°) faults. The rolling-hinge and lower-crustal flow models instead propose that slip on detachment faults starts at high angles (>30°) and that the faults subsequently rotate to lower angles. The evolving shear zone model suggests that all depositional basins would form and fill coevally. In contrast, the rolling-hinge model requires that depositional basins form and fill sequentially with time, becoming younger in the direction of relative upper plate offset. The rolling hinge model also implies that fanning dips are found in syntectonic rocks everywhere except immediately above the down-dip segments of the detachment fault. A feature common to the evolving shear zone model and the rolling hinge model is that the raised topography of the metamorpic core compexes resulted from isostatic rebound.

Sedimentologic evidence and isotopic ages from syntectonic Miocene deposits in the CRec show that basin formation and localized volcanism began approximately synchronously throughout the region. An initial episode of detachment faulting at 20 Ma and a second one between 18 Ma and ca. 17.5 Ma each spanned time intervals too short for resolution by isotopic dating. Upwarping and exposure of the lower plate in the core complexs occurred after ca. 14 Ma during a third episode of detachment faulting. Lower Miocene strata in fault blocks near the up-dip (west) margin of the CRec have uniform moderate dips, whereas coeval strata above the down-dip (east) side are steeply tilted and have fanning dips, demonstrating differences in tilting histories. Depositional basins formed only on the down-dip (eastern) side of the

Beratan, K. K., and Nielson, J. E., 1996, Tests of detachment fault models using Miocene syntectonic strata, Colorado River extensional corridor, southeastern California and west-central Arizona, *in* Beratan, K. K., ed., Reconstructing the History of Basin and Range Extension Using Sedimentology and Stratigraphy: Boulder, Colorado, Geological Society of America Special Paper 303.

terrane. These observations favor aspects of the evolving shear-zone model over the rolling-hinge model.

The most intense phase of extension, marked by uplift of mylonitic rocks to near-surface levels and voluminous volcanism, was separated from final uplift and unroofing of the range cores by 4 m.y. The last extension-related deformation occurred even later. Isostatic rebound should occur very rapidly, perhaps on the order of thousands of years. The delay between intense extension and uplift in the CRec is more consistent with uplift caused by transfer of lower-crustal material from relatively unextended areas to highly extended areas, as postulated by the lower-crustal flow model.

On the basis of observations from synextensional strata, we favor a model of detachment fault evolution that incorporates elements of both the evolving shear-zone and lower-crustal flow models. Rapid extension occurred on a fault initially dipping 30° or less. This extension was accompained by isostatic rebound, which brought midcrustal rocks to near-surface depths. Subsequent uplift of the range cores was accomplished by lower-crustal flow. Observations are incompatible with the rolling hinge model.

INTRODUCTION

The Colorado River extensional corridor (CRec) of Howard and John (1987), southeastern California and west-central Arizona (Fig. 1), is part of a belt of Cordilleran metamorphic core complexes that stretches from Canada to Mexico (Coney, 1980). Metamorphic core complexes are domiform upwarps flanked by regional-scale normal faults (detachment faults) that undulate but generally dip at low angles. The detachment faults in the CRec separate a lower plate, commonly of ductilely deformed mylonitic gneiss, from an upper plate of Tertiary synextensional sedimentary and volcanic rocks that nonconformably overlie Mesozoic or older rocks. The upper-plate rocks are distended by generally northwest-striking high-angle normal faults; thus the core complexes resulted from extreme crustal extension in a northeast-southwest (N60°E) direction.

The Whipple detachment fault crops out in the central Whipple Mountains of California (Fig. 1). It has been correlated with major low-angle faults exposed in the Chemehuevi and Sacramento Mountains of California to the north (Spencer and Turner, 1983; John, 1987; Simpson et al., 1991), and in the Buckskin and Rawhide Mountains of Arizona to the south (Davis et al., 1980; Spencer and Reynolds, 1989). The western headwall region of the Whipple detachment fault is thought to lie between the Old Woman Mountains and the Turtle Mountains and Mopah Range in California, west of the Whipple Mountains (Fig. 1; Davis et al., 1980; Howard et al., 1982). In Arizona, the detachment fault underlies the Mohave Mountains and dips to the east toward the unextended Colorado Plateau (Howard et al., 1995). Tentative matches of rock units and dike swarms in rocks above and below the Whipple and Chemehuevi detachment faults have produced estimates of 100% extension or more (e.g., Davis et al., 1980; Reynolds and Spencer, 1985; Howard and John, 1987).

Controversy surrounds the mechanics and evolution of detachment faults. A key element in this debate is the orientation of the faults at inception and during activity. The continuity of lower plate slip directions, the angle between the fault and bedding of tilted synextensional sedimentary deposits, and the apparent absence of large changes in metamorphic grade in the down-dip direction within footwall rocks suggest that detachment faults formed and moved at low (<15°) angles (Howard and John, 1987; Davis and Lister, 1988; Lister and Davis, 1989). However, modern faults that produce large earthquakes and have normal slip appear to be restricted to dip angles between 30° and 60° in continental crust (Jackson and McKenzie, 1983; Arbasz and Julander, 1986; Jackson and White, 1989). In addition, principles of classic rock mechanics indicate that high-angle faults are favored in an extensional setting (Anderson, 1951). Such theoretical considerations, and the field observations of Proffett (1977) at Yerington, Nevada, led to the argument that detachment faults must initiate at high angles and rotate to lower dips after most of the extension is accomplished (Gans et al., 1985; Gans, 1987; Buck, 1988; Hamilton, 1988; Jackson and White, 1989).

Current models that explain the origin and evolution of detachment faults include (1) the *evolving shear zone* model of Lister and Davis (1989), a modification of earlier shear-zone models such as Davis (1983), (2) the *rolling-hinge* model (Hamilton, 1988; Wernicke and Axen, 1988; Bartley et al., 1990), and (3) the *lower-crustal flow* model (Gans, 1987; Block and Royden, 1990; D. McKenzie, 1992, written communication). These models allow predictions about the progression and timing of structural events that affect the upper plates of detachment faults. In this chapter, we describe the structural evolution of the upper plate of the CRec, as determined from stratigraphic and structural relations and the sedimentary facies patterns of synextensional deposits. We then compare these observations and inferences to predictions derived from the three models of detachment fault evolution.

OBSERVATIONS

The history of detachment fault evolution that we present is based on a synthesis of our own stratigraphic data and map-

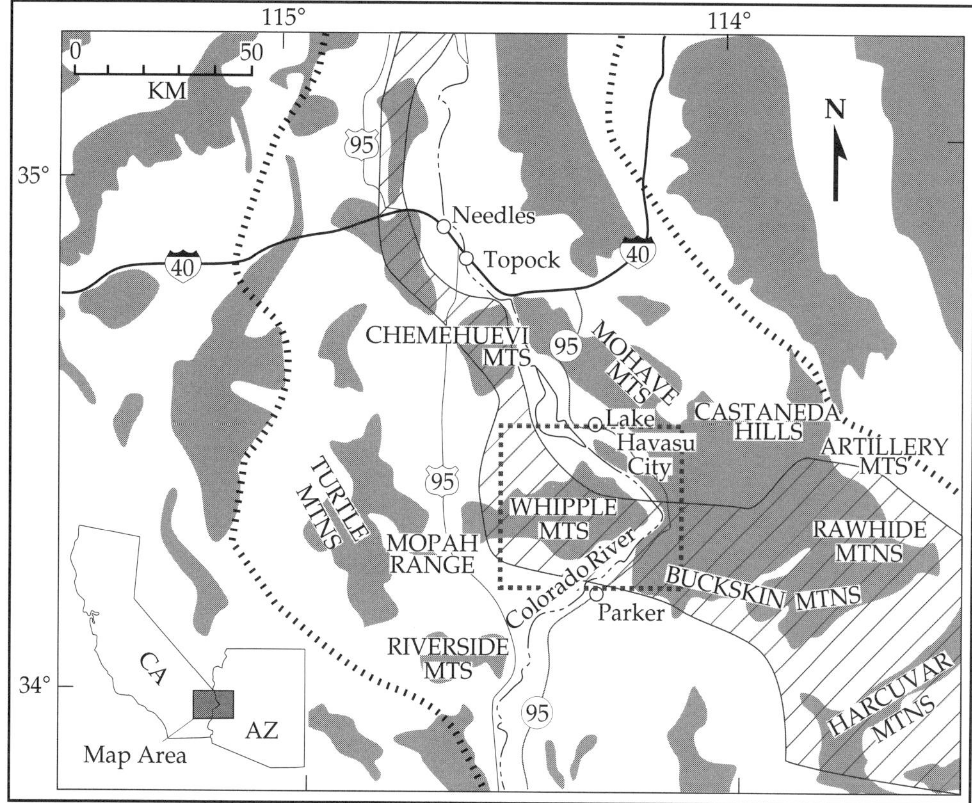

Figure 1. Location map, showing the major mountain ranges (shaded) of the Colorado River extensional corridor. The thick broken lines show the limit of extended upper plate crustal blocks; crustal blocks outside this zone display either minor or no tilting of Tertiary strata. The area filled by diagonal lines corresponds to the zone containing metamorphic core complexes, which are exposures of lower and middle crust in blister-like upwarps that mark the focii of greatest extension.

ping as well as the data of many colleagues. Stratigraphic sections are presented in Sherrod and Nielson (1993). Nielson and Beratan (1994, 1995) synthesized these data and proposed a correlation of Miocene syntectonic rocks throughout the CRec. Following is review of the key observations.

CRec structural domains

The primary expressions of detachment faulting in the upper plate of the CRec include volcanism, basin formation, segmentation of the upper plate by faults, rotation of fault-bounded blocks, and tilting of syntectonic deposits. Spencer and Reynolds (1989) recognized a pattern in the spatial distribution of these effects, and they distinguished three structural belts within the "Whipple tilt-block domain": a western "synformal keel of distended upper-plate rocks above warped regional detachment faults," a central belt of "archlike uplifts" that form metamorphic core complexes, and an eastern belt of "moderately to highly tilted and extended upper-plate rocks." Nielson and Beratan (1994, 1995) further subdivide the upper plate into six structural domains (Fig. 2), characterized by differences in tilting history. The *Mopah* domain is identical to the western belt of Spencer and Reynolds (1989); all synextensional rocks of early Miocene age in this domain have uniform, moderate tilts (Fig. 2). The *Topock*, *Crossman*, *Aubrey*, *Parker Dam,* and *Buckskin-Rawhide* domains lie east of the zone of core complexes, collectively corresponding to Spencer and Reynolds's (1989) eastern belt; all these domains contain steeply tilted lower Miocene rocks (Fig. 2).

Volcanism occurred independently of basin formation in the CRec; a chain of volcanoes dominated the Mopah domain while fault-bounded basins formed in the central and eastern belts. Correlation of lower Miocene rocks (Nielson and Beratan, 1994; 1995) show that sedimentary strata in the eastern belt basins were deposited both during periods of block subsidence and of block rotation. Uniform tilts of lower Miocene strata indicate deposition during block subsidence prior to block tilting, whereas fanning dips indicate their deposition during rapid block rotation. Tilting occurred as discrete episodes and created unconformities. The limited areal extent of individual unconformities indicates that the upper plate segmented during the initial stage of detachment faulting, creating structural domains that responded semi-independently to faulting events (Nielson and Beratan, 1994, 1995).

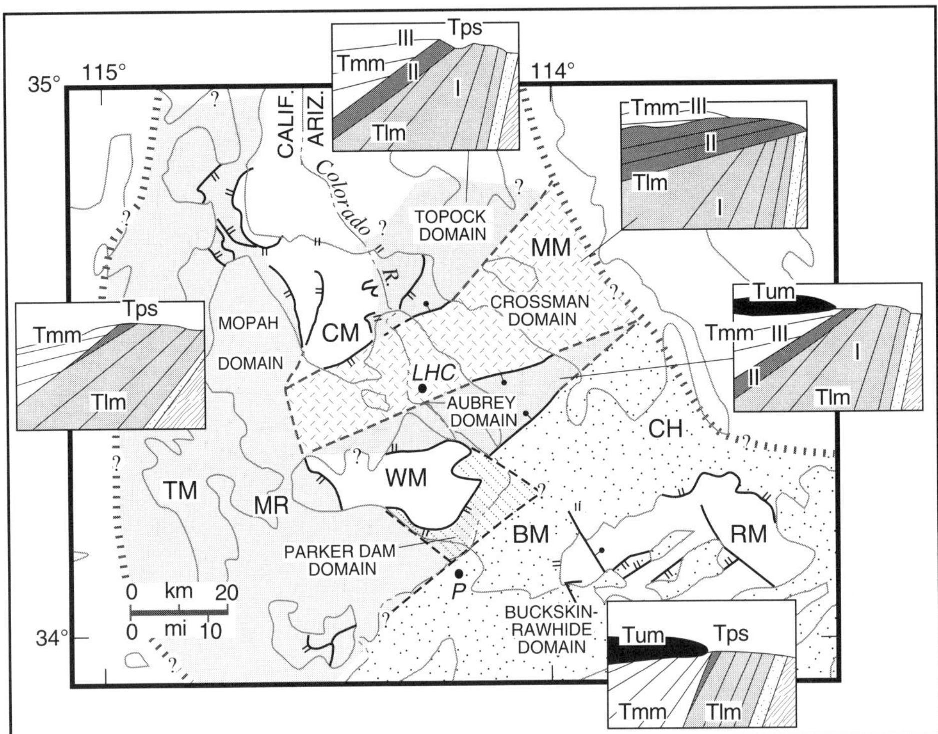

Figure 2. Upper-plate tilting domains of the Colorado River extensional corridor (CRec). Areal limits of the domains (mostly inferred, queried where highly uncertain) may be coincident with faults, shown by heavy dashed lines. Unpatterned map areas within the CRec either lack data for domain assignments or are exposures of lower plate windows in the Whipple, Chemehuevi, and Rawhide detachment faults. Cross sections depict generalized structural relations for five domains; relations in the Parker Dam domain are probably similar to those found in the Aubrey domain. Symbols in cross sections: wavy diagonal lines = pre-Tertiary rocks; small dots = basal arkose; medium shading = lower Miocene strata (Tlm) underlying unconformity U1L, or underlying the Miocene Peach Springs Tuff (Tps) in domains lacking unconformity U1L (sequence I of Nielson, 1986); heavy shading = lower Miocene strata lying between unconformities U1L and U1 or the Peach Springs Tuff (sequence II of Nielson, 1986); unpatterned = strata lying between unconformities U1 and U2 (sequence III of Nielson, 1986); black = middle (Tmm) and upper (Tum) Miocene strata overlying unconformity U2 (sequence IV of Nielson, 1986). BM = Buckshin Mountains; CH = Castaneda Hills; CM = Chemehuevi Mountains; LHC = Lake Havasu City; MM = Mohave Mountains; MR = Mopah Range; P = Parker; RM = Rawhide Mountains; TM = Turtle Mountains; WM = Whipple Mountains.

Miocene deposits and structural relations

The lowest unit in many sections is oxidized arkosic conglomerate and sandstone that grades downward into grus and overlies pre-Tertiary rocks that provided most of the clasts in the unit. The grus zone represents a deeply weathered regional surface. This surface was mantled by the arkose, consisting of compositionally and texturally immature sediment deposited by unconfined sheet floods, debris flows, and braided streams in poorly integrated drainage systems. Local lenses of finer-grained siliciclastic strata and limestone were deposited in ponds or small playas (Nielson and Beratan, 1990; Beratan, 1991). This basal arkose unit probably developed prior to the onset of Tertiary extension. The subdued Oligocene to earliest Miocene landscape was altered when volcanic centers and fault-bounded half-graben basins formed in the early Miocene.

Throughout the CRec, the conformably overlying synextensional Miocene deposits can be related to three major intervals of deposition on the basis of isotopic ages: the oldest represents volcanic eruptions and (or) deposition in basins during the early Miocene (ca. 22–18 Ma), the middle interval took place in the later early to middle Miocene (ca. 18–13 Ma), and the youngest represents deposition during the later middle to late Miocene (ca. 13–8 Ma). Within the Mohave Mountains of Arizona, Nielson (1986) recognized as many as four unconformity-bounded sequences, numbered I–IV (from oldest to youngest) that can be equated to these regionwide depositional intervals. Nielson and Beratan (1995) designated the intervening unconformities (from oldest to youngest), U1L, U1, U2, and U3 (Fig. 2). Where it is found, U1L underlies the 18.5 Ma Peach Springs Tuff (Young and Brennan, 1974) or age-equivalent strata (Nielson and Beratan, 1990; Nielson et al., 1990), and unconformity U1 overlies the Peach Springs Tuff. Unconformity U2 formed between 12 Ma and 14 Ma, perhaps diachronously, and U3 formed between 7–9 Ma.

Older Miocene Deposits (Pre-U1). Miocene strata deposited prior to development of unconformity U1 consist of volcanic rocks and volcaniclastic and epiclastic sedimentary rocks with aggregate thickness of 1–2 km. Basins were depositionally isolated and small, no more than 15 km on a side (Nielson and Beratan, 1990). Sedimentary facies patterns in the southern and eastern Whipple Mountains suggest that the basins were bounded on their northwest and/or southeast sides by high-angle faults with northeast strike, parallel to the direction of extension. These northeast-striking faults have been interpreted as transfer faults as defined by Etheridge et al., (1988) (Nielson and Beratan, 1990; Beratan, 1991). Very coarse grained sedimentary materials shed from these basin margins include rock-avalanche and debris-flow deposits. The presence of these units indicates that steep slopes or scarps were associated with the faults.

The Mopah domain contains several volcanic centers; stratigraphic sections are composed predominantly of volcanic rocks, locally interspersed with deposits of disrupted streams (Hazlett, 1986). Volcanic activity also characterized the northern part of the eastern belt, as the Topock and Crossman domains were dominated by volcanic rocks. The volcanic component decreases southward through the Aubrey domain, with sedimentary rocks much more abundant than volcanic rocks in the Parker Dam and Buckskin-Rawhide domains (Nielson and Beratan, 1990; Lucchitta and Suneson, 1993b; Yarnold, 1993; Spencer and Reynolds, 1993). Clasts in lower Miocene sedimentary units were derived entirely from upper-plate sources.

The Mopah domain contains a single moderately tilted lower Miocene sequence that is capped locally by the Peach Springs Tuff and bounded above by unconformity U1, which formed between 18.1 Ma and 17.5 Ma (Howard et al., 1993; Nielson and Beratan, 1995). In contrast, the Topock and Aubrey domains contain two lower Miocene sequences separated by an unconformity (U1L) with about 10° of angular discordance. Basal units in the lower sequence (sequence I of Nielson, 1986) have dips that fan from 70° to 90° at the basal nonconformity to 40° or 56° upsection. The Peach Springs Tuff immediately overlies U1L in most exposures, providing a minimum age for development of the unconformity. The Peach Springs Tuff is within the second lower Miocene sequence (sequence II of Nielson, 1986), which is overlain by unconformity U1. The lower Miocene sequence of the Parker Dam domain consists almost entirely of sedimentary rocks but appears to have an identical tilt history to the Topock and Aubrey domains, although no angular discordance between the Peach Springs Tuff and underlying sedimentary units has been demonstrated.

The Crossman domain displays two sequences that are stratigraphically equivalent to sequences I and II in the flanking Topock and Aubrey domains (Fig. 2), and similar fanning dips characterize the basal part of sequence I. However, in the area of Crossman Peak, in Arizona, the unconformity between sequences I and II (U1L) has an angular discordance of 20° to as much as 40°, and it formed earlier than unconformity U1L in the Topock and Aubrey domains. Rocks in sequence I yielded ages of 21.5 Ma to ca. 20 Ma, and the lowest units above unconformity U1L give closely similar ages (19.9 Ma and 19.7 Ma: Nielson and Beratan, 1990, 1995; Faulds et al., 1994). Thus, the U1L hiatus appears to represent an interval of <1 m.y. Sequence II in this area lacks the Peach Springs Tuff, but is composed mostly of volcanic rocks that bracket the Peach Springs Tuff in age.

Sections in the Buckskin-Rawhide domain, which includes the Castaneda Hills, do not contain angular unconformities equivalent to U1L or U1. Sections consist of interbedded sedimentary and volcanic rocks with steep, uniform dips (Lucchitta and Suneson, 1993a; Sherrod, 1988). The Peach Springs Tuff crops out locally and is conformably overlain by sedimentary strata.

Middle Miocene deposits (between U1 and U2). Throughout the CRec, the deposits that began to accumulate after eruption of the Peach Springs Tuff contain a higher proportion of sedimentary strata than do underlying units. Sedimentary facies patterns suggest that basins were still internally drained, although with more stable and better integrated drainage systems than in the early Miocene. The middle Miocene strata commonly lap onto former topographic highs along the basin margins, suggesting that the basins were filling with sediments during a time of relative tectonic quiescence.

The middle sequence in the Mopah domain is mostly composed of bimodal volcanic rocks (ages between 18.1 Ma and 14 Ma) and minor sedimentary units; this contrasts with the older Miocene deposits, which lacked sedimentary interbeds. The sequence overlies unconformity U1 and generally dips around 10°, but locally as much as 30° (Howard et al., 1993). The Topock, Aubrey, and Parker Dam domains contain a sedimentary rock-dominated sequence (sequence III of Nielson, 1986). Because of diachroneity of faulting, units in the upper part of this sequence in these domains are equivalent in age to untilted (post-unconformity U2) deposits elsewhere in the CRec. Rocks equivalent in age to sequence III are minor in the Crossman domain but form a substantial succession in the Buckskin-Rawhide domain, where dips of units stratigraphically above the Peach Springs Tuff fan from steep to horizontal.

Paleocurrent and sedimentary facies patterns in middle Miocene strata of the Parker Dam domain suggest that a moderate topographic high existed in the approximate location of the high part of the present-day Whipple Mountains during deposition of the middle Miocene sequence (Beratan, 1991). Clast compositions indicate that the topographic high was entirely mantled by upper-plate rocks. In most parts of the CRec, clasts in middle Miocene strata were derived exclusively from upper-plate sources. Clasts from the detachment fault zones or the lower plate appear in the upper part of the middle sequence in the eastern Chemehuevi Mountains, Aubrey Hills, and northeastern Whipple Mountains. These lower-plate fragments are associated with megabreccia deposits, interpreted as large rock avalanches (landslides), that generally are found east of the core-complex upwarps. For example, mylonitic clasts from the lower plate of the Whipple detachment fault appear in strata above a landslide megabreccia in the Aubrey Hills and northeastern Whipple Mountains that was derived from a non-mylonitic source (Dunn, 1986; Nielson and Beratan, 1990; Beratan, 1990; Howard et al., 1995; Yin and Dunn, 1992). This megabreccia deposit overlies siltstone interbedded with basalt flows that erupted at ca. 14.1 Ma (Beratan, 1990; Nielson and Beratan, 1990; Howard et al., 1995).

Younger Miocene Deposits (Between U2 and U3). The middle to upper Miocene deposits (sequence IV of Nielson, 1986), which overlie unconformity U2, record continued sediment deposition, widespread eruption of thin basalt flows, and localized bimodal volcanic centers such as the Castaneda Hills, between ca. 13 Ma and 8.2 Ma (Nielson and Beratan, 1990, 1995; Lucchitta and Suneson, 1993a, b). Locally, the strata are slightly tilted, indicating continued activity on the high-angle normal faults. Basins are poorly delineated; ages are difficult to obtain because volcanic units generally are rare, and sedimentary strata are difficult to separate from lithologically and texturally similar younger, postextensional units. Where units of this age can be identified, their spatial distribution indicates that topographic boundaries that separated older basins locally were overtopped. For example, basalt flows and rhyolitic volcanic rocks that overlie tilted middle Miocene rocks in the Castaneda Hills can be traced continuously into the nearby area of Standard Wash, where they overlie tilted strata that were deposited in a different basin.

Throughout the CRec, clasts derived from lower plate sources are common in the younger Miocene conglomerates, owing to widespread surface exposure of the Whipple and Chemehuevi detachment faults. In many areas, this transition in clast types occurrs abruptly across unconformity U2. The relatively abrupt transition in clast type over such a large area suggests that the unroofing was caused by an episode of detachment fault-related deformation rather than by progressive erosion through a continuous upper-plate cover (Beratan and Nielson, 1993). In the Whipple Mountains, this episode resulted in sufficient displacement on the Whipple detachment fault to juxtapose post–Peach Springs sedimentary strata against lower-plate mylonitic gneiss, and to "strand" blocks of upper-plate rocks on top of the progressively exhumed detachment fault. Structural "stranding" of blocks is considered more likely than subsequent erosional isolation for two reasons; sedimentary facies patterns do not suggest that a significant part of the basin is missing, and reletively few intraclasts derived from middle sequence sedimentary strata are found in younger strata.

DISCUSSION

Detachment fault evolution

Cooling ages between 18 Ma and 20 Ma on mylonitic gneiss from the core of the Whipple Mountains (Dokka and Lingrey, 1979; E. DeWitt and J. F. Sutter, cited in Davis and Lister, 1988) provided the first evidence of rapid extension in the early Miocene. This was also the time of most voluminous volcanism in the western part of the CRec (Nielson and Beratan, 1990) and emplacement of the Chambers Well and Crossman dike swarms (Howard et al., 1982; Howard and John, 1987; Nielson et al., 1993). During this time, sequences of volcanic and sedimentary strata as much as 2 km thick accumulated in rapidly subsiding basins above the down-dip segment of the regional detachment faults.

Three distinct episodes of detachment faulting are recorded in the Tertiary strata (Nielson and Beratan, 1994, 1995). The first episode at ca. 20 Ma is expressed by volcanism in the Mopah domain; segmentation of the upper plate to form the structural domains and further subdivision of the domains by northwest-striking faults; formation of basins by subsidence of the smaller upper-plate blocks in the eastern domains; and rotation of blocks beneath the Topock, Aubrey, Parker Dam, and Crossman domains that created fanning dips close to the basal nonconformity. The Crossman Peak area (Crossman block) of the Crossman domain underwent extreme tilting in this episode. In contrast, upper-plate blocks west and southeast of the Whipple Mountains did not rotate until later. These observations suggest that the zone of maximum extension was located in the eastern half of the CRec, probably beneath the original position of the Crossman block.

A second episode of fault motion, occurring after 18.5 Ma, tilted upper-plate blocks of the Mopah domain a moderate amount in a short interval between 18.1 Ma and 17.5 Ma. In the Buckskin-Rawhide domain, fanning of dips in strata above the Peach Springs Tuff record rapid rotation of upper-plate blocks during this episode. In contrast, all other domains of the eastern belt experienced only moderate tilting and do not exhibit fanning dips. This localization of rapid tilting suggests a transfer of extension from the Whipple detachment fault to the Rawhide detachment fault. The Parker Dam basin expanded southwestward at this time and captured a small part of the Mopah domain when a new high-angle normal fault began to control tilting (Beratan, 1991).

A third, apparently diachronous episode of detachment-fault motion after 14 Ma is recorded in every domain of the CRec. This event was synchronous with exposure of detachment faults, fragmentation of earlier basins by more closely spaced normal faults, and localized bimodal volcanism in the area of the Rawhide detachment fault. Elsewhere, a widespread pattern of isolated fissure-fed basaltic eruptions developed. The lower plate of the Whipple Mountains detachment system appears to have been widely exposed to erosion in the interval after 14.1 Ma and before 12.7 Ma (Beratan and Nielson, 1993). The most dramatic expressions of this event are found on the east side of the CRec, close to exposures of the detachment faults.

The youngest strata in the CRec that were affected by detachment-related deformation and volcanism formed between 11 Ma and ca. 8 Ma. Centralized volcanism was limited to the

Buckskin-Rawhide domain. Rocks of this age record only a small degree of tilting, generally less than 15°.

Detachment fault models

Three main models of extension along detachment faults in the upper crust have been proposed in recent years. These are (1) the evolving shear-zone model, which incorporates field evidence for primary low-angle normal faults; (2) the rolling-hinge model; and (3) the lower-crustal flow model. The latter two postulate mechanisms that rotate originally high-angle faults to low angles, in an attempt to reconcile field evidence that favors originally low-angle faults with evidence against initially low-angle faults from continental seismic records and theoretical considerations. In the following section, we include brief descriptions of each model and specific predictions about the surface expression of detachment faulting derived from the model. In each case, we test these models for the CRec by comparing predictions to observations on synextensional rocks.

Evolving shear-zone model. Davis and Lister (1988) postulated that detachment faults are zones of evolving low-angle shear that probably root into the middle crust or lower parts of the upper crust. The detachment zones propagate upward and can either reach the surface directly as low-angle faults or terminate at shallow depths in pull-apart complexes of closely spaced normal faults. Metamorphic core complexes form as the lower plate is bowed upward by isostatic rebound to adjust for differential extension (Fig. 3A); the shape of uplifts is modified by local isostatic adjustments caused by intrusions of granite into the middle crust (Lister and Davis , 1989). This model is an elaboration of earlier simple-shear models (e.g., Wernicke, 1981; Davis, 1983).

Several structural and chronologic consequences are implied by this model:

(1) Because the fault is active at a low angle, nearly the entire width of the extended zone should experience extension-related deformation at the same time, or over a restricted time interval.

(2) Isostatic rebound will occur in the most highly deformed part of the detachment terrane. Rebound should isolate basins up-dip of the uplift while fault motion continues down-dip, resulting in younger ages for the end of deformation in the down-dip direction.

(3) Isostatic rebound should occur after tectonic unloading, with a relatively short response time (essentially immediate).

Observations relevant to these predictions include:

(1) Compatible with the first prediction, volcanism and/or basin formation began throughout the CRec at roughly the same time.

(2) Tilting of rocks deposited after early Miocene time is less variable in the Mopah domain (up-dip), compared with units in the eastern domains (down-dip), an observation that is partly compatible with the prediction.

(3) Essentially immediate simple isostatic rebound may have occurred during the first and second episodes of extension. In the first episode, a substantial amount of magmatism and volcanism, extreme block rotation, and erosion occupied a time interval so short that it has yet to be resolved by isotopic dating. Sedimentologic and age data, which show positive relief at the current position of the Whipple Mountains core just after this initial event, are compatible with the third prediction. However, detachment faulting, core uplift, and tectonic unloading appear to have continued for some time afterward. The post–14 Ma extensional event, which exposed exhumed detachment faults and later created the major topographic elements of present-day topography, is separated by about 4 m.y. from the end of major extension in the middle Miocene, suggesting that isostatic rebound may not have been the only process contributing to core uplift.

Rolling hinge model. The rolling-hinge model (Hamilton, 1988; Bartley et al., 1990) postulates that detachment faults originate as master high-angle normal faults; the footwalls rise as they become tectonically denuded by hanging-wall slip, with the result that blocks of the hanging wall are stranded on the inactivated parts of the faults. The dip of the detachment fault decreases where the fault intersects the surface and then becomes inactive, resulting in surface exposures that display low dip angles (Fig. 3B). The model provides a mechanism by which detachment faults are seismically active only along their steeply dipping parts. Related models were proposed by Howard et al., (1982), Spencer (1984), Wernicke and Axen (1988), and Buck (1988).

Predictions about surface events derived from this model include:

(1) Active faulting migrates in space and time; depositional basins should form sequentially in the down-dip direction. Only the basin closest to the moving part of the fault is active at any given time, and thus the site of detachment-related sedimentation also must migrate (Hamilton, 1988).

(2) Pieces of the hanging wall are sliced off and carried along on top of the detachment fault "conveyer belt." Range-size blocks become stranded on top of flattened up-dip sectors of the detachment fault while slip continues on more steeply dipping down-dip sectors (Hamilton, 1988; Bartley et al., 1990).

(3) Pieces of upper plate are detached and quickly rotate as they migrate away from the hinge; only blocks near the hinge are not highly rotated. The lowermost units deposited on all blocks that migrated away from the hinge should show fanning dips.

(4) Major extension terminates at progressively later time in the direction of initial dip of the detachment fault as the fault flattens and becomes inactive progressively in the down-dip direction (Hamilton, 1988).

(5) Isostatic rebound should occur immediately following tectonic unloading, and the location of uplift caused by isostatic rebound should migrate in the down-dip direction.

Observations relevant to these predictions include:

(1) No migration of the time of detachment faulting can be discerned from the inception of volcanism or time of basin-formation in the CRec. Moreover, variable dips of lower Miocene rocks show segmentation of the upper plate in the first

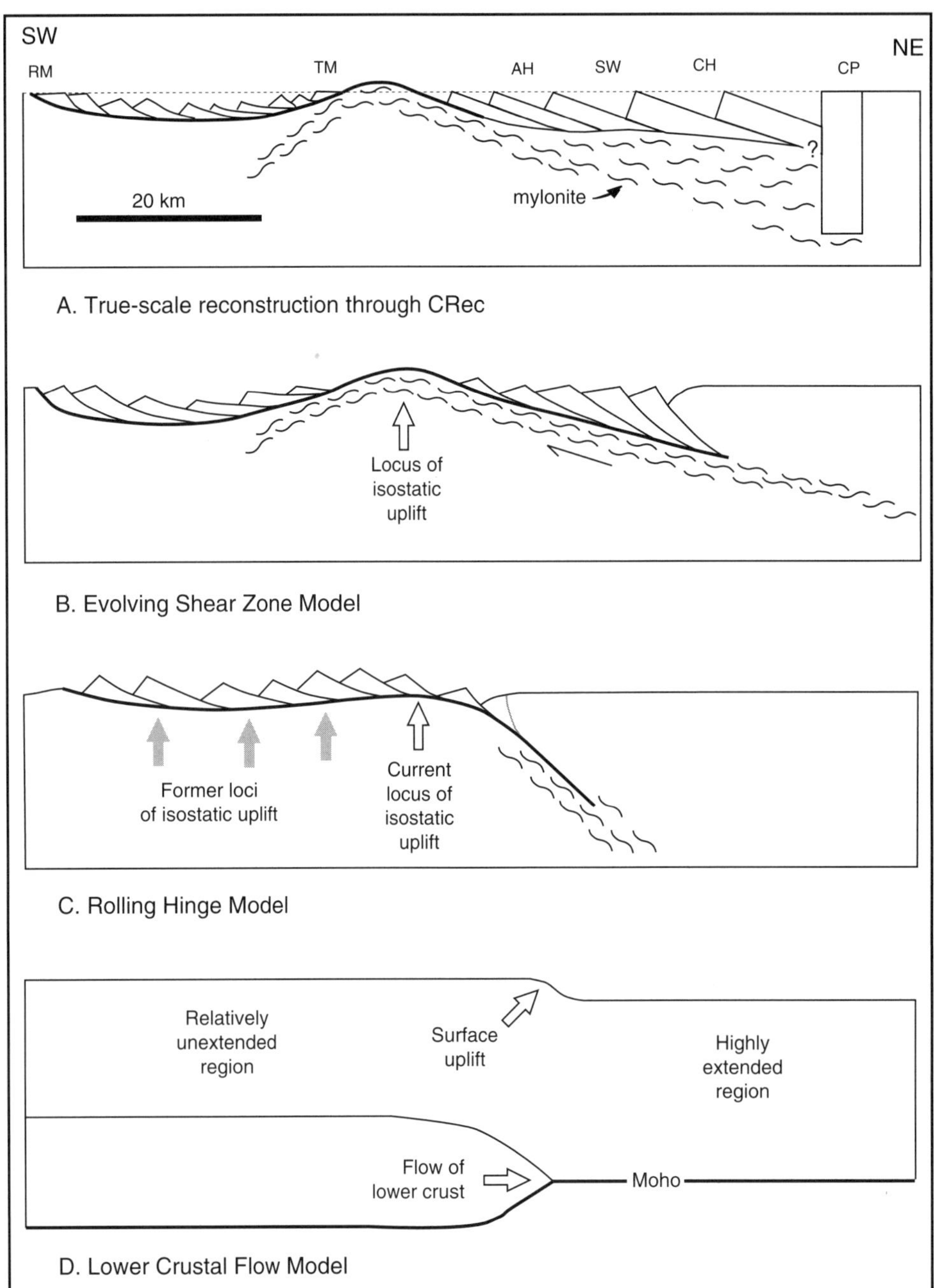

Figure 3. Schematic cross sections showing the major features of the three extensional models discussed in the chapter. (A) True-scale reconstruction through the CRec, passing through the center of the Whipple Mountain's core, parallel to extension direction. No vertical exaggeration. Based on measured distances and tilt angles. Note the low tilt angles of blocks at the southwest edge and the high tilt angles of blocks east of the range core. RM = Riverside Mountains; TM = Turks Mine area, southwestern Whipple Mountains; AH = Aubrey Hills; SW = Standard Wash; CH = just north of the Castaneda Hills; CP = Colorado Plateau. (B) Evolving shear-zone model (Lister and Davis, 1989). Extension in the upper crust occurs along a primary low-angle (<30°) normal fault which roots a mid-crustal levels. the doming that is responsible for the formation of metamorphic core complexes results from isostatic rebound following crustal thinning. tilt angles of blocks are not specified in this model, but the model is permissive of the observed tilt distribution. (C) Rolling-hinge model (Hamilton, 1988; Wernicke and Axen, 1988; Bartley et al., 1990). Extension occurs along high-angle (>30°) normal faults, which "roll over" or flatten at the hinge, where they intersect the surface. Sedimentary basins from as pieces of the upper plate are sliced off, and thus there is a systematic younging of basin-formation and deposition in the down-dip direction. Doming results from isostatic rebound, and the locus of isostatic uplift migrates in the down-dip direction. This model requires that the northeasternmost block(s) be less steeply tilted than blocks to the southwest, contrary to observed relations. (D) Lower-crustal flow model (Gans, 1987; Block and Royden, 1990; D. McKenzie, 1982, written communication). Doming results from flow of ductile lower-crustal material following extension from relatively unextended regions to highly extended regions, ultimately producing a flat Moho. This model does not specify tilt relations.

extensional episode. These observations are incompatible with the first prediction.

(2) Compatible with the second prediction, upper-plate fault blocks are cut by the Whipple detachment fault and some are stranded on exposures of lower-plate rocks. However, in the CRec, such relations are exposed only in the narrow central belt, and most of the terrane constitutes nearly contiguous, variably rotated upper plate blocks.

(3) Incompatible with and opposite to relations implied by the third prediction, variable tilting relations are exhibited by coeval lower Miocene rocks in the CRec; extreme tilting and fanning dips are found only above the down-dip parts of detachment faults, whereas coeval units near the breakaway are only moderately tilted.

(4) Centers of volcanism migrated southeastward rather than northeastward in the relative direction of upper-plate transport, shifting from the western and northern parts of the CRec in early Miocene time to the Buckskin-Rawhide domain in late Miocene time. Between 20 Ma and 18 Ma the focus of substantial tilting migrated south, from the Whipple detachment fault to the Buckskin-Rawhide detachment fault. Additional sites of sedimentation developed throughout the CRec after the early Miocene, and basins began to coalesce in middle Miocene time. These observations do not support patterns of active detachment migration suggested by the fourth prediction.

(5) As noted above, the timing of initial range-core uplift is compatible with simple isostatic rebound in the initial phase of extension. Exhumation of the lower plate occurred 4 m.y. after the initial episode of detachment faulting, and the relatively few stranded blocks are localized in the central belt of the CRec. There also is no evidence from sedimentation patterns or from stratigraphic correlations in the CRec to suggest progressive migration of the position of the topographic high that marks the lower-plate upwarp. Therefore, data from syntectonic deposits are incompatible with the fifth prediction.

Lower-crustal flow model. The lower-crustal flow model (e.g., Block and Royden, 1990; Kruse et al., 1991; D. McKenzie, 1992, written communication) postulates that regional-scale, lateral flow in the lower crust causes the uplift of range cores (Fig. 3C) and induces rotation of upper-crustal high-angle faults to low angles. Material flows from sites beneath unextended blocks and toward zones of extended upper crust, resulting in a relatively flat Moho, as shown in seismic studies of the region (Henyey et al., 1986; McCarthy et al., 1991). Such flow can cause large (up to several kilometers) vertical motions over scales of a few tens of kilometers. Calculations suggest that lower-crustal material of low viscosity at the base of relatively thick (unextended) crust will tend to spread out as a layer with a steep front (D. McKenzie, 1992, written communication). As the low-viscosity layer spreads, it adds to the crustal thickness of the extended crust, resulting in isostatic uplift and formation of a step at the Earth's surface. Flow may begin during extension; the effects of this lower-crustal flow will be seen at upper-crustal levels in a time frame on the order of 1–10 m.y. In the region of the step above the flow front, there is likely to be tilting in the direction of lower-crustal flow, followed by back-tilting after the flow front has passed (D. McKenzie, 1992, written communication). This model has been suggested as a mechanism to rotate detachment faults to shallower dips (e.g., Buck, 1988; Wernicke and Axen, 1988; Spencer and Reynolds, 1989; and Block and Royden, 1990). However, because the lower-crustal flow occurs in *response* to extension, the model is independent of the extensional mechanism.

Predictions based on this model are:

(1) Areas that have undergone large amounts of extension will experience late-stage uplift as lower-crustal material is transferred from less highly extended regions.

(2) Tilting directions during lower-crustal flow will be in the direction of flow, followed by back-tilting after the front in the low viscosity layer has passed.

(3) The duration of crustal adjustments to detachment faulting should be on the order of 1–10 m.y.

Observations relevant to these predictions include:

(1) A topographic high was present in the area now occupied by the high part of the Whipple Mountains during deposition of the middle sequence of strata. This feature may have resulted from isostatic rebound that accompanied the first episode of extension. However, uplift continued and those older basins were fragmented by exposure of the lower plate after 14 Ma, during the third extensional episode (responsible for development of unconformity U2). The length of time between the second extensional episode and earliest exposure of the lower plate (~4 m.y.) suggests that the core expanded slowly, an observation compatible with the first prediction.

(2) Strata throughout the region tilted to the southwest throughout detachment faulting, except in areas of minor folding; no reversal of tilting direction has been noted. These observations are incompatible with late-stage back-tilting.

(3) Extension in the CRec spanned the early and middle Miocene and ended at ca. 8 Ma, an interval of 10–12 m.y. About 4–5 m.y. separated the end of the most intense phase of extension (20–18 Ma) and the third extensional event (14–12.7 Ma), which was marked by full development of the Whipple Mountains metamorphic core complex and initiation of present-day topography, a time scale that is compatible with the third prediction.

CONCLUSIONS

This work represents the first attempt to apply stratigraphic data on a regional scale as a test of detachment fault models. Current data are sufficient to provide significant constraints for an evaluation of detachment fault models; however, some patterns are allowed by all the models, and thus are inconclusive constraints. For example, progressively later cessation of volcanism in the down-dip direction is compatible with all three models.

Observations in the CRec support aspects of the evolving shear zone and lower-crustal flow models, but are incompatible with the rolling hinge model. Although the rolling hinge model explains stranded blocks, such features are minor in the CRec. The rolling hinge model also conflicts with evidence that the initial event of detachment faulting occurred throughout the region simultaneously. Our correlation of syntectonic deposits shows no down-dip migration of the time of basin formation, sedimentation, or deformation related to detachment faulting. Another argument against the rolling hinge model is that steepest dips of early syntectonic strata are not found adjacent to the breakaway, but formed above the down-dip segment of the detachment fault and remained there. Analyses of structural relations in the Chemehuevi Mountains (John and Foster, 1993) and Castaneda Hills (Lucchitta and Suneson, 1993a) also do not support the rolling-hinge model.

The evolving shear-zone and rolling-hinge models both imply that isostatic uplift was responsible for formation of the range cores. Isostatic adjustment takes place on a time scale of about 10^4 years (England and Molnar, 1990), a time compatible with two short detachment fault episodes that produced rapid rise of the lower plate between 20 Ma and 18 Ma, as predicted by the evolving shear zone model. The continued development of range cores during at least 4 m.y. following the major extensional phase and the continued surface displacements until ca. 9 Ma are more consistent with the lower-crustal flow model than with the other models.

The evolving shear-zone model successfully predicts that all basins should be actively forming and filling during the same time period, but by itself does not adequately explain the late-stage tectonic event that created the major topographic elements forming present-day topography after 8 Ma. The lower-crustal flow model successfully predicts the timing of the late-stage event, but does not constrain the initial fault geometry. Therefore, the data favor a model in which rapid extension is accomplished along a primary low-angle fault and is followed immediately by isostatic rebound, as in the evolving shear zone model, with the resulting geometry modified over time by inflow of lower-crustal material.

ACKNOWLEDGMENTS

The stratigraphic data upon which this paper is based is the result of more than a decade of detailed studies of CRec stratigraphy by numerous researchers (cited in Nielson and Beratan, 1995). The authors' ideas regarding structural models were particularly influenced by discussions with Greg Davis, Warren Hamilton, Keith Howard, Rick Hazlett, Barbara John, Ivo Lucchitta, and Julia Miller. Beratan benefited from discussions with D. McKenzie in the field; his fortitude in the face of rising floodwaters was also appreciated. R. Blom at the Jet Propulsion Laboratory provided Beratan with support and encouragement during the initial phase of manuscript preparation. We thank D. Sherrod and R. Christiansen for their very helpful reviews of the paper.

REFERENCES CITED

Anderson, E. M., 1951, The dynamics of faulting (second edition): Edinburgh, Scotland, Oliver and Boyd, 191 p.

Arbasz, W. J., and Julander, D. R., 1986, Geometry of seismically active faults and crustal deformation within the Basin and Range–Colorado Plateau transition, *in* Meyer, L. T., ed., Extensional tectonics of the southwestern United States: A perspective on processes and kinematics: Boulder, Colorado, Geological Society of America Special Paper 208, p. 43–74.

Bartley, J. M., Fletcher, J. M., and Glazner, A. F., 1990, Tertiary extension and contraction of lower-plate rocks in the central Mojave metamorphic core complex, southern California: Tectonics, v. 9, p. 521–534.

Beratan, K. K., 1990, Basin development during Miocene detachment faulting, Whipple Mountains, southeastern California [Ph.D. dissert.]: Los Angeles, University of Southern California, Ph.D. dissertation, 267 p.

Beratan, K. K., 1991, Miocene synextension sedimentation patterns, Whipple Mountains, southeastern California: Implications for the geometry of the Whipple detachment system: Journal of Geophysical Research, v. 96, p. 12,425–12,442.

Beratan, K. K., and Nielson, J. E., 1993, Evidence for late movement on the Whipple detachment fault, southeastern California: Geological Society of America Abstracts with Programs, v. 25, no. 5, p. A8.

Block, L., and Royden, L. H., 1990, Core complex geometries and regional scale flow in the lower crust: Tectonics, v. 9, p. 557–567.

Buck, W. R., 1988, Flexural rotation of normal faults: Tectonics, v. 7, p. 959–973.

Coney, P. F., 1980, Cordilleran metamorphic core complexes: An overview, *in* Crittenden, M. D., Jr., Coney, P. F., and Davis, G. H., eds., Cordilleran metamorphic core complexes: Boulder, Colorado, Geological Society of America Memoir 153, p. 7–31.

Davis, G. A., and Lister, G. S., 1988, Detachment faulting in continental extension; perspectives from the southwestern Cordillera, *in* Clark, S. P., Jr., Processes in continental lithospheric deformation: Boulder, Colorado, Geological Society of America Special Paper 218, p. 133–159.

Davis, G. A., Anderson, J. L., Frost, E. G., and Shackelford, T. J., 1980, Mylonitization and detachment faulting in the Whipple-Buckskin-Rawhide Mountains terrane, southeastern California and western Arizona, *in* Crittenden, M. D., Coney, P. J., and Davis, G. H., eds., Cordilleran metamorphic core complexes: Boulder, Colorado, Geological Society of America Memoir 153, p. 79–129.

Davis, G. H., 1983, Shear-zone model for the origin of metamorphic core complexes: Geology, v. 11, p. 342–347.

Dokka, R. K., and Lingrey, S. H., 1979, Fission track evidence for a Miocene cooling event, Whipple Mountains, southeastern California, *in* Armentrout, J. M., Cole, M. R., and TerBest, H., eds., Cenozoic paleogeography of the western United States: Los Angeles, Society of Economic Paleontologists and Minerologists, Pacific Section, p. 141–146.

Dunn, J. F., 1986, The structural geology of northeastern Whipple Mountains detachment terrane [M.S. thesis]: Los Angeles, University of Southern California, 113 p.

England, P., and Molnar, P., 1990, Surface uplift, uplift of rocks, and exhumation of rocks: Geology, v. 18, p. 1,173–1,177.

Etheridge, M. A., Symonds, P. A., and Powell, T. G., 1988, Application of the detachment fault model for continental extension to hydrocarbon exploration in extensional basins: Australian Petroleum Exploration Association, v. 28, p. 167–187.

Faulds, J. E., Gans. P. B, and Smith, E. I., 1994, Spatial and temporal patterns of extension in the northern Colorado River extensional corridor, northwestern Arizona and southern Nevada: Geological Society of America Abstracts with Programs, v. 26, no. 2, p. 51.

Gans, P. B., 1987, An open-system two-layer crustal stretching model for the eastern Great Basin: Tectonics, v. 6, p. 1–12.

Gans, P. B., Miller, E. L., McCarthy, J., and Ouldcott, M. L., 1985, Tertiary extensional faulting and evolving ductile-brittle transition zones in the northern Snake Range and vicinity: New insights from seismic data: Geology, v. 13, p. 189–193.

Hamilton, W. B., 1988, Detachment faulting in the Death Valley region, California and Nevada: U.S. Geological Survey Bulletin 1790, p. 51–85.

Hazlett, R. W., 1986, Geology of a Tertiary volcanic center, Mopah Range, San Bernardino County, California [Ph.D. dissert.]: Los Angeles, CA, University of Southern California, 303 p.

Henyey, T. E., Okaya, D. A., Frost, E. G., and McEvilly, T. V., 1986, CAL-CRUST (1985) seismic reflection survey, Whipple Mountains detachment terrane, California: An overview: Geophysical Journal of the Royal Astronomical Society, v. 89, p. 111–118.

Howard, K. A., and John, B. E., 1987, Crustal extension along a rooted system of imbricate low-angle faults: Colorado River extensional corridor, California and Arizona, *in* Coward, M. P., Dewey, J. F., and Hancock, P. L., eds., Continental extensional tectonics: London, Geological Society of London Special Paper 28, p. 299–311.

Howard, K. A., Stone, P., Pernokas, M. A., and Marvin, R. F., 1982, Geologic and geochronologic reconnaissance of the Turtle Mountains area, California: West border of the Whipple Mountains detachment terrane, *in* Frost, E. G., and Martin, D. L., eds., Mesozoic-Cenozoic tectonic evolution of the Colorado River region, California, Arizona, and Nevada (Anderson-Hamilton Volume): San Diego, California, Cordilleran Publishers, p. 341–354.

Howard, K. A. Christiansen, P. P., and John, B. E., 1993, Cenozoic stratigraphy of northern Chemehuevi Valley and flanking Stepladder Mountains and Sawtooth Range, southeastern Calif.: U.S. Geological Survey Bulletin 2053, p. 123–126.

Howard, K. A., Nielson, J. E., Wilshire, H. G., Nakata, J. K., Goodge, J. W., Reneau, S. L., John, B. E., and Hansen, V. L. 1995, Preliminary geologic map of the Mohave Mountains area Mohave County, western Arizona: U.S. Geological Survey Miscellaneous Investigations Map I-2308, scale 1:48,000, in press.

Jackson, J. A., and McKenzie, D., 1983, The geometrical evolution of normal fault systems: Journal of Structural Geology, v. 5, p. 471–482.

Jackson, J. A., and White, N. J., 1989, Normal faulting in the upper continental crust: Journal of Structural Geology, v. 11, p. 15–36.

John, B. E., 1987, Geometry and evolution of a mid-crustal extensional fault system: Chemehuevi Mountains, southeastern California, *in* Coward, M. P., Dewey, J. F., and Hancock, P. L., eds., Continental extensional tectonics: London, Geological Society of London Special Publication 28, p. 312–333.

John, B. E., and Foster, D. A., 1993, Structural and thermal constraints on the initiation angle of detachment faulting in the southern Basin and Range: The Chemehuevi Mountains case study: Geological Society of America Bulletin, v. 105, p. 1,091–1,108.

Kruse, S., McNutt, M., Phipps-Morgan, J., and Royden, L., 1991, Lithospheric extension near Lake Mead, Nevada: A model for ductile flow in the lower crust: Journal of Geophysical Research, v. 96, p. 4,435–4,456.

Lister, G. S., and Davis, G. A., 1989, The origin of metamorphic core complexes and detachment faults formed during Tertiary continental extension in the northern Colorado River region, U.S.A.: Journal of Structural Geology, v. 11, p. 65–94.

Lucchitta, I., and Suneson, N. H., 1993a, Dips and extension: Geological Society of America Bulletin, v. 105, p. 1,346–1,356.

Lucchitta, I., and Suneson, N. H., 1993b, Stratigraphic section of the Castaneda Hills–Signal area, Ariz.: U.S. Geological Survey Bulletin 2053, 137–144.

McCarthy, J., Larkin, S. P., Fuis, G. S., Simpson, R. W., and Howard, K. A., 1991, Anatomy of a metamorphic core complex: Seismic refraction/wide-angle reflection profiling in southeastern California and western Arizona: Journal of Geophysical Research, v. 96, p. 12,259–12,292.

Nielson, J. E., 1986, Miocene stratigraphy of the Mohave Mountains, Arizona, and correlation with adjacent ranges, *in* Geological Society of America, Cenozoic stratigraphy, structure and mineralization in the Mojave Desert: Los Angeles, Geological Society of America, Cordilleran Section, 82nd Annual Meeting Guidebook. Field Trip nos. 5 and 6: Los Angeles, California State University, p. 15–24.

Nielson, J. E., and Beratan, K. K., 1990, Tertiary basin development and tectonic implications, Whipple detachment system, Colorado River extensional corridor, California and Arizona: Journal of Geophysical Research, v. 95, p. 599–614.

Nielson, J. E., and Beratan, K. K., 1994, Stratigraphic record of detachment fault evolution, Colorado River extensional corridor: Geological Society of America Abstracts with Programs, v. 26, no. 2, p 77.

Nielson, J. E., and Beratan, K. K., 1995, Stratigraphic and structural synthesis of a Miocene extensional terrane, southeast California and west-central Arizona: Geological Society of America Bulletin, v. 107, p. 241–252.

Nielson, J. E., Lux, D. R., Dalrymple, G. B., and Glazner, A. F., 1990, Age of the Peach Springs Tuff, southeastern California and western Arizona: Journal of Geophysical Research, v. 95, p. 571–580.

Nielson, J. E., Pease, V. L., and Nakata, J. K., 1993, Paleomagnetic reversals in Miocene dikes, and tectonic evolution of the Crossman block, Mohave Mountains, Arizona: Geological Society of America Abstracts with Programs, v. 25, no. 5, p. 127.

Proffett, J. M., Jr., 1977, Cenozoic geology of the Yerington district, Nevada, and implications for the nature and origin of Basin and Range faulting: Geological Society of America Bulletin, v. 88, p. 247–266.

Reynolds, S. J., and Spencer, J., 1985, Evidence for large-scale transport on the Bullard detachment fault, west-central Arizona: Geology, v. 13, p. 353–356.

Sherrod, D. R., 1988, Preliminary geologic map of the Monkeys Head quadrangle, Mohave and La Paz Counties, Arizona: U.S. Geological Survey Open-File Report 88-597, scale 1:24,000.

Sherrod, D. R., and Nielson, J. E., 1993, Tertiary stratigraphy of highly extended terranes, California, Arizona, and Nevada; Proceedings of a workshop held at the Desert Research Center, Soda Springs (Zzyzx) California, February 9–12, 1990: U.S. Geological Survey Bulletin 2053, 250 p.

Simpson, C., Schweitzer, J., and Howard, K. A., 1991, A reinterpretation of the timing, position, and significance of the Sacramento Mountains detachment fault, southeastern California: Geological Society of America Bulletin, v. 103, p. 751–761.

Spencer, J. E., 1984, Role of tectonic denudation in warping an uplift of low-angle normal faults: Geology, v. 12, p. 95–98.

Spencer, J. E., and Reynolds, S. J., 1989, Tertiary structure, stratigraphy, and tectonics of the Buckskin Mountains, *in* Spencer, J. E., and Reynolds, S. J., eds., Geology and mineral resources of the Buckskin and Rawhide Mountains, west-central Arizona: Arizona Geological Survey Bulletin 198, p. 103–167.

Spencer, J. E., and Reynolds, S. J., 1993, Stratigraphy of middle Tertiary rocks in the central and eastern Buckskin Mountains, west-central Ariz.: U.S. Geological Survey Bulletin 2053, p. 149–150.

Spencer, J. E., and Turner, R. D., 1983, Geologic map of part of the northwestern Sacramento Mountains, southeastern California: U.S. Geological Survey Open-File Report 83-614, scale 1:24,000.

Wernicke, B., 1981, Low-angle normal faults in the Basin and Range province; nappe tectonics in an extending orogen: Nature, v. 291, p. 645–648.

Wernicke, B., and Axen, G., 1988, On the role of isostacy in the evolution of normal fault systems: Geology, v. 16, p. 848–851.

Yarnold, J. C., 1993, Middle Tertiary stratigraphy of the northern Rawhide Mountains and Artillery Mountains, Ariz.: U.S. Geological Survey Bulletin 2053, p. 145–148.

Yin, A., and Dunn, J. F., 1992, Structural and stratigraphic development of the Whipple-Chemehuevi detachment fault system, southeastern California: Implications for the geometrical evolution of domal and basinal low-angle normal faults: Geological Society of America Bulletin, v. 104, p. 659–674.

Young, R. A., and Brennan, W. J., 1974, Peach Springs Tuff: Its bearing on structural evolution of the Colorado Plateau and development of Cenozoic drainage in Mohave County, Arizona: Geological Society of America Bulletin, v. 85, p. 83–90.

Manuscript Accepted by the Society April 21, 1995

Printed in U.S.A.

Geological Society of America
Special Paper 303
1996

Stratigraphic effects and tectonic implications of the growth of normal faults and extensional basins

Roy W. Schlische
Department of Geological Sciences, Rutgers University, Busch Campus, Piscataway, New Jersey 08855-1179
Mark H. Anders
Department of Geological Sciences and Lamont-Doherty Earth Observatory, Columbia University, Palisades, New York 10964

ABSTRACT

Recent research on normal faults has established that (1) cumulative displacement is highest near the fault center and decreases toward the tips, and (2) faults lengthen as cumulative displacement increases. Half-graben-type basins are a fundamental manifestation of displacement on large normal fault systems, and thus are expected to be deepest near their centers and to grow in depth, width, and length through time. Basin growth models predict that progressively younger synextensional strata will onlap basement rocks, especially if sedimentation keeps pace with increasing basin capacity. Under certain circumstances, the models predict a transition from conditions in which sediment supply exceeds capacity (predominantly open basin) to one in which the basin is underfilled (predominantly closed basin) if sedimentation cannot keep pace with increasing basin capacity.

Some basins evolve through the merger of originally isolated subbasins whose border faults grew toward one another. The oldest strata form restricted sequences in each subbasin. Strata deposited shortly after consolidation thin toward the intrabasin high that forms in the merger zone, a region of short-term displacement deficit. For two merging coplanar faults, the merger zone is located at the center of the combined fault system, and displacement must increase to conform to the typical displacement profile and the scaling law between fault length and displacement. Thus, the youngest strata thicken toward the former location of the intrabasin high. If the growing faults overlap in the extension direction, displacement is distributed on multiple splay faults. The intrabasin high then has long-term expression, even though the summed fault displacement within the high is equal to or slightly greater than that of the deeper, flanking subbasins.

Oblique-slip accommodation zones form in the overlap zone of basins whose propagating faults dip in opposite directions. In general, faults growing in length obviate the need for transfer faults at the fault tips. Transfer faults may form if both fault tips cannot propagate. If only one tip is fixed, the other tip propagates away from the fixed tip, and the depocenter migrates in the same direction.

The fault and basin growth models described above provide a useful framework for interpreting the stratigraphic record of extensional basins and extracting their tectonic development, as demonstrated by examples from the Basin and Range and Mesozoic rifts of eastern North America.

Schlische, R. W., and Anders, M. H., 1996, Stratigraphic effects and tectonic implications of the growth of normal faults and extensional basins, *in* Beratan, K. K., ed., Reconstructing the History of Basin and Range Extension Using Sedimentology and Stratigraphy: Boulder, Colorado, Geological Society of America Special Paper 303.

INTRODUCTION

Considerable progress has been made in recent years in defining the architecture of extensional basins, which are, in nearly all cases, asymmetrical features (Bally, 1982; Wernicke and Burchfiel, 1982; Anderson et al., 1983; Jackson and McKenzie, 1983; Gibbs, 1984; Leeder and Gawthorpe, 1987; Rosendahl, 1987; Rosendahl et al., 1992; Schlische, 1993). Half graben are bounded on one margin by the border fault system (BFS) consisting of a network of mostly normal-slip faults; the opposite ramping margin is inclined at a shallower angle toward the BFS. Along-strike changes in basin geometry are known in less detail, but it is clear that basin asymmetry may alternate polarity in some rift systems (e.g., Rosendahl, 1987) or may remain unchanged for considerable distances in others (e.g., Schlische, 1993). In addition, there is a growing appreciation that extensional basins and their bounding fault systems are segmented (e.g., Gawthorpe and Hurst, 1993) and that basin evolution involves fault-tip propagation, basin growth, and linkage of smaller subbasins (Chapin, 1979; Bosworth, 1985; Ebinger, 1989; Gibson et al., 1989; Morley et al., 1990; Schlische, 1991, 1992, 1993; Nelson et al., 1992; Klitgord and Ager, 1993; Anders and Schlische, 1994). Nonetheless, details of extensional basin evolution remain sketchy.

A parallel research area in extensional tectonics concerns the three-dimensional geometry of normal faults and their growth through time. There are three important points: (1) Displacement is generally greatest at the center of a fault and decreases toward the fault tips (Fig. 1) (e.g., Barnett et al., 1987). (2) A positive correlation between displacement and fault length (Fig. 2) indicates that faults apparently grow in length as displacement increases (e.g., Watterson, 1986). (3) Faults are segmented at a variety of scales (Fig. 3) (e.g., Jackson and White, 1989; Peacock and Sanderson, 1991). In some cases, the segmented faults form a kinematically linked array (e.g., Walsh and Watterson, 1991).

In this chapter, we apply the results of research on normal fault displacement geometry, fault segmentation, and fault growth to extensional basin evolution. In particular, we explore the first-order stratigraphic effects of the growth of extensional basins in length, width, and depth through time as well as the consolidation of originally isolated subbasins. The models presented in this report are based purely on the geometric evolution of faults and the associated basins. Consequently, the models ignore the isostatic effects of sediment loading and erosion of uplifted footwall blocks, fail to account for sediment compaction, and do not incorporate changes in climate. Despite these obvious limitations, it is our hope that these basin growth models will provide a framework for interpreting the stratigraphic record of extensional basins in order to unravel the tectonic controls on sedimentation as well as the evolution of these basins and their bounding fault systems.

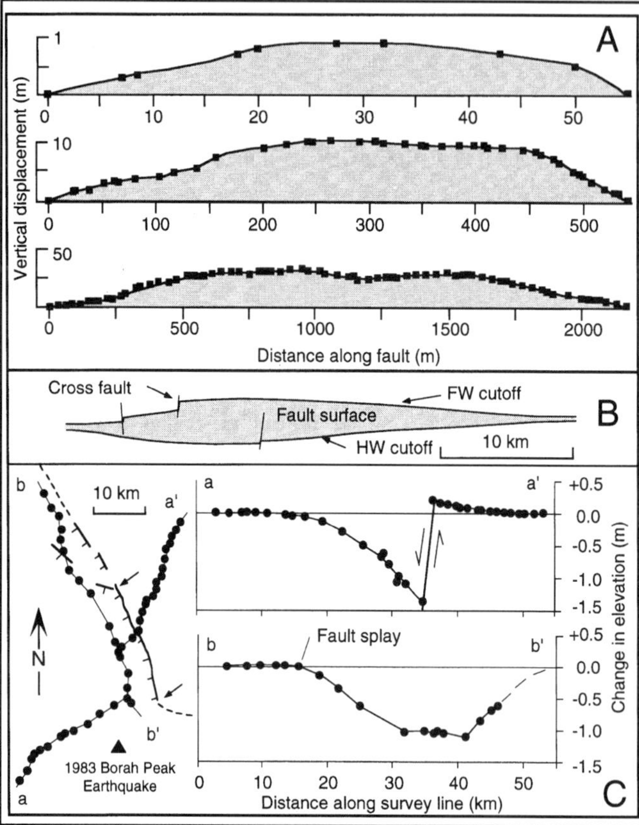

Figure 1. Scale invariance of displacement geometry of normal faults. (A) Variations in scarp height (a measure of displacement) on normal faults within the Volcanic Tablelands, California. Note the different horizontal and vertical scales. Modified from Dawers et al. (1993). (B) Along-strike variations in fault displacement shown by horizon separation diagram for Tiim Phonolite on the Saimo normal fault, Kenya rift valley. Modified from Chapman et al. (1978). (C) Changes in ground surface elevation following the 1983 Borah Peak earthquake on the Lost River normal fault, Idaho (epicenter indicated by triangle). Black circles indicate survey points; arrows delineate segment boundaries of Crone and Haller (1991); tick marks indicate parts of fault that developed prominent scarps following the 1983 earthquake. Modified from Barrientos et al. (1987).

GEOMETRY AND GROWTH OF NORMAL FAULTS

Detailed studies of individual normal faults have shown that maximum displacement occurs at or near the center of the fault and decreases to zero at the tips (Fig. 1) (Chapman et al., 1978; Muroaka and Kamata, 1983; Barnett et al., 1987; Walsh and Watterson, 1987; Scholz et al., 1993; Dawers et al., 1993). These variations in fault displacement are scale-invariant, having been reported on faults ranging in length from a few centimeters (microfaults cutting lacustrine strata in Mesozoic rift basins in eastern North America; Young et al., 1995) to tens and hundreds of meters (faults cutting the Bishop Tuff in the Volcanic Tablelands, California; Fig. 1A) to tens of kilometers (normal fault

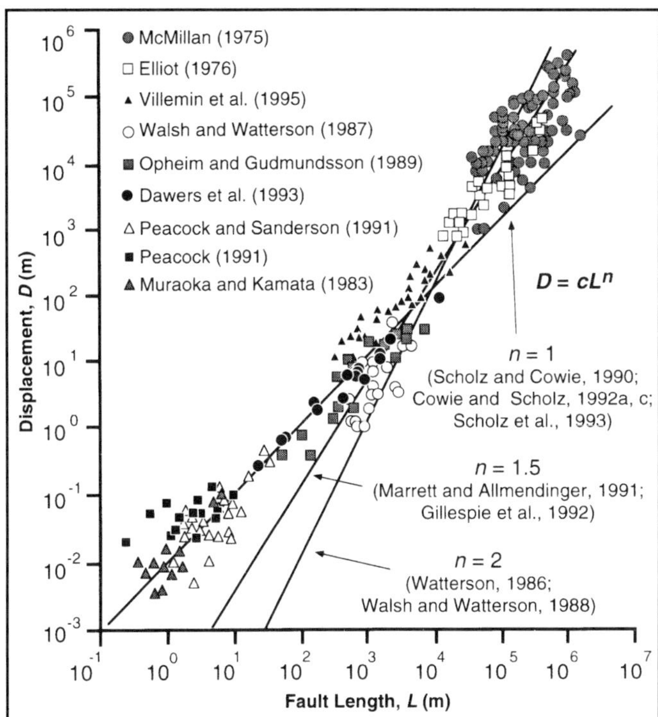

Figure 2. Log-log plot of fault length versus displacement. The positive correlation suggests faults grow in length as displacement accrues. General fault-growth scaling law and proposed values of n are also indicated (see text for discussion). Figure compiled mostly from Cowie and Scholz (1992a).

within the Kenya rift; Fig. 1B). As is amply demonstrated by the faults in the Volcanic Tablelands and Kenya rift, the along-strike displacement profile resembles a bell-shaped curve, a geometry that is self-similar over a wide range of fault length scales. As a result of these along-strike variations in fault displacement, hanging-wall depressions produced by the faulting resemble elongate synclines in longitudinal profile (parallel to the fault).

Along-strike variations in fault displacement are also evident in the patterns of footwall uplift associated with large range-bounding normal faults (Zandt and Owens, 1980; Jackson and McKenzie, 1983; King and Ellis, 1990). The magnitude of the footwall uplift commonly reaches a maximum near the center of the fault system and tapers off to zero near the ends (e.g., Zandt and Owens, 1980; Jackson and McKenzie, 1983; Anders and Schlische, 1994). Uplift results from elastic rebound during seismic slip events (Fig. 1C; Stein and Barrientos, 1985) and isostatic adjustment to unloading of the footwall block (Jackson and McKenzie, 1983).

Displacement also varies with distance measured normal to the fault: maximum displacement occurs at the fault surface and exponentially decreases away from the fault in both the hanging wall and the footwall (Barnett et al., 1987). The term "reverse-drag" or "rollover" is commonly applied to this type of hanging-wall geometry, which is historically associated with movement on listric normal faults (e.g., Hamblin, 1965; Gibbs, 1984). As noted by Barnett et al. (1987) and Roberts and Yielding (1994), this geometry is also associated with planar faults and cannot be exclusively used to infer the presence of listric normal faults.

The displacement geometry parallel and normal to fault surfaces described above also applies to the slip distribution following earthquakes on normal faults, such as the 1983 Borah Peak earthquake on the Lost River fault, Idaho (Fig. 1C; Stein and Barrientos, 1985; Barrientos et al., 1987). Note how longitudinal profile b–b′ resembles an elongate syncline; furthermore, transverse profile a–a′ exhibits a classic half-graben geometry. These geometric similarities strongly suggest that half-graben basins associated with large normal faults formed as a result of repeated seismic slip events (Stein et al., 1988).

In addition to the characteristic displacement geometry, fault populations exhibit a relationship between length and displacement (Fig. 2):

$$D = cL^n \quad \text{(Equation 1)}$$

where D is displacement, L is length, c is related to rock properties, and n is some exponent. This relation is interpreted to indicate that short faults with small displacements grow into longer faults with larger displacements over time (e.g., Watterson, 1986). Cowie and Scholz (1992c) presented a quantitative model in which fault growth occurs because additional displacement on a fault of finite length increases the displacement gradient, which controls the stress concentration at the fault tips; when this stress exceeds the strength of the rock, fracturing occurs at the fault tips, which then propagate through the fractured rock.

Considerable controversy surrounds the exact value of n in the scaling relation in Equation 1. Watterson (1986) and Walsh and Watterson (1988) originally proposed that $n = 2$. Using larger data sets and by combining data, Marrett and Allmendinger (1991), Gillespie et al., (1992), and Walsh and Watterson (1992) argued that $n \approx 1.5$ best fits the data. Scholz and Cowie (1990) and Cowie and Scholz (1992a,c) suggested that $n \approx 1$ for individual data sets; furthermore, Cowie and Scholz (1992a) argued that fault populations from different tectonic regions and rock types would likely have different values of c and therefore should not be combined. For $n = 1$, the growth process is self-similar (Cowie and Scholz, 1992a). Dawers et al. (1993) recently examined a population of normal faults from the Volcanic Tablelands, California. Fault lengths span three orders of magnitude and exhibit a linear relation with fault displacement ($n = 1$) (see Fig. 2). The value of n does not affect the qualitative predictions of the basin growth models presented in a following section, although it does influence basin geometry and stratigraphy to some extent.

Normal fault systems at all scales are commonly segmented (Schwartz and Coppersmith, 1984; Jackson and White, 1989;

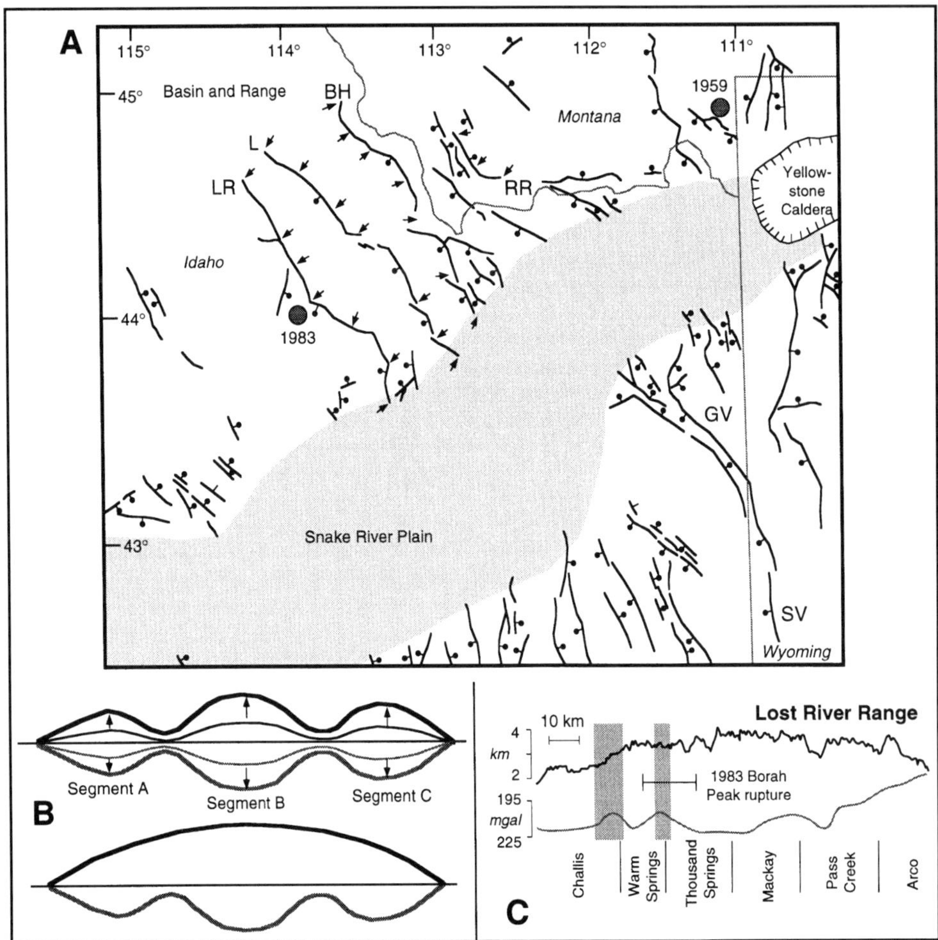

Figure 3. Segmented normal fault systems. (A) Normal faults of the northern Basin and Range. Arrows delineate fault segment boundaries defined by Crone and Haller (1991). Abbreviations are as follows: BH, Beaverhead fault system (see Fig. 12A); GV, Grand Valley fault system; L, Lemhi fault system; LR, Lost River fault system (see Fig. 12B), RR, Red Rock fault system; and SV, Star Valley fault system (see Fig. 12C). Modified from Machette et al. (1991). (B) Idealized footwall uplift and hanging-wall subsidence profiles (drawn parallel to fault surface) for segmented faults. In the upper diagram, fault segment boundaries are sites of persistent displacement deficit. In the lower diagram, segment boundaries are marked by overlapping faults that distribute displacement, reducing net basin subsidence, forming intrabasin highs. Modified from Anders and Schlische (1994). (C) Plot of footwall elevation and Bouguer gravity anomaly for the hanging wall of the Lost River fault. Gravity highs correlate with basement highs. Named fault segments are derived from Crone and Haller (1991). Shading indicates segment boundaries marked by fault overlaps, which do not correspond to footwall elevation lows. Modified from Anders and Schlische (1994).

Crone and Haller, 1991; Machette et al., 1991; Peacock and Sanderson, 1991; Roberts and Jackson, 1991; Stewart and Hancock, 1991; Zhang et al., 1991; Gawthorpe and Hurst, 1993; Anders and Schlische, 1994; Davison, 1994; Dawers and Anders, 1995; Jackson and Leeder, 1994; Trudgill and Cartwright, 1994; Wu and Bruhn, 1994). Fault-segment boundaries are recognized by bends in the fault trace, overlaps, offsets, apparent deficits in hanging-wall subsidence, and differences in the age of faulting on either side of the segment boundary. For basin-bounding faults in the Basin and Range, the typical segment length is 20–25 km. For some larger fault systems, centrally located segments (with higher displacements) are longer than distal segments (Fig. 3A; Crone and Haller, 1991; Machette et al., 1991; Wu and Bruhn, 1994). We suggest that the longer segments may have formed through the consolidation of smaller segments.

Neotectonicists consider many segment boundaries to be barriers to seismic ruptures and thus regions of displacement deficit (Schwartz and Coppersmith, 1984; Wheeler, 1987; Crone and Haller, 1989; Zhang et al., 1991). Anders and Schlische (1994) argued that some segment boundaries within the northern Basin and Range—except those at the ends of large fault systems—are not associated with footwall elevation lows, which would be the case if there were a displacement deficit (see Figs.

3B and 3C). However, some segment boundaries coincide with (1) regions of overlapping normal faults that distribute displacement and subsidence among the multiple splays and (2) regions in which fault tips have propagated toward one another. Specific examples are given in a following section.

Fault segments within a larger array are commonly kinematically linked (Larsen, 1988; Morley et al., 1990; Peacock and Sanderson, 1991; Walsh and Watterson, 1991; Davison, 1994; Dawers and Anders, 1995; Trudgill and Cartwright, 1994). In perhaps the most common linkage involving overlapping faults, the displacement on one fault decreases as that on the other increases; relay ramps form in the region of overlap. Individual faults within the array display along-strike variations in fault displacement similar to those mentioned above for isolated faults; however, displacement gradients are higher in the vicinity of overlapping fault segments (Peacock and Sanderson, 1991; Trudgill and Cartwright, 1994).

MODELS FOR THE GROWTH OF EXTENSIONAL BASINS

Basins bounded by a single fault

A half graben is simply a fault-bounded basin. Therefore, as the border fault increases in length during accumulation of displacement, the basin increases in length, width, and depth (Fig. 4; Gibson et al., 1989; Schlische, 1991). Along-strike variations in displacement on the border fault are responsible for the broadly synclinal geometry of the basin in longitudinal section (Fig. 4). Because fault growth requires maintaining a critical displacement geometry, which is governed by the cohesive strength of the surrounding material (Cowie and Scholz, 1992c), the geometry of displacement variations remains fairly constant during the growth process (Fig. 4). Note that the fault need not lengthen following each slip event; furthermore, the entire fault need not rupture during a given slip event. Once a sequence of slip events has created a critical displacement profile, the fault tips will propagate (Cowie and Scholz, 1992c).

Basin expansion affects the thickness and geometry of synextensional stratal units. These units are thickest adjacent to the boundary normal fault at or near the center of the fault and thin in all directions toward the edges of the basin. Assuming that sedimentation can keep pace with subsidence, younger sediments are deposited over a progressively larger depositional surface area. Consequently, strata progressively onlap pre-extensional rocks along the entire hanging-wall margin of the basin. Stratal onlap should be observable in transverse and longitudinal sections (Fig. 4). Transverse onlap in continental extensional basins is generally interpreted to reflect the filling of a basin that is widening through time (Leeder and Gawthorpe, 1987; Gibson et al., 1989; Schlische and Olsen, 1990; Schlische, 1991). In basins affected by erosion, longitudinal onlap may also be expressed in map view. As shown in stage 4 of Figure 4, the second oldest stratal unit pinches out along strike;

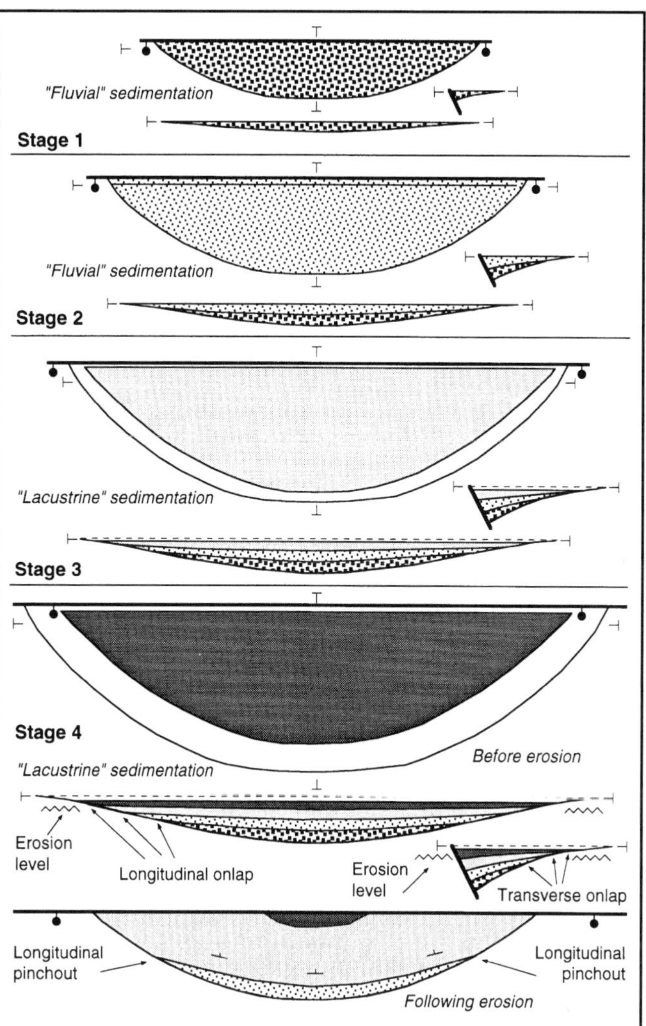

Figure 4. Model for the growth of a half-graben basin associated with a single border fault. Note that basin growth results in an ever-increasing depositional surface area manifested by onlap geometries and a transition from "fluvial" to "lacustrine" deposition as the enlarging basin becomes sediment starved. In stage 4, the map view of the basin shows the distribution of stratal units following erosional truncation. The oldest unit has no surface expression; the second oldest unit pinches out along strike (longitudinal pinchout).

Schlische (1993) termed this geometry "longitudinal pinchout." Note also that the oldest stratigraphic unit does not crop out at the surface.

The above-stated assumption that sedimentation can always keep pace with subsidence is unlikely in many circumstances. If the amount of hanging-wall subsidence relative to footwall uplift implied by Figure 1C is correct, then the footwall could not supply enough sediment to fill the basin, even if erosion could keep up with uplift. However, other data (Stein et al., 1988) suggest that the ratio of hanging-wall subsidence to footwall uplift is 2:1 to 1:1. In the latter case, the footwall could theoretically supply enough sediment to keep the basin filled. The hanging-wall block is also an important source of sediment. In fact, in the East African rift system, much of the sediment entering the basins is derived from the hanging-wall block or from axial sources (e.g., Cohen, 1990). Of course, in areas of distributed extension, such as the Basin and Range, the distinction between hanging-wall and footwall blocks blurs: one basin's footwall is another's hanging wall.

The large number of modern basins that are at present underfilled with sediments (e.g., basins containing lakes in East Africa, Baikal rift system, some basins in the Basin and Range) also indicates that sedimentation cannot always keep up with subsidence. On the other hand, the stratigraphic record of many ancient extensional basins (specific examples are discussed in a following section) indicates that at least initially sedimentation matched or exceeded subsidence. This condition is more likely to be met in basins that are fed by regional drainage systems. A modern example of this is provided by the Ruzizi River, which has infilled the northernmost half-graben unit within the Tanganyika rift system (Lambiase, 1990).

For basins in which there is initially an excess of sediment, basin growth (the progressive increase in the length, width, and depth of the basin) suggests that these basins may undergo a transition from being overfilled to underfilled (sediment starved). If the rate at which basin capacity increases is much greater than the rate of sediment infilling, younger strata may no longer onlap pre-extensional rocks but instead pinch out against older synextensional strata. Once tectonic activity ceases, the basin will fill with sediment. In summary, basin growth and filling models predict a tripartite basin evolution: (1) initially, sediment supply exceeds basin capacity; drainage is predominantly external; (2) the basin becomes underfilled with predominantly internal drainage; and (3) the basin fills and drainage becomes external after faulting ceases. Note that the stage 1 may not always be developed (depending on relative rates of sediment supply and increases in basin capacity) but is especially likely in the initial stages of basin development when the basin is short, narrow, and shallow. Episodic extension may lead to multiple tectonostratigraphic packages (see Lambiase, 1990). External changes in drainage systems and climate may complicate the stratigraphy. For example, erosional downcutting of basin outlet may result in transition from an underfilled closed basin to an overfilled open basin, and increased precipitation may lead to an increase in sediment supply.

According to Schlische and Olsen (1990), depositional environments in basins in which sediment supply exceeds capacity consist predominantly of basinwide fluvial systems, although local ponding of water can occur. The basin is externally drained. In the remainder of this chapter, we will refer to this depositional environment simply as "fluvial." If the basin is underfilled, the excess capacity of the basin may be occupied by a lake, provided sufficient water is available. Fluvial and deltaic sedimentation occurs around the margins of the lake. The relative amounts of lacustrine and fluvial/deltaic sedimentation depend on the size of the lake (a function of climate) with respect to the excess capacity of the basin. Under most circumstances, the underfilled basin will be internally drained, unless the level of the lake reaches the lowest outlet of the basin. Despite the wide range of depositional environments represented by the underfilled state, we will refer to them simply as "lacustrine." The above discussion applies to basins in terrestrial settings. Postma and Drinia (1993) applied some of these concepts to marine half graben.

The value of the exponent n (Equation 1) in the fault-growth scaling law is important in influencing extensional basin geometry and stratigraphy. As shown in Figure 5, for equivalent displacements and values of the scaling parameter c, the basin for which $n = 1$ is considerably larger than the basins for which $n > 1$. This suggests that the relationships among basin length, width, and depth can be used to constrain n. Furthermore, as n decreases and *all other parameters are held constant*, the transition from sediment excess to sediment-starvation should occur at earlier and earlier stages in the extensional history. However, it may be unrealistic to assume that sediment supply rates will remain constant if growing basins capture additional drainages.

The value of n also affects trends in sediment accumulation rates in growing extensional basins. Accumulation rate data shown in Figures 6A ($n = 1$) and 6B ($n = 2$) were obtained from basin growth and filling models similar to those in Figure 4 in which the total volume of sediment supplied to the basin per unit

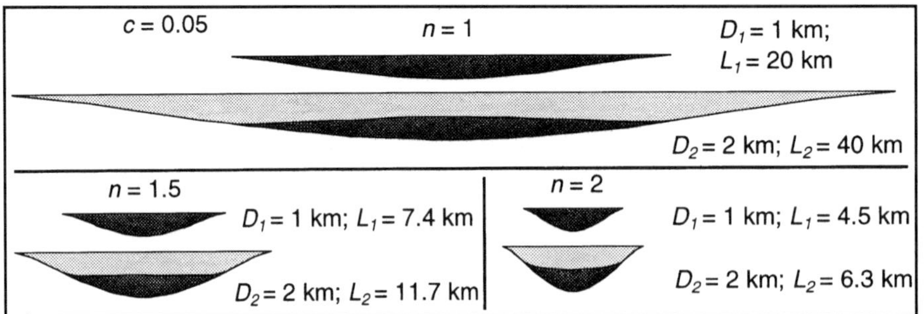

Figure 5. Effect of different values of n (Equation 1) on basin geometry (parallel to border fault system) and geometry of synextensional units (assuming basin always filled to lowest outlet) as seen in longitudinal section. The depth of the basins in both stages is the same in the three cases.

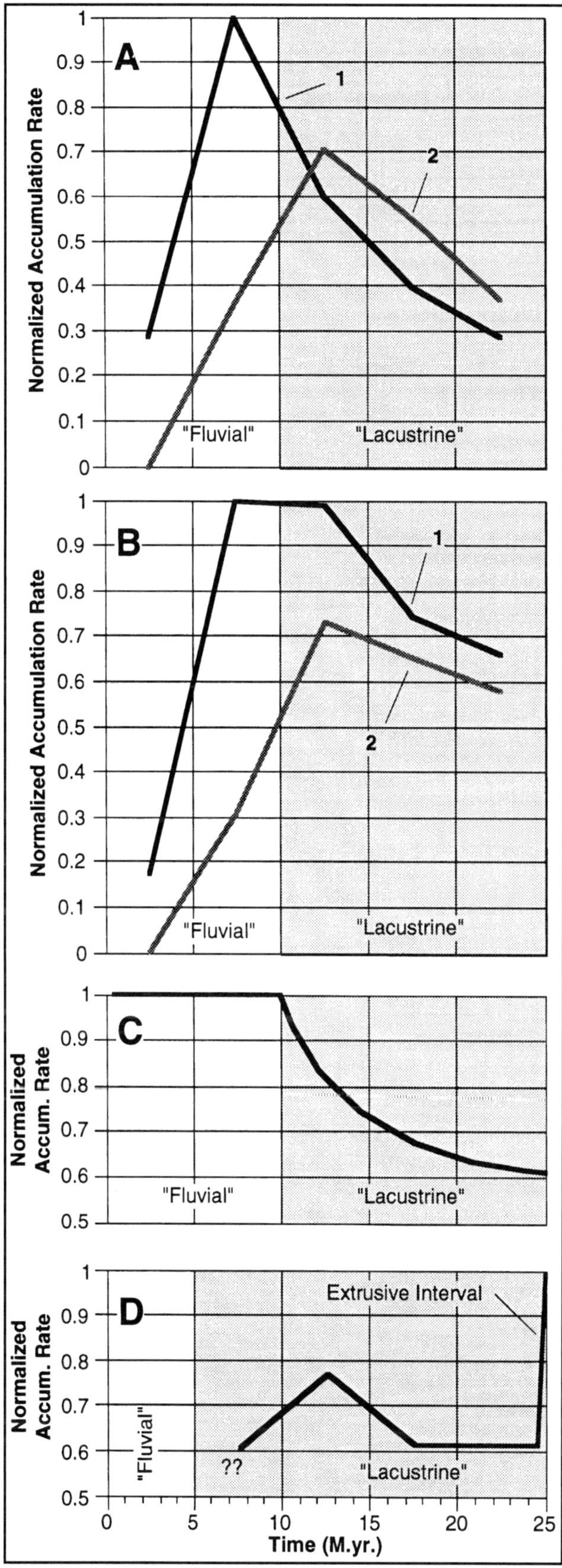

time was constant (Schlische, 1991). Accumulation rates (normalized to permit easier comparison) in both models first rapidly increase, reach a maximum near the "fluvial"-"lacustrine" transition, and then decline. The rate of decline is faster in the basin for which $n = 1$. The models also indicate that trends in accumulation rate depend on location within the basin: curve 1 in Figures 6A and 6B is based on data from near the center of the basin, whereas curve 2 was derived from measurements taken nearer the edge of the basin. Note that the form of the accumulation rate curves for the basin growth models is substantially different from the full-graben model (Fig. 6C) in which the length of the basin is fixed and the width and depth increase through time (Schlische and Olsen, 1990).

The trends in accumulation rates are considerably different if the basin is always filled to capacity. In a growing basin for which $n = 1$, accumulation rates remain constant at the center of the basin (as long as the fault displacement rate is constant) but increase through time at more edgeward localities. For the $n = 2$ case, accumulation rates increase everywhere within the basin through time (Gibson et al., 1989). In the full-graben model, accumulation rates everywhere within the basin are equal to the subsidence rate.

The models shown in Figure 4 assume that both fault tips are free to propagate along strike. Even if one fault tip is pinned (perhaps it has propagated into a region where the cohesive strength of the rock is locally higher), the other tip may still propagate, and the locus of maximum subsidence shifts with it (Fig. 7A). Thus the depocenter of the basin migrates in the same direction as the propagating fault tip. Like all features associated with fault and basin growth, this process should leave a characteristic signature in the stratigraphic record. Note that in Figure 7A the position of the depocenter is offset somewhat from the region of maximum depth to basement. Transfer faults may form to accommodate strain if both fault tips are inhibited from propagating (Fig. 7B).

Basins bounded by multiple fault segments

As noted above, normal fault systems commonly consist of multiple fault segments, and thus the model shown in Figure 4 is an oversimplification in some cases. The geometry of these mul-

Figure 6. Normalized accumulation rates plotted against time since initiation of extension. (A) Data derived from half-graben growth and filling model (similar to Fig. 4) for which the volume of sediment supplied to the basin per unit time is constant; half-graben growth governed by Equation 1, where $n = 1$ (data based on Fig. 7 of Schlische, 1991). Curve 1 represents data derived from location near the center of the basin; data in curve 2 were derived from a more edgeward location. (B) Same as A, but $n = 2$ (data based on Fig. 6 of Schlische, 1991). (C) Accumulation rates based on full-graben basin filling model of Schlische and Olsen (1990). (D) Accumulation rates derived from core data in the Newark basin (Olsen et al., 1995). Accumulation rates were averaged over 5 m.y. long time increments (except for extrusive interval).

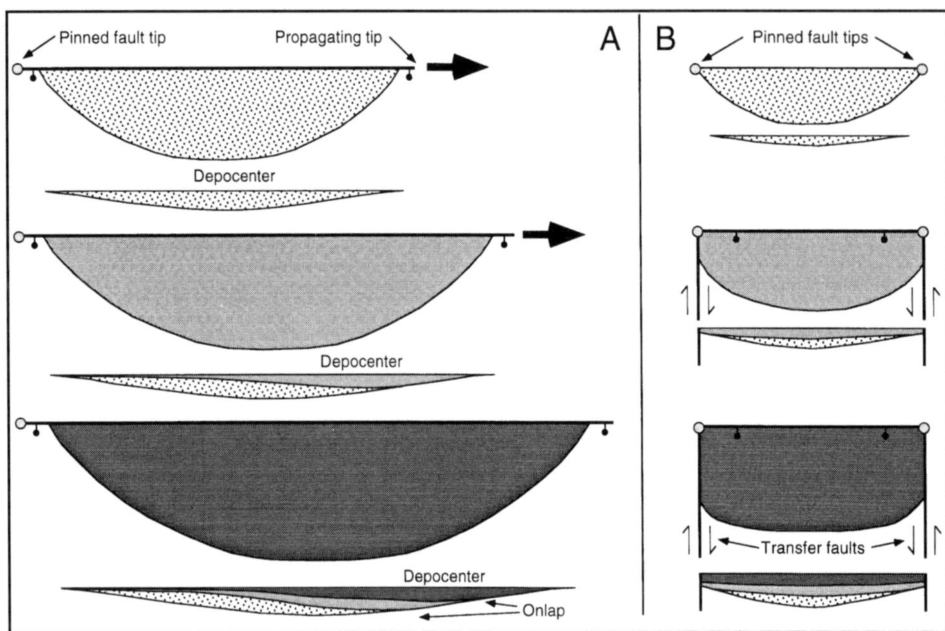

Figure 7. (A) Basin growth and filling model for a fault in which one segment boundary is pinned and the other is capable of lateral propagation. Note the migration of the basin depocenter in the direction of fault-tip propagation. (B) Basin growth and filling model for a normal fault in which both tips are pinned. Increase in displacement gradient requires the formation of transfer faults to accommodate strain.

tiple fault segments may be classified on the basis of the amount of overlap of adjacent fault segments (Anders and Schlische, 1994). Case 1 involves faults that do not overlap in map view (Fig. 8). In Case 2 , the fault segments overlap but are closely spaced relative to the size of composite basin (Fig. 9). Case 3 concerns widely spaced overlapping faults (Fig. 10). Next we explore the stratigraphic implications of the growth of segmented fault systems.

Case 1: No overlap. In the first stages, each fault segment is associated with its own isolated sedimentary basin (Fig. 8, stage 1). Each subbasin resembles a simple basin bounded by a single border fault (Fig. 4). The oldest stratigraphic units are restricted to each subbasin and thicken from the edges of the subbasin toward its center. As displacement increases on the individual fault segments, the fault tips propagate toward one another.

After the faults have merged, stratigraphic units are deposited basinwide but nonetheless thin toward the region where the fault segments merged, for this is an area of displacement deficit known as an intrabasin high (Fig. 8, stage 2; Anders and Schlische, 1994). In order for the fault system to maintain the proper scaling relationship between length and displacement (Equation 1; Fig. 2), displacement must increase in the zone of merger to compensate for the greatly increased length of the fault zone following merger. As displacement increases in the zone of merger, we expect little increase in the length of the fault zone; this is because the displacement profile is not in the critical state required for fault tip propagation (Cowie and Scholz, 1992c). After the displacement deficit is eliminated and a critical displacement profile is reestablished, the fault system can begin to lengthen again. Stratigraphic units deposited during this stage of basin evolution thicken toward the center of the composite basin (Fig. 8, stage 3).

The model in Figure 8 assumes that the fault segments merge in the same plane; this is reasonable if basin development occurs as a result of localized partial reactivation of preexisting faults. In other cases, it is more likely that the fault segments will overlap to some extent as the fault tips propagate toward and past one another.

Case 2: Closely spaced overlapping faults. The first stages of the evolution of a basin bounded by closely spaced overlapping faults are similar to that of a Case 1 basin: each fault segment is associated with an isolated subbasin containing a restricted stratigraphic wedge (Fig. 9, stage 1). While displacement accrues on the individual fault segments, the fault tips propagate; eventually the fault segments overlap in map view. At this stage, the originally isolated basins merge, and stratal units are deposited basinwide. Once again, stratigraphic units are expected to thin toward the intrabasin high marking the region of merger and its attendant temporary fault displacement deficit (Fig. 9, stage 2). The geometry of the basin and stratigraphic units deposited after this stage depends on the number of fault segments that remain active. If the basin develops a single, continuous (albeit somewhat curved) border fault system, the displacement deficit in the merger zone is overcome, and the

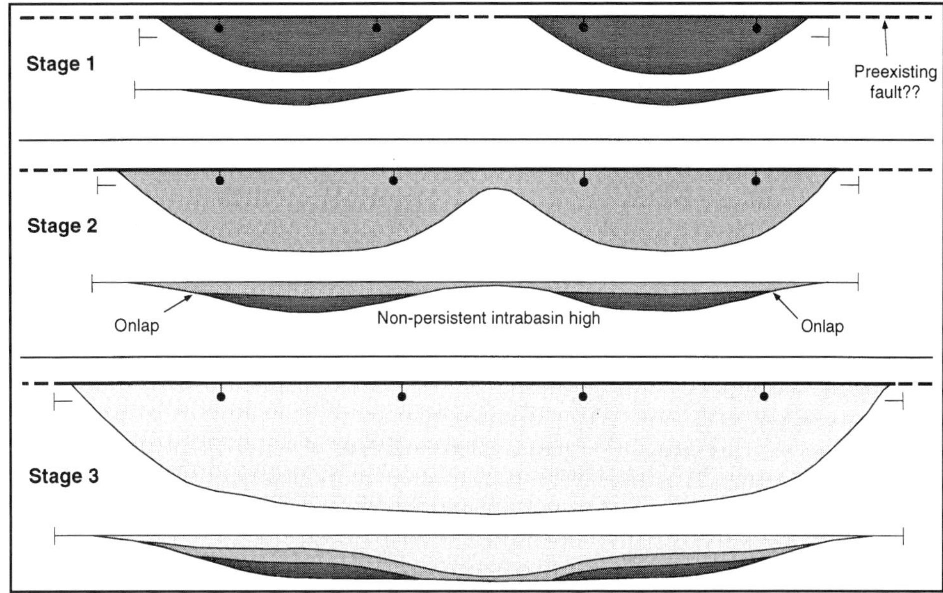

Figure 8. Basin growth and filling model for two fault segments that lie in the same plane. An intrabasin high initially forms in the area where the two fault segments merged but becomes buried as displacement accrues on the composite fault system.

geometry of stratigraphic units in longitudinal section closely resembles that of a Case 1 basin in stage 3 of its evolution (see Fig. 8). If multiple overlapping fault segments remain penecontemporaneously active, total fault system displacement is distributed among the various fault segments, such that the depth of the basin in the region of merger is reduced compared with those parts of the basin bounded by a single fault (Fig. 9, stage 3). Thus, even those units deposited after the original displacement deficit is overcome in the merger zone still thin toward this region, which may be termed a persistent intrabasin high (R. V. Ackermann, 1993, personal communication). The intrabasin high developed in the Case 1 basin is not persistent.

According to Gawthorpe and Hurst (1993), the zone of overlap in stage 1 of Figure 9 is classified as an interbasin transfer zone, whereas it is called an intrabasin transfer zone in stages 2 and 3. In stage 1, footwall uplift is minimal in the zone of overlap, which could therefore serve as a region in which drainages enter the basins (Gawthorpe and Hurst, 1993) and deposit coarse clastic material. By stage 3, footwall uplift is at its maximum in the zone of the overlap (see Fig. 7b in Anders and Schlische, 1994), thus limiting drainage into the basin in this region. Sediments deposited on and around the intrabasin high should be finer grained than those that accumulated during stages 1 and 2.

Case 3: Widely spaced overlapping faults. In this type of basin geometry, the fault segments are so widely separated that the individual basins may never merge (Fig. 10). Therefore, the geometry of stratigraphic units in each basin resembles that of the isolated basin in Figure 4. Although separate from one another in map view, the overlapping faults may be still kinematically linked. Linkage would be indicated if the total displacement profile (obtained by summing the displacement on the multiple segments in the direction of extension) is identical to that of an isolated fault. For all three stages in Figure 10, the zone of overlap is classified as an interbasin transfer zone by Gawthorpe and Hurst (1993). Footwall uplift is minimal in the zone of overlap throughout the evolution of the fault system; drainages are therefore likely to exploit these lows (Gawthorpe and Hurst, 1993).

The three cases of overlapping faults described above involve faults that dip in the same general direction. Figure 11 shows the development of a composite basin in which the border faults dip toward one another. Consequently, as the fault tips approach, they begin to interfere with one another. Interference may be avoided if the fault tips curve (Fig. 11, stage 3). Because these sections of the faults are no longer perpendicular to the extension direction, they experience a higher component of strike slip. The higher strike-slip component and reduced displacement toward the fault tips result in a lower amount of basin subsidence in this zone, forming a high-relief accommodation zone (nomenclature of Rosendahl, 1987), also referred to as isolation accommodation zones (Versfelt and Rosendahl, 1989) and antithetic interbasin highs (Gawthorpe and Hurst, 1993). Stratigraphic units should thin toward this region.

The geometries of the stratigraphic wedges shown in Figures 7–11 were constructed assuming that basins or subbasins were always filled to the lowest outlet of the basin. This corresponds to a "fluvial" depositional environment, as described above. Assuming a finite supply of sediment, a transition from "fluvial" deposition to "lacustrine" sedimentation could be produced at any time

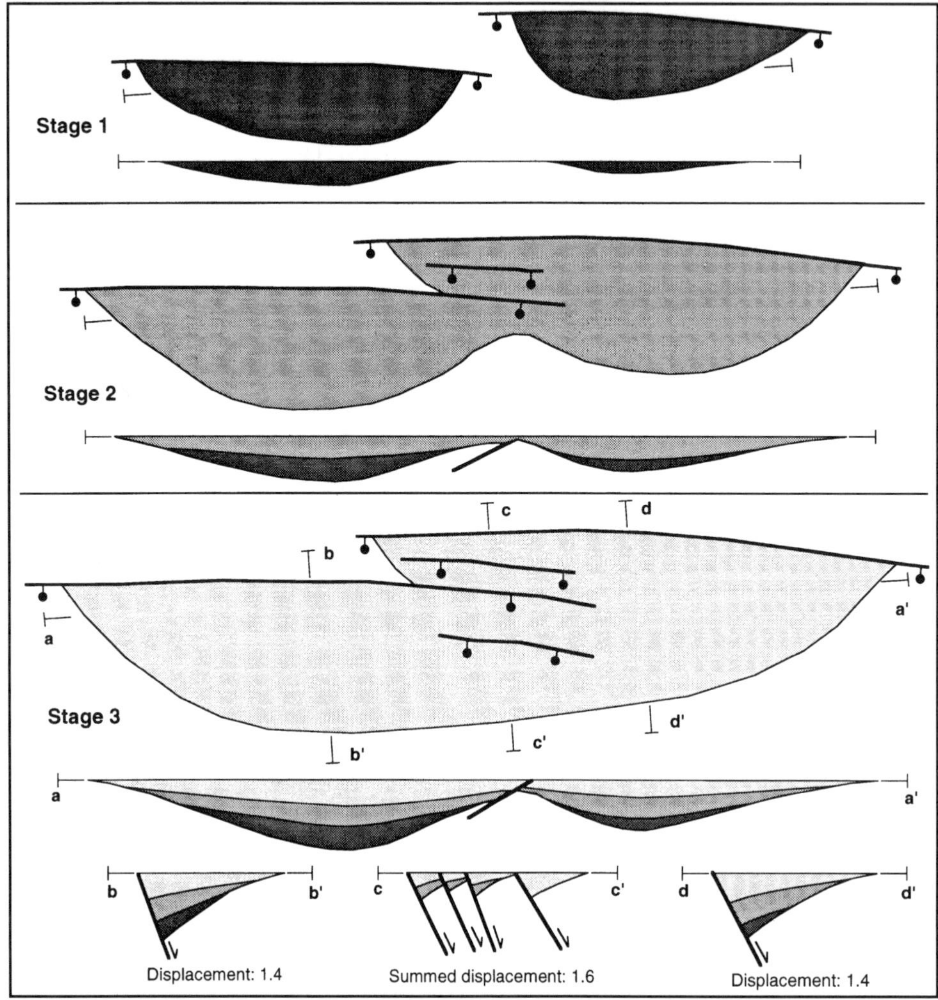

Figure 9. Basin growth and filling model for two closely overlapping fault segments. An intrabasin high forms in the region of fault overlap. The intrabasin high persists into stage 3 because displacement is distributed on multiple faults. Note that the summed displacement in the zone of overlap is slightly greater than that of the faults bounding the flanking subbasins while total basin depth is considerably reduced in the zone of overlap, an intrabasin high.

during the enlargement of the basin or subbasin. A switchover to "lacustrine" sedimentation may be more likely in the period during and following subbasin merger, especially if the width of the basin increases to scale with newly increased length. Again, as noted previously, the timing of any "fluvial"-"lacustrine" transition may be influenced by the growing basin capturing additional drainages. The geometry of stratal units deposited in a sediment-starved basin will not differ too greatly from those depicted in Figures 7–11. At the edges of the composite basins, younger stratal units may pinch out against older synextensional units instead of onlapping "basement" rocks.

EXAMPLES OF FAULT AND BASIN GROWTH

The fault and basin growth models predict specific fault, basin (or subbasin), and stratal geometries as well as major facies transitions. In the following sections, we document examples of these features from the Basin and Range province and the Mesozoic rift system of eastern North America.

Basin and Range

As discussed previously, many of the fault systems within the Basin and Range are segmented. According to Crone and Haller (1991), the 120 km long Beaverhead fault system of eastern Idaho consists of six fault segments (Fig. 3A). The boundaries of the southernmost three segments are marked by regions of overlapping faults and intrabasin highs (Fig. 12A; Anders and Schlische, 1994). The southern intrabasin high occurs near Blue Dome and is characterized by exposures of Tertiary and Paleozoic rocks flanked by Quaternary alluvium of the adjacent subbasins of Upper and Lower Birch Creek Valleys. The Blue

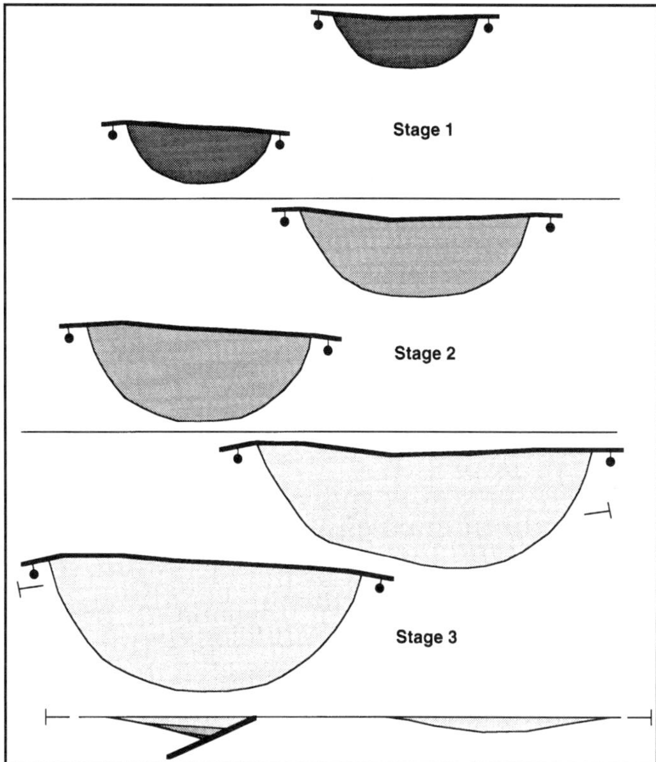

Figure 10. Basin growth and filling model for widely spaced overlapping faults.

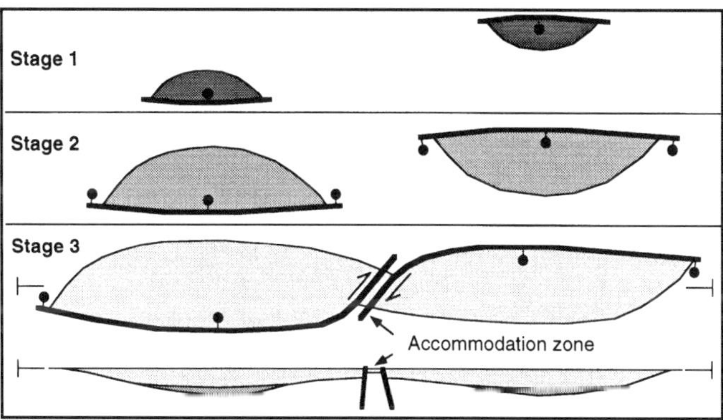

Figure 11. Basin growth and filling model for faults dipping in opposite directions. An accommodation zone forms in the region where the faults overlap. Based partly on Rosendahl (1987).

Dome intrabasin high is also characterized by several overlapping intrabasinal splay faults. Offsets of key stratigraphic marker horizons (the late Miocene [10 Ma] Medicine Lodge Limestone, basalt flows [~8 Ma], and the 6.6 Ma ash-flow tuff of Blacktail) within the Blue Dome region indicate that the summed fault displacement on all faults shown in cross section X–X′ is approximately 2.5 km (Fig. 12A; Anders and Schlische, 1994). Displacements on the main range-bounding faults to the north and south of the intrabasin high are both approximately 3 km (Anders et al., 1993; Anders and Schlische, 1994). Therefore, although the displacements are comparable in the subbasins and intrabasin high, maximum basin depth in the intrabasin high (~0.5 km) is considerably less than that of the flanking subbasins because displacement is distributed on multiple overlapping faults. The northern intrabasin high known as Middle Ridge is also characterized by multiple overlapping faults and reduced hanging-wall subsidence as evidenced by outcrops of Tertiary strata (Fig. 12A; Anders and Schlische, 1994). Neither intrabasin high is associated with footwall elevation lows, further suggesting that these are not sites of long-term displacement deficit (Anders and Schlische, 1994).

The stratigraphic sequence within the Blue Dome intrabasin high is tilted uniformly (Rodgers and Anders, 1990). Note that two locations in the Blue Dome intrabasin high reported to have tilts of 4° ± 4° and 5° ± 4° (Rodgers and Anders, 1990) were subsequently determined to be on a large landslide block. A similar stratigraphic sequence to the south of Blue Dome exhibits progressively gentler tectonic tilts. Anders and Schlische (1994) concluded that the southernmost segment of the Beaverhead fault system (Fig. 12A) was active prior to the splay faults within the intrabasin high. These splay faults subsequently formed as the tips of the fault segments bounding Upper and Lower Birch Creek Valleys propagated toward one another, similar to the basin evolution shown in Figure 9. Anders and Schlische (1994) suggested a similar origin for the Middle Ridge intrabasin high. The difference in strike between the intrabasinal splay faults and the main range-bounding fault segments of the Beaverhead fault system is attributed to a change in the extension direction during fault and basin growth (Anders and Schlische, 1994).

Six fault segments comprise the Lost River fault system of east-central Idaho (Figs. 3A and 3C; Crone and Haller, 1991), two of which ruptured in 1983 during the $M_s = 7.3$ Borah Peak earthquake (Richins et al., 1987; Crone et al., 1987). As with the Beaverhead fault system, the hanging wall of the northern Lost River fault system is characterized by elevation lows (Round Valley, Antelope Flat, Thousand Springs Valley) covered mostly by Quaternary alluvium separated by intrabasin highs (Antelope Hills and Willow Creek Hills) marked by outcrops of Paleozoic rocks and multiple overlapping splay faults (Fig. 12B). Furthermore, the intrabasin highs are not associated with marked footwall elevation lows (Fig. 3C), suggesting no long-term displacement deficit in these regions. It is our contention that displacement is distributed on the overlapping faults within the intrabasin highs, the sum of which is similar to the displacement on the main range-bounding faults adjoining the Antelope Flat and Thousand Springs Valley subbasins. However, a lack of appropriate exposures precludes more precise quantitative estimates of fault displacement. Nonetheless, the similarity of the geometry of the intrabasin highs along the Lost River fault to those along the Beaverhead fault suggest that the intrabasin highs are the sites where fault segments propagated toward one another.

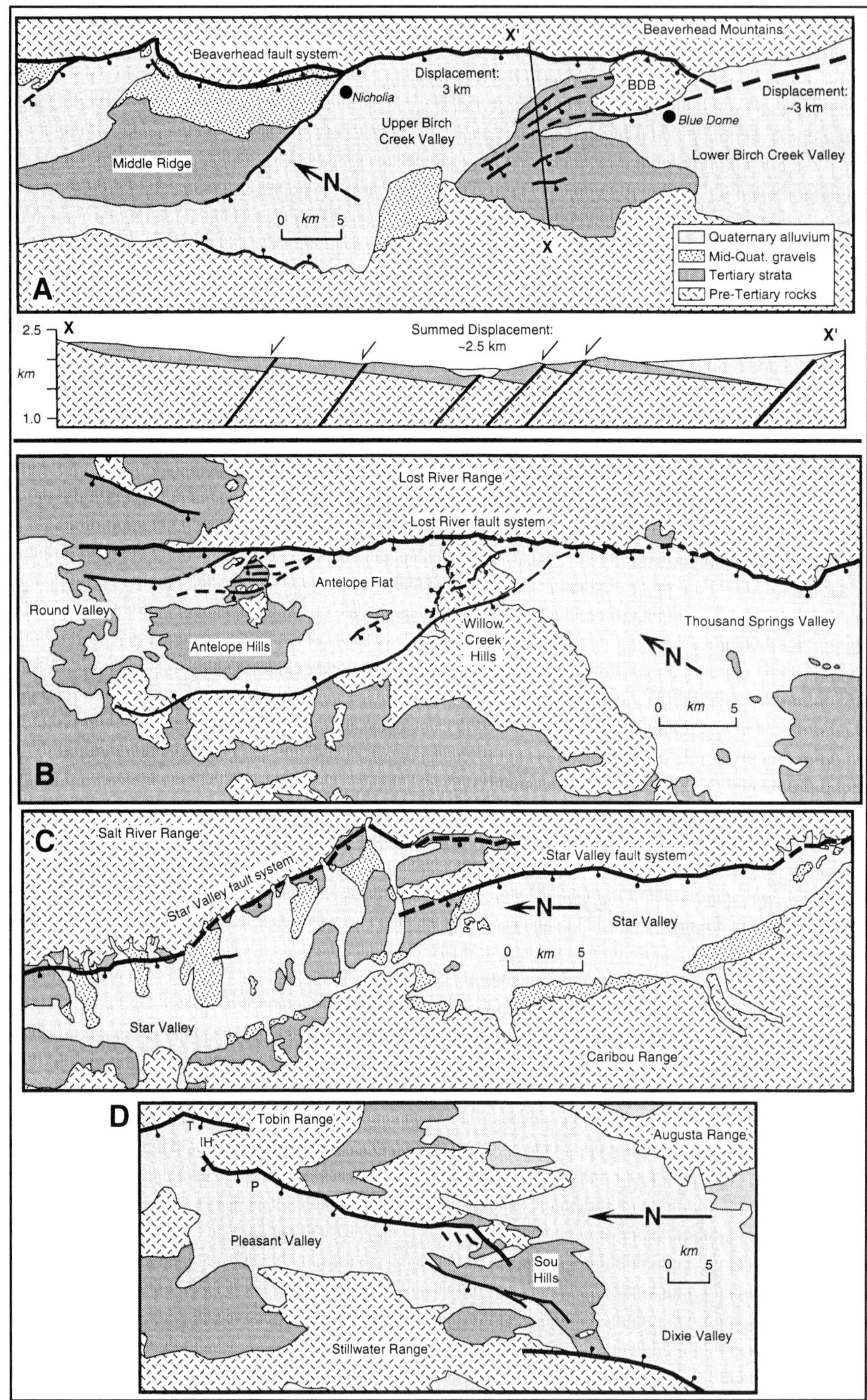

Figure 12. Segmented fault systems and associated intrabasin highs of the northern Basin and Range (see Fig. 3A for locations). Range-bounding fault systems shown with heavier lines. (A) Generalized geologic map of two intrabasin highs along the southern part of the Beaverhead fault system. Cross section passes through Blue Dome intrabasin high. BDB = Blue Dome block. Modified from Anders and Schlische (1994). (B) Generalized geologic map of two intrabasin highs along the northern half of the Lost River fault system. Based on Crone et al. (1987), Crone and Haller (1989), Fisher et al. (1983), and mapping by Mark H. Anders and Noel Howe. (C) Generalized geologic map of Star Valley showing intrabasin high where two segments of the Star Valley fault system overlap. Modified from Anders et al. (1989). (D) Generalized geologic map of interbasin high (Sou Hills) and intrabasin high (IH) along the Pleasant Valley–Dixie Valley fault system, Nevada. P and T are the Pierce and Tobin segments of the Pleasant Valley fault system, respectively. Modified from Fonseca (1988) and Zhang et al. (1991).

Two strands of the 140 km long Grand Valley/Star Valley fault system, Wyoming (Fig. 3A), overlap each other over a distance of 10 km, forming an intrabasin high between them (Fig. 12C). Studies of latest Quaternary rupturing history of the Star Valley fault system (Piety et al., 1986, 1992; Anders et al., 1989; McCalpin et al., 1990) showed that coeval ruptures occurred on both sides of the intrabasin high, suggesting that ruptures transfer from one segment to the other. In the zone of overlap, the basin depth for each fault strand is greater than 1 km (proprietary Chevron Research seismic reflection data). Seismic reflection studies (Dixon, 1982) and proprietary seismic reflection data (Chevron Research) also show that the maximum basin depth is about 3.5 km. Fault displacement is highest near the centers of both the northern and southern fault segments and tapers to zero at the segment ends. The displacement profiles of these segments are therefore similar to those of isolated normal faults (Barnett et al., 1987; Walsh and Watterson, 1987). This strongly suggests that the intrabasin high formed where the two segments linked together to form the larger Star Valley fault system. As with the Blue Dome intrabasin high along the Beaverhead fault, the total displacement in the region of the intrabasin high represents a large percentage of the total displacement for the deepest parts of the adjoining basins.

Wu and Bruhn (1994) concluded that the South Oquirrh Mountains fault system of Utah resulted from the growth and linkage of four fault segments. On the basis of geomorphological data, Jackson and Leeder (1994) inferred that the northern end of the Pierce segment of the Pleasant Valley fault system in Nevada (Fig. 12D) grew by ~50 m per earthquake slip event. Other examples of fault segmentation, overlapping faults, and associated structures in the Basin and Range are discussed by Schwartz and Coppersmith (1984), Jackson and White (1989), Crone and Haller (1991), Machette et al. (1991), Zhang et al. (1991), and Gawthorpe and Hurst (1993).

In the specific examples discussed thus far, the fault segments dip in the same direction. The Sou Hills is a bedrock high (Fonseca, 1988) exposed in the overlap zone between the oppositely dipping Dixie Valley and Pleasant Valley fault systems (Fig. 12D). This overlap zone is classified as antithetic interbasin by Gawthorpe and Hurst (1993). Unlike the intrabasin highs in the northern Basin and Range described above, the Sou Hills bedrock high (an interbasin high) corresponds to a low in the elevation of the footwall blocks. In this case, a fault segment boundary does correlate with a footwall elevation low and a region of displacement deficit.

As discussed in the basin growth and filling models above, hanging-wall onlap is diagnostic of basin growth. Transverse onlap has been reported in the following Cenozoic basins of the Basin and Range: Dixie Valley, northern Fallon basin, and Diamond Valley basin of Nevada (Anderson et al., 1983, their figs. 4–6) as well as the Railroad Valley basin of Nevada (Fig. 13A; Vreeland and Berrong, 1979). The onlap suggests that these basins widened as they filled.

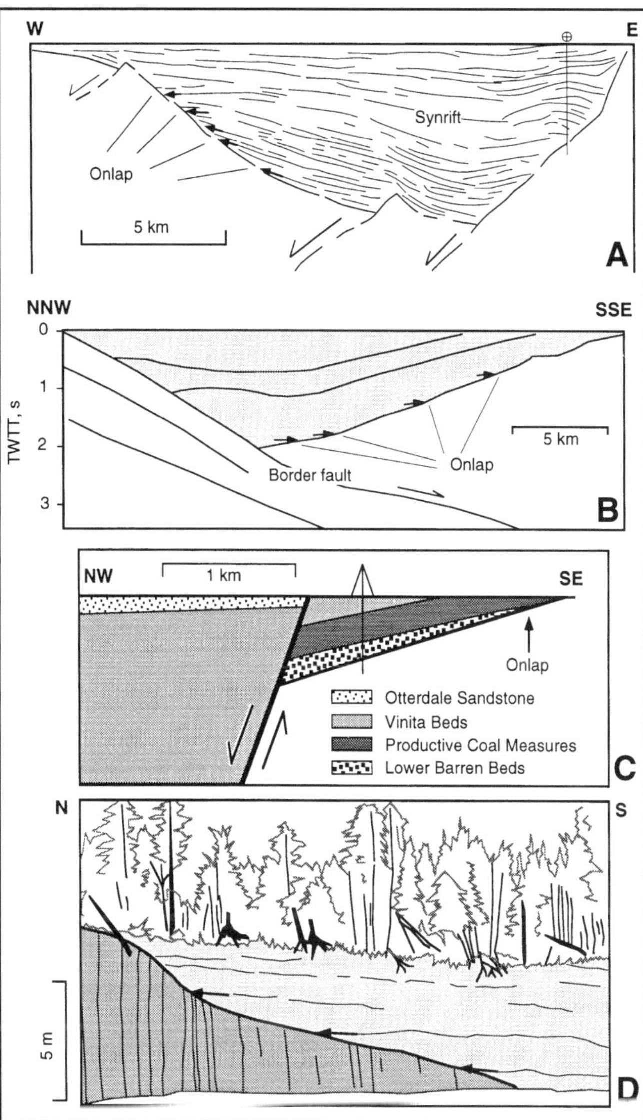

Figure 13. Transverse hanging-wall onlap. (A) Transverse onlap interpreted from seismic profile of the Railroad Valley basin, Nevada (Vreeland and Berrong, 1979). (B) Onlap of fluvial Stockton Formation onto basement rocks observed on seismic profile across the Newark basin, Pennsylvania (Schlische, 1992). (C) Enlargement of cross section M–M' in Figure 15, showing onlap along the southeastern margin of the Richmond basin based on outcrop and drill hole data (Olsen et al., 1989). (D) Sketch of sea cliff at Tennycape, Nova Scotia, along southeastern margin of Fundy basin showing gently dipping fluvial strata of Triassic Wolfville Formation onlapping subvertical lacustrine strata of the Mississippian Horton Group. View is to the south. For location, see arrow in inset for Figure 14A. Note that both stratal thickening toward the border fault and hanging-wall onlap occur in A, B, and C.

Basin growth and filling models also predict a tripartite stratigraphy, which has been reported by DiGiuseppi and Bartley (1991) for the White River Valley basin, Nevada. The middle lacustrine unit accumulated during closed basin sedimentation during active faulting. The overlying fluvial unit represents the post-faulting infilling of the basin. The present-day open drainage of the basin is a result of downcutting by tributaries to the Colorado River.

The simple tripartite stratigraphy predicted by the models need not be present in all Basin and Range basins for the following reasons: (1) Given the regionally distributed nature of extension, not all basins may have started out with a sediment excess corresponding to "fluvial" sedimentation. (2) A large part of the basin fill consists of igneous material, which the basin filling models did not consider. In addition, lava flows can create dams that raise the outlets of basins. (3) Downcutting by streams can lower the elevation of the basin's outlet. (4) As much of the basin fill is buried by the present-day depositional surface, considerable uncertainty surrounds the recognition of basin-scale changes in depositional environment through time.

Triassic-Jurassic basins of eastern North America

The early Mesozoic breakup of Pangea resulted in the formation of numerous rift basins along the east coast of North America (Fig. 14A, inset) (e.g., Manspeizer, 1988). The majority of these basins are half graben (see Fig. 15) bounded along one margin by predominantly normal-slip border fault systems (BFS). These faults were syndepositionally active as indicated by an increase in thickness of synextensional units toward the BFS and a decrease in dip in younger units (Fig. 15; Schlische, 1993, and references therein). The BFS generally parallels the grain of older Paleozoic structures; consequently, most of the border faults are known or inferred to be reactivated structures (e.g., Swanson, 1986). The BFS is also commonly segmented, with segment boundaries marked by changes in fault strike or fault overlap. Relay ramps and rider blocks occur between overlapping fault segments (Schlische, 1993).

Basin-scale morphology in longitudinal section may be inferred from the map trace of the contact between prerift rocks and synrift strata (Fig. 14A). The distance of this contact from the BFS (the width of the basin) is a proxy for the amount of extension and hence fault displacement (Schlische, 1993). Where the trace of this contact is concave toward the BFS, the basin is syncline shaped. Intrabasin highs occur where the trace of the contact is convex, typically at fault segment boundaries. The Pomperaug, Richmond, Culpeper, and Newark basins each consist of a single plunging elongated syncline, although this geometry is complicated by intrabasinal faults. The Hartford-Deerfield, Dan River, and Deep River basins consist of multiple broad synclines separated by narrower intrabasin highs (Fig. 14A). The synclinal geometry of the Newark basin is corroborated by stratigraphic thickness relations (Schlische, 1992); for example, the width of the Lockatong Formation (shown in black on Fig. 14A) increases appreciably from the ends of the basin toward its center. The synclinal geometry of basins or subbasins reflects the characteristic displacement geometry of normal faults (Fig. 1), a feature reproduced in the models in Figures 4 and 7–11.

Maximum basin or subbasin length is plotted against maximum basin or subbasin width in Figure 14B. It is clear that shorter basins or subbasins are narrower than longer basins or subbasins and that the relation is linear; maximum width is approximately one-fifth of maximum length. If basin width is linearly related to the amount of BFS displacement as suggested by Gibson et al. (1989), then this linear plot may be indicative of a linear relationship between fault length and displacement, suggesting that $n = 1$ in Equation 1. A plot of maximum basin or subbasin length versus maximum inferred basin depth (Fig. 14C) shows considerably more scatter. This is not entirely surprising because basin depth is a function of both fault displacement and dip angle (Morley, 1988).

The typical Triassic stratigraphy consists of a basal fluvial unit overlain by lacustrine strata (Fig. 16; Olsen et al., 1989; Schlische and Olsen, 1990; Schlische, 1991, 1993; Smoot, 1991). The deepest lake facies occur near the base of the lacustrine succession and then generally shoal upward. Where the fluvial-lacustrine transition is known in detail (e.g., from cores of the Newark basin; Olsen et al., 1995), it appears that the change is not abrupt. In some basins, the shallow-water lacustrine deposits are capped by fluvial strata. In the northern basins, this Triassic sequence is overlain by an Early Jurassic-age package of lava flows and intercalated lacustrine (commonly deep water) strata overlain by shallow lacustrine and, in some cases, fluvial strata (Fig. 16). Similar stratigraphic sequences occur in numerous rift basins of varied age, although the basal "fluvial" sequence is not universally present (Table 1; Lambiase, 1990).

The switchover from initial "fluvial" to subsequent "lacustrine" deposition in these basins represents a transition from an excess supply of sediments to sediment starvation. Schlische (1991, 1993) attributed this transition to basin growth: the increasing length, width, and depth of the basin resulted in a positive rate of increase in basin capacity (see Fig. 4). As noted by Schlische (1993), "fluvial"-"lacustrine" transitions may also be produced by a decrease in the sediment-supply rate. However, we consider it more likely that sediment supply would actually increase through time as the expanding basin captures more drainage systems, although regional downcutting could take drainages away from the basin (e.g., DiGiuseppi and Bartley, 1991). Another possibility is that some tectonic event deepened the basins, allowing lacustrine sedimentation to occur. However, the gradual "fluvial"-"lacustrine" transition in the Newark basin is not compatible with abrupt tectonic deepening of the basin.

Lacustrine deposits in the Mesozoic basins generally shoal upward. Superimposed on this long-term trend are periodic and hierarchical Milankovitch-type climatic cycles (e.g., Olsen et al., 1989). Schlische and Olsen (1990) attributed the long-term

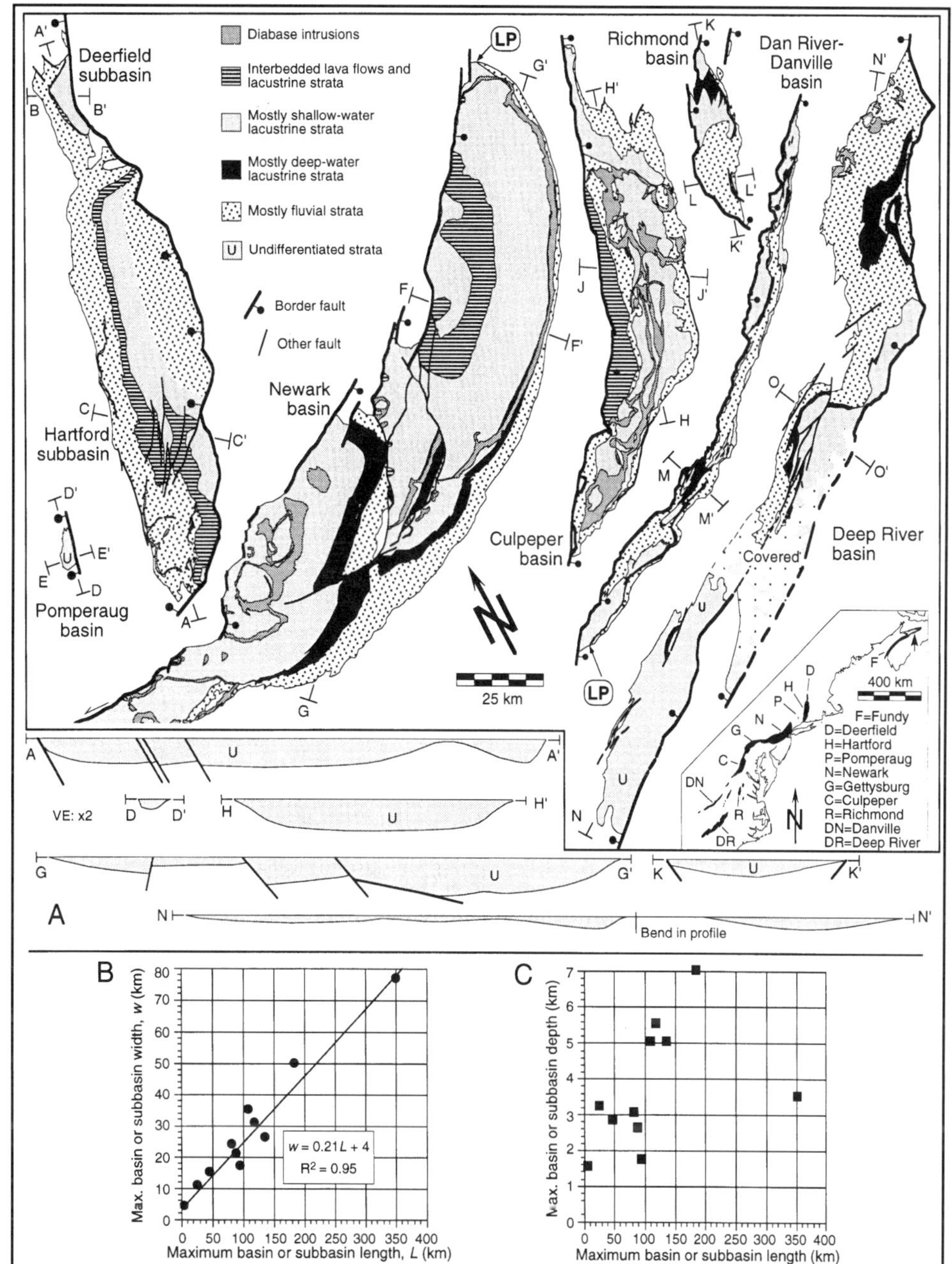

Figure 14. (A) Geologic maps and longitudinal cross sections of representative Triassic-Jurassic rift basins, eastern North America (see inset for locations). Transverse cross sections given in Figure 15. LP refers to longitudinal pinchout in the Newark and Dan River basins. Maps and cross sections simplified from Schlische (1993). (B) Approximately linear relationship between maximum basin or subbasin length and maximum basin or subbasin width. (C) Plot of length versus depth.

upward shoaling to the effects of basin growth, i.e., available water was spread out over a larger and larger region. Another possibility is that the long-term upward shoaling of the lakes represents a regional climatic change.

As we noted earlier, long-term trends in sediment accumulation rate may be influenced by basin growth. Normalized accumulation rates for the Newark basin (Fig. 6D) are based on thicknesses of fixed-period Milankovitch-type lake-level cycles measured in seven long drill cores (Schlische et al., 1991; Silvestri and Schlische, 1992; Olsen et al., 1995). Accumulation rates are not known in the fluvial interval because Milankovitch cycles are not developed. Within the lacustrine part of the Newark basin section, accumulation rates first increase, reach a maximum approximately 10 m.y. after the fluvial-lacustrine transition, decline to their initial values, and then remain unchanged through the close of the Triassic. This pattern is in general agreement with

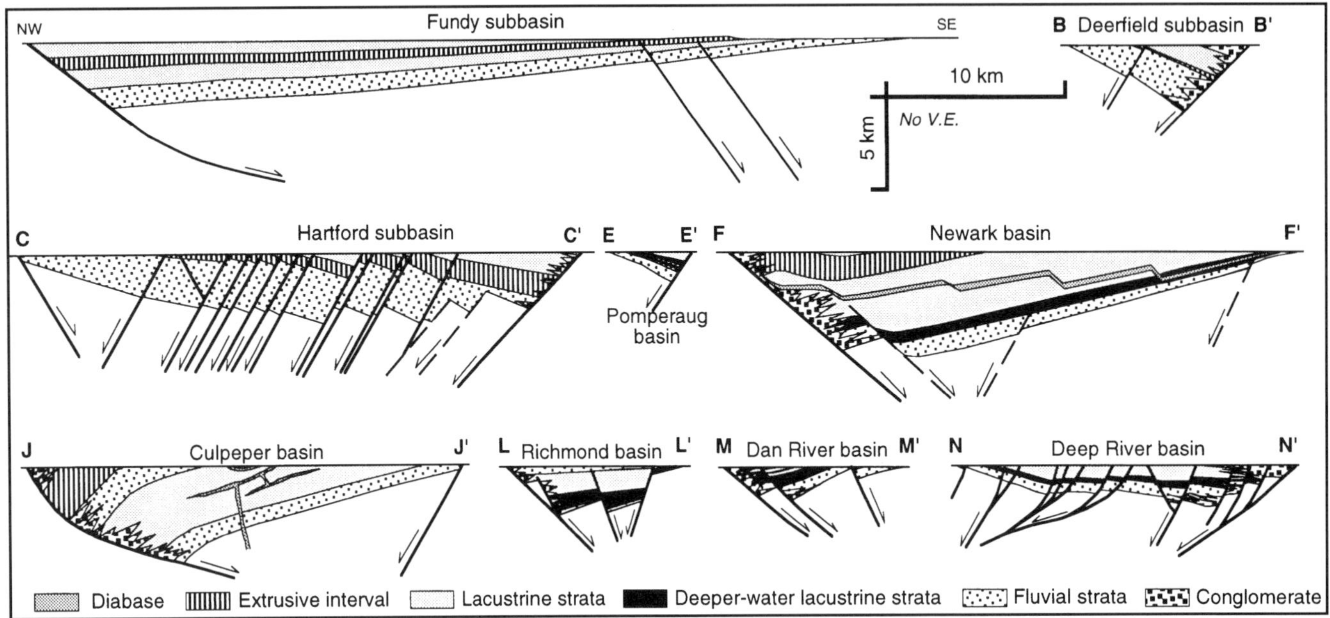

Figure 15. Transverse cross sections of Mesozoic rift basins, eastern North America. Note the asymmetrical nature of most basins and increase in stratal thickness toward growth border fault systems. Modified from Schlische (1993).

predictions of the basin growth models shown in Figure 6; however, maximum accumulation rates occur considerably later in the rifting history of the Newark basin than in the models and none of the models reproduces the uniform accumulation rates observed in latest Triassic strata. The Newark basin data are clearly not consistent with the predictions of the full-graben model of Schlische and Olsen (1990).

Interestingly, accumulation rates for the lower part of the lacustrine section are virtually identical to those for the upper part, which in general was deposited under much shallower water (Figs. 6 and 16). This indicates that over the long term, accumulation rates are not linked with the wetness or dryness of the lacustrine facies. On the other hand, Silvestri and Schlische (1992) demonstrated that accumulation rates are affected by Milankovitch-type climatic cycles (higher accumulation rates during wetter periods). This suggests that the long-term upward shoaling is not a manifestation of climatic change but rather reflects basin growth.

Accumulation rates determined for earliest Jurassic lacustrine strata intercalated with extrusive rocks of the Newark basin are significantly higher than for Triassic strata (see Fig. 6D). These markedly higher accumulation rates in the Newark basin are associated with a period of apparent increase in fault displacement rates (Fig. 17). Similar increases in fault heave are associated with the extrusive intervals within the Fundy, Deerfield, Hartford, and Culpeper basins. Schlische and Olsen (1990) suggested that this increase in accumulation rates reflected asymmetric basin deepening associated with an increased extension rate. An alternative interpretation is that the loading of large volumes of basalt in a short period of time (600 k.y.; Olsen and Fedosh, 1988; Olsen et al., 1989) caused additional hanging-wall subsidence.

Transverse onlap has been observed in sections transverse to the BFS in the Fundy, Newark, and Richmond basins (Fig. 13). Onlap is difficult to observe in longitudinal section but may be inferred from longitudinal pinchouts that occur in map view (see Fig. 4, stage 4). Longitudinal pinchout occurs at the northern end of the Newark basin and the southern end of the Dan River/Danville basin (LP in Fig. 14A). Schlische (1993) interpreted longitudinal pinchout as evidence that the Mesozoic basins lengthened through time.

SUMMARY, DISCUSSION, AND IMPLICATIONS

The scaling relationship between fault length and displacement (Fig. 2) and theoretical models (e.g., Cowie and Scholz, 1992c) strongly suggest that faults grow in length as displacement accrues. Additional research indicates that many normal fault systems are segmented on a variety of scales (e.g., Jackson and White, 1989; Peacock and Sanderson, 1991) and that longer fault systems form as a consequence of the linkage of shorter fault segments (e.g., Anders and Schlische, 1994; Dawers and Anders, 1995; Wu and Bruhn, 1994). Although earthquake ruptures are commonly reported to be confined to discrete fault segments separated by barriers to rupture propagation (e.g., Schwartz and Coppersmith, 1984), the concept of faults lengthening through time suggests that the locations of rupture terminations cannot stay fixed over the long term (Cowie and Scholz, 1992b). During the 1983 Borah Peak earthquake, for example, the rupture propagated through what was thought to be a long-

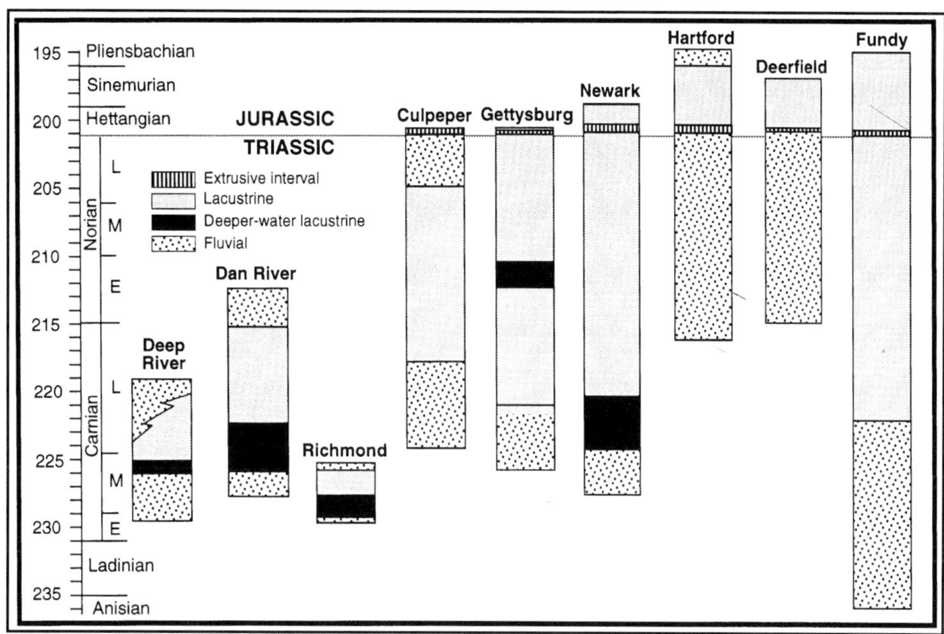

Figure 16. Chrono- and lithostratigraphy of Mesozoic rift basins, eastern North America. All basins contain a basal fluvial unit invariably overlain by lacustrine strata or interbedded lava flows and lacustrine strata. Similar stratigraphic sections occur in a number of terrestrial rift basins (Table 1). Time scale and figure based on Olsen et al. (1989) and Schlische and Olsen (1990).

TABLE 1. STRATIGRAPHY OF SELECTED CONTINENTAL EXTENSIONAL BASINS*

Basin	Location	Age	Stratigraphic Succession
Keweenawan	USA/Canada	Precambrian	Fluvial—lacustrine—fluvial
Morondava	Madagascar	Permian-Triassic	Fluvio-deltaic—deep lacustrine—fluvio-deltaic—fluvial
Mombasa	Kenya	Permian-Triassic	Lacustrine—fluvio-deltaic
Reconcavo	Brazil	Cretaceous	Deep lacustrine—deltaic—fluvial
West African†	Gabon/Angola	Cretaceous	Fluvial—deep lacustrine—deltaic; shallow lacustrine—deltaic
Sudan†	Southern Sudan	Cretaceous	Fluvio-lacustrine—deep lacustrine—fluvial; lacustrine—deltaic—fluvial
Central Sumatra	Sumatra	Paleogene	Shallow lacustrine—deep lacustrine—fluvio-deltaic

*Summarized from Lambiase, 1990.
†Indicates basin that experienced multiple tectonic cycles.

term rupture barrier (Fig. 12B; Crone and Haller, 1991). Thus, as discussed by Anders and Schlische (1994), the rupture pattern of faults defined by neotectonic features is likely a transient phenomenon.

In light of the transience of some segment boundaries imposed by the concept of fault growth, we suggest that the geometry of extensional basins may be a better indicator of longer-term fault segmentation. Thus syncline-shaped subbasins should be present along long-lived fault segments and intrabasin highs should lie adjacent to segment boundaries (see Figs. 3B and 4). However, as noted in this chapter, intrabasin highs do not necessarily reflect regions of displacement deficit if they are marked by multiple overlapping faults that distribute displacement and therefore reduce overall basin depth. The geometry of the uplifted footwall block may also provide useful information on fault segmentation (Crone and Haller, 1991; Simpson and Anders, 1992; Anders and Schlische, 1994) provided that erosion and differences in erosional resistance have not significantly altered the geometry.

Several authors have stressed the role of transfer faults in extensional tectonics (e.g., Gibbs, 1984; Lister et al., 1986). Transfer faults are generally depicted as strike-slip faults ori-

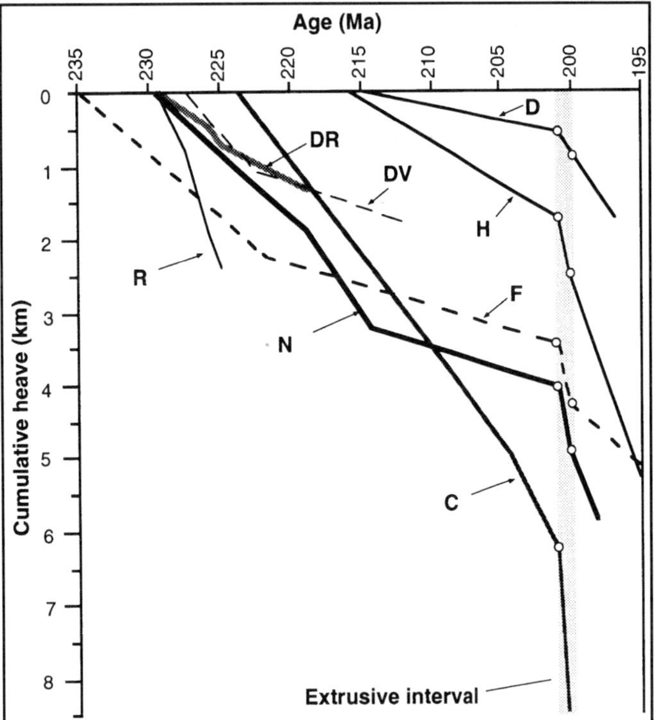

Figure 17. Plot of cumulative fault heave against geologic time based on cross sections in Figure 15 and chronostratigraphy in Figure 16. All of the northern basins experienced higher extension rates in the earliest Jurassic in association with the emplacement of lava flows. The southern basins ceased subsiding in the Late Triassic. Abbreviations are the following: C, Culpeper basin; D, Deerfield subbasin; DR, Dan River basin; DV, Danville basin; H, Hartford subbasin; F, Fundy subbasin; N, Newark basin; R, Richmond basin.

ented parallel to the extension direction that link normal faults that may or may not dip in opposite directions. Discrete transfer faults as defined above generally do not link border fault segments in the East African rift system (Morley et al., 1990), the Aegean Sea (Roberts and Jackson, 1991), and the early Mesozoic basins of eastern North America (Schlische, 1993); such faults also are absent from the range-bounding normal fault systems in the Basin and Range province discussed in this report. Similar observations were reported by Gawthorpe and Hurst (1993) (but see Milani and Davison, 1988). (For smaller-scale faults, transfer-type faults may link closely overlapping normal fault segments during breaching of relay ramps [Peacock and Sanderson, 1991; Trudgill and Cartwright, 1994].) The dearth of transfer faults is not surprising given that (1) fault tips generally propagate as displacement accumulates and (2) displacement generally decreases to zero at the fault tips. Given that normal faults are generally elliptical (although the top and bottom of the ellipse may be truncated by the Earth's surface and the base of the zone of brittle faulting, respectively) with the long axes oriented normal to the slip direction (e.g., Barnett et al., 1987; Walsh and Watterson, 1987), it is unlikely that the extreme lateral fault tips are exposed at the surface. If for some reason both fault tips cannot propagate as displacement increases, some form of transfer fault or strain accommodation structure will be kinematically required. As noted above, if only one fault tip is pinned, the other should be free to propagate, leaving a distinctive stratigraphic signature in the associated basin (Fig. 7A).

Despite recent advances in our understanding of fault growth, details of the actual growth and linkage process are limited, for geologists can study only the present-day geometry. On the other hand, the growth of large normal fault systems involves the formation and filling of extensional basins. Furthermore, the geometry and facies transitions recorded in synextensional strata ought to reflect the evolving geometry of the basin. Given sufficient seismic-stratigraphic or well control tied into a precise chronostratigraphy, it should be possible to determine the three-dimensional geometry of basin-filling units through time and thus constrain not only how the fault zone and its segments evolved but also the rate at which this fault system grew.

The evolutionary models presented in this chapter suggest that the following features result from basin growth: (1) a positive relationship between basin length and width and possibly between length and depth, (2) transverse onlap, (3) longitudinal onlap, (4) longitudinal pinchout; and (5) a transition from initial basinwide "fluvial" sedimentation to "lacustrine" deposition. Basin growth also involves the consolidation of subbasins associated with a segmented fault system; this is manifested in the stratigraphic record by older units that form restricted sequences in the original subbasins and younger units that are deposited basinwide but may still thin toward intrabasin highs that form in the regions where fault tips propagate toward one another.

The features listed above need not always be indicative of basin growth, however. Progressive hanging-wall onlap, for example, can be produced simply by the infilling of a preexisting depression, although evidence of synfaulting sedimentation obviates this interpretation (see Fig. 13). A transition from areally restricted sedimentation to basinwide deposition may reflect progressive infilling of differential relief. As noted earlier, the transition from "fluvial" to "lacustrine" sedimentation may be attributed to a decrease in sediment-supply rate rather than a progressive increase in basin capacity. Taken together, however, these features constitute strong evidence in support of basin growth. With the exception of longitudinal onlap, all features have been described in the Mesozoic rift basins of eastern North America and Cenozoic basins of the Basin and Range. The similarity in stratigraphic architecture of so many continental extensional basins (Fig. 16 and Table 1) strongly suggests some fundamental tectonic control on sedimentation; we propose that fault and basin growth is primarily responsible for this phenomenon.

We attribute the lack of evidence for longitudinal onlap to a dearth of published seismic reflection profiles shot parallel to basin-bounding fault systems. Seismic profiles oriented normal to the boundary faults are far more numerous, and it is perhaps not surprising that there are many well-documented examples of transverse onlap (e.g., Fig. 13). Progressive hanging-wall onlap has important implications for ancient extensional basins: even though these basins are commonly deeply eroded, the oldest and most tilted synextensional deposits will not crop out at the Earth's surface (see Fig. 4, stage 4). Consequently, dating of the basin fill based on outcrop samples underestimates the age of extension and basin formation. Rift basins in eastern North America are likely to predate significantly the current estimates of ca. 230–235 Ma (Olsen et al., 1989).

Finally, we acknowledge that the basin growth and filling models described in this chapter are overly simplistic. In all models except that shown in Figure 4, the basins were always filled to the lowest outlet of the basin. Furthermore, the models fail to take into account sediment compaction, the isostatic effects of sediment loading and erosion of uplifted footwall blocks, the mechanisms by which sediment is transported into and within the basin (see two-dimensional models of Travis and Nunn, 1994), syntectonic igneous activity, and any climatically driven fluctuations in lake level during lacustrine sedimentation. Thus, the actual stratigraphy of extensional basins will be considerably more complicated than that shown in the models. It is therefore not surprising that actual basin stratigraphy exhibits considerable deviations from the model predictions (e.g., Fig. 6). Nonetheless, we hope that these models will provide a framework for interpreting the stratigraphic record of extensional basins in light of the concepts of fault and basin growth and serve as springboard for developing more realistic models.

ACKNOWLEDGMENTS

Discussions with Rolf Ackermann, Nancye Dawers, David McCormick, Paul Olsen, and Chris Scholz significantly influenced our understanding of fault and basin growth. We also appreciate the helpful reviews of Rolf Ackermann, Kathi Beratan, and two anonymous reviewers. Research was funded by grants from the National Science Foundation (EAR-90-17785 to Schlische and EAR-90-04534 to Anders).

REFERENCES CITED

Anders, M. H., and Schlische, R. W., 1994, Overlapping faults, intrabasin highs, and the growth of normal faults: Journal of Geology, v. 102, p. 165–180.

Anders, M. H., Geissman, J. W., Piety, L. A., and Sullivan, J. T., 1989, Parabolic distribution of circumeastern Snake River Plain seismicity and latest Quaternary faulting: Migratory pattern and association with the Yellowstone hotspot: Journal of Geophysical Research, v. 94, p. 1589–1621.

Anders, M. H., Spiegelmann, M., Rodgers, D. W., and Hagstrum, J. T., 1993, The growth of fault-bounded tilt blocks: Tectonics, v. 12, p. 1451–1459.

Anderson, R. E., Zoback, M. L., and Thompson, G. A., 1983, Implications of selected subsurface data on the structural form and evolution of some basins in the northern Basin and Range province, Nevada and Utah: Geological Society of America Bulletin, v. 94, p. 1055–1072.

Bally, A. W., 1982, Musings over sedimentary basin evolution: Royal Society of London, Philosophical Transactions, ser. A, v. A305, p. 325–338.

Barnett, J. A. M., Mortimer, J., Rippon, J. H., Walsh, J. J., and Watterson, J., 1987, Displacement geometry in the volume containing a single normal fault: American Association of Petroleum Geologists Bulletin, v. 71, p. 925–937.

Barrientos, S. E., Stein, R. S., and Ward, S. N., 1987, Comparison of the 1959 Hebgen Lake, Montana, and the 1983 Borah Peak, Idaho, earthquakes from geodetic observations: Seismological Society of America Bulletin, v. 77, p. 784–808.

Bosworth, W., 1985, Geometry of propagating rifts: Nature, v. 316, p. 625–627.

Chapin, C. E., 1979, Evolution of the Rio Grande rift—A summary, *in* Riecker, R. E., ed., Rio Grande rift —Tectonics and magmatism: Washington, D.C., American Geophysical Union, p. 1–5.

Chapman, G. R., Lippard, S. J., and Martyn, J. E., 1978, The stratigraphy and structure of the Kamasia Range, Kenya rift valley: Geological Society of London Journal, v. 135, p. 265–281.

Cohen, A. S., 1990, Tectono-stratigraphic model for sedimentation in Lake Tanganyika, Africa, *in* Katz, B. J., ed., Lacustrine basin exploration—Case studies and modern analogues: Tulsa, Oklahoma, American Association of Petroleum Geologists Memoir 50, p. 137–150.

Cowie, P. A., and Scholz, C. H., 1992a, Displacement-length scaling relationship for faults: Data synthesis and discussion: Journal of Structural Geology, v. 14, p. 1149–1156.

Cowie, P. A., and Scholz, C. H., 1992b, Growth of faults by accumulation of seismic slip: Journal of Geophysical Research, v. 97, p. 11,085–11,095.

Cowie, P. A., and Scholz, C. H., 1992c, Physical explanation for displacement-length relationship of faults using a post-yield fracture mechanics model: Journal of Structural Geology, v. 14, p. 1133–1148.

Crone, A. J., and Haller, K. M., 1989, Segmentation of Basin and Range normal faults: Examples from east-central Idaho and southwestern Montana, *in* Schwartz, D. P., and Sibson, R. H., eds., Proceedings of Workshop XLV, Fault Segmentation and Controls of Rupture Initiation and Termination: U.S. Geological Survey Open File Report 89-315, p. 110–130.

Crone, A. J., and Haller, K. M., 1991, Segmentation and the coseismic behavior of Basin and Range normal faults: Examples from east-central Idaho and southwestern Montana, U.S.A.: Journal of Structural Geology, v. 13, p. 151–164.

Crone, A. J., Machette, M. N., Bonilla, M. G., Lienkaemper, J. J., Pierce, K. L., Scott, W. E., and Bucknam, R. C., 1987, Surface faulting accompanying the Borah Peak earthquake and segmentation of the Lost River fault, central Idaho: Seismological Society of America Bulletin, v. 77, p. 739–770.

Davison, I., 1994, Linked faults systems; extensional, strike-slip and contractional, *in* Hancock, P. L., ed., Continental deformation: New York, Pergamon, p. 121–142.

Dawers, N. H., and Anders, M. H., 1995, Displacement-length scaling and fault linkage: Journal of Structural Geology, v. 17, p. 607–614.

Dawers, N. H., Anders, M. H., and Scholz, C. H., 1993, Fault length and displacement: Scaling laws: Geology, v. 21, p. 1107–1110.

DiGiuseppi, W. H., and Bartley, J. M., 1991, Stratigraphic effects of a change from internal to external drainage in an extending basin, southeastern Nevada: Geological Society of America Bulletin, v. 103, p. 48–55.

Dixon, J. S., 1982, Regional structural synthesis, Wyoming salient of western Overthrust Belt: American Association of Petroleum Geologists Bulletin, v. 66, p. 1560–1580.

Ebinger, C. J., 1989, Tectonic development of the western branch of the East African rift system: Geological Society of America Bulletin, v. 101, p. 885–903.

Elliot, D., 1976, Energy balance and deformation mechanisms of thrust sheets: Royal Society of London, Philosophical Transactions, ser. A, v. A283, p. 289–312.

Fisher, F. S., McIntyre, D. H., and Johnson, K. M., 1983, Geologic map of the Challis 1° x 2° quadrangle, Idaho: U.S. Geological Survey Open File Report 83-523, 60 p.

Fonseca, J., 1988, The Sou Hills: A barrier to faulting in the central Nevada seismic belt: Journal of Geophysical Research, v. 93, p. 475–489.

Gawthorpe, R. L., and Hurst, J. M., 1993, Transfer zones in extensional basins: their structural style and influence on drainage development and stratigraphy: Geological Society of London Journal, v. 150, p. 1137–1152.

Gibbs, A. D., 1984, Structural evolution of extensional basin margins: Geological Society of London Journal, v. 141, p. 609–620.

Gibson, J. R., Walsh, J. J., and Watterson, J., 1989, Modelling of bed contours and cross-sections adjacent to planar normal faults: Journal of Structural Geology, v. 11, p. 317–328.

Gillespie, P. A., Walsh, J. J., and Watterson, J., 1992, Limitations of dimension and displacement data from single faults and the consequences for data analysis and interpretation: Journal of Structural Geology, v. 14, p. 1157–1172.

Hamblin, W. K., 1965, Origin of "reverse drag" on the downthrown sides of normal faults: Geological Society of America Bulletin, v. 76, p. 1145–1164.

Jackson, J., and Leeder, M., 1994, Drainage systems and the development of normal faults: An example from Pleasant Valley, Nevada: Journal of Structural Geology, v. 16, p. 1041-1059.

Jackson, J., and McKenzie, D., 1983, The geometrical evolution of normal fault systems: Journal of Structural Geology, v. 5, p. 471–482.

Jackson, J. A., and White, N. J., 1989, Normal faulting in the upper continental crust: Observations from regions of active extension: Journal of Structural Geology, v. 11, p. 15–36.

King, G. C. P., and Ellis, M., 1990, The origin of large local uplift in extensional regions: Nature, v. 348, p. 20–27.

Klitgord, K. D., and Agar, S. M., 1993, Propagating border faults and rift flank segmentation: Examples from the Baikal rift: Geological Society of America Abstracts with Programs, v. 25, no. 6, p. A409–A410.

Lambiase, J. J., 1990, A model for tectonic control of lacustrine stratigraphic sequences in continental rift basins, *in* Katz, B. J., ed., Lacustrine exploration: Case studies and modern analogues: American Association of Petroleum Geologists Memoir 50, p. 265–276.

Larsen, P.-H., 1988, Relay structures in a Lower Permian basement-involved extension system, East Greenland: Journal of Structural Geology, v. 10, p. 3–8.

Leeder, M. R., and Gawthorpe, R. L., 1987, Sedimentary models for extensional tilt-block/half-graben basins, *in* Coward, M. P., Dewey, J. F., and Hancock, P. L., eds., Continental extensional tectonics: London, Geological Society of London Special Publication 28, p. 139–152.

Lister, G. S., Etheridge, M. A., and Symonds, P. A., 1986, Detachment faulting and the evolution of passive continental margins: Geology, v. 14, p. 246–250.

Machette, M. N., Personius, S. F., Nelson, A. R., Schwartz, D. P., and Lund, W. R., 1991, The Wasatch fault zone, Utah—Segmentation and history of Holocene earthquakes: Journal of Structural Geology, v. 13, p. 137–149.

Manspeizer, W., 1988, Triassic-Jurassic rifting and opening of the Atlantic: An overview, *in* Manspeizer, W., ed., Triassic-Jurassic rifting, continental breakup, and the origin of the Atlantic Ocean and passive margins: New York, Elsevier, p. 41–79.

Marrett, R., and Allmendinger, R. W., 1991, Estimates of strain due to brittle faulting: Sampling of fault populations: Journal of Structural Geology, v. 13, p. 735–738.

McCalpin, J. P., Piety, L. A., and Anders, M. H., 1990, Latest Quaternary faulting and structural evolution of Star Valley, Wyoming, *in* Roberts, S., ed., Geologic field tours of western Wyoming and parts of adjacent Idaho, Montana and Utah: Geological Survey of Wyoming Public Information Circular 29, p. 4–12.

McMillan, R. A., 1975, The orientation and sense of displacement of strike-slip faults in continental crust [B.S. thesis]: Ottawa, Ontario, Carleton University.

Milani, E. J., and Davison, I., 1988, Basement control and transfer tectonics in the Reconcavo-Tucano-Jatoba rift, northeast Brazil: Tectonophysics, v. 154, p. 41–70.

Morley, C. K., 1988, Variable extension in Lake Tanganyika: Tectonics, v. 7, p. 785–801.

Morley, C. K., Nelson, R. A., Patton, T. L., and Munn, S. G., 1990, Transfer zones in the East African rift system and their relevance to hydrocarbon exploration in rifts: American Association of Petroleum Geologists Bulletin, v. 74, p. 1234–1253.

Muraoka, H., and Kamata, H., 1983, Displacement distribution along minor fault traces: Journal of Structural Geology, v. 5, p. 483–495.

Nelson, R. A., Patton, T. L., and Morley, C. K., 1992, Rift-segment interaction and its relation to hydrocarbon exploration in continental rift systems: American Association of Petroleum Geologists Bulletin, v. 76, p. 1153–1169.

Olsen, P. E., and Fedosh, M., 1988, Duration of early Mesozoic extrusive igneous episode in eastern North America determined by use of Milankovitch-type lake cycles: Geological Society of America Abstracts with Programs, v. 20, no. 1, p. 59.

Olsen, P. E., Schlische, R. W., and Gore, P. J. W., eds., 1989, Tectonic, depositional, and paleoecological history of early Mesozoic rift basins of eastern North America, *in* International Geological Congress, Field Trip Guidebook T-351: Washington, D.C., American Geophysical Union, 174 p.

Olsen, P. E., Kent, D. V., Cornet, B., Witte, W. K., and Schlische, R. W., 1995, High-resolution stratigraphy of the Newark rift basin (early Mesozoic, eastern North America): Results of the Newark Basin Coring Project: Geological Society of America Bulletin, (in press).

Opheim, J. A., and Gudmundsson, A., 1989, Formation and geometry of fractures, and related volcanism, of the Krafla fissure swarm, northeast Iceland: Geological Society of America Bulletin, v. 101, p. 1608–1622.

Peacock, D. C. P., 1991, Displacements and segment linkage in strike-slip fault zones: Journal of Structural Geology, v. 13, p. 1025–1035.

Peacock, D. C. P., and Sanderson, D. J., 1991, Displacements, segment linkage and relay ramps in normal fault zones: Journal of Structural Geology, v. 13, p. 721-733.

Piety, L. A., Wood, C. K., Gilbert, J. D., Sullivan, J. T., and Anders, M. H., 1986, Seismotectonic study for Palisades Dam and Reservoir, Palisades Project, Idaho: U.S. Bureau of Reclamation Seismotectonic Report 86-3, 355 p.

Piety, L. A., Sullivan, J. T., and Anders, M. H., 1992, Segmentation and earthquake potential of the Grand Valley fault, Idaho and Wyoming: Boulder, Colorado, Geological Society of America Memoir 179, p. 155–182.

Postma, G., and Drinia, H., 1993, Architecture and sedimentary facies evolution of a marine, expanding half-graben (Crete, Late Miocene): Basin Research, v. 5, p. 103–124.

Richins, R. D., Pechmann, J. C., Smith, R. B., Langer, C. J., Goter, S. K., Zollweg, J. E., and King, J. J., 1987, The 1983 Borah Peak, Idaho, earthquake and its aftershocks: Bulletin of the Seismological Society of America, v. 77, p. 694–723.

Roberts, A., and Yielding, G., 1994, Continental extensional tectonics, *in* Hancock, P. L., ed., Continental deformation: New York, Pergamon, p. 223–250.

Roberts, S., and Jackson, J., 1991, Active normal faulting in central Greece: An overview, *in* Roberts, A. M., Yielding, G., and Freeman, B., eds., The geometry of normal faults: London, Geological Society of London Special Publication 56, p. 125–142.

Rodgers, D. W., and Anders, M. H., 1990, Neogene evolution of Birch Creek Valley near Lone Pine, Idaho, *in* Roberts, S., ed., Geologic field tours of western Wyoming and parts of adjacent Idaho, Montana and Utah: Geological Survey of Wyoming Public Information Circular 29, p. 26–38.

Rosendahl, B. R., 1987, Architecture of continental rifts with special reference to East Africa: Annual Review of Earth and Planetary Science, v. 15, p. 445–503.

Rosendahl, B. R., Kilembe, E., and Kaczmarick, K., 1992, Comparison of the Tanganyika, Malawi, Rukwa, and Turkana rift zones from analyses of seismic reflection data: Tectonophysics, v. 213, p. 235–256.

Schlische, R. W., 1991, Half-graben filling models: Implications for the evolution of continental extensional basins: Basin Research, v. 3, p. 123–141.

Schlische, R. W., 1992, Structural and stratigraphic development of the Newark extensional basin, eastern North America: Implications for the growth of the basin and its bounding structures: Geological Society of America Bulletin, v. 104, p. 1246–1263.

Schlische, R. W., 1993, Anatomy and evolution of the Triassic-Jurassic continental rift system, eastern North America: Tectonics, v. 12, p. 1026–1042.

Schlische, R. W., and Olsen, P. E., 1990, Quantitative filling model for continental extensional basins with applications to early Mesozoic rifts of eastern North America: Journal of Geology, v. 98, p. 135–155.

Schlische, R. W., Olsen, P. E., Cornet, B., and Silvestri, S. M., 1991, Preliminary analysis of Newark basin drilling project cores: Implications for basin tectonics: Geological Society of America Abstracts with Programs, v. 23, no. 5, p. A251.

Scholz, C. H., and Cowie, P. A., 1990, Determination of total strain from faulting using slip measurements: Nature, v. 346, p. 837–838.

Scholz, C. H., Dawers, N. H., Yu, J.-J., Anders, M. H., and Cowie, P. A., 1993, Fault growth and scaling laws: Preliminary results: Journal of Geophysical Research, v. 98, p. 21,951–21,961.

Schwartz, D. P., and Coppersmith, K. J., 1984, Fault behavior and characteristic earthquakes: Examples from the Wasatch and San Andreas fault zones: Journal of Geophysical Research, v. 89, p. 5681–5698.

Silvestri, S. M., and Schlische, R. W., 1992, Analysis of Newark basin drill cores and outcrop sections: Tectonic and climatic controls on lacustrine facies architecture: Geological Society of America Abstracts with Programs, v. 24, no. 3, p. 76.

Simpson, D. W., and Anders, M. H., 1992, Tectonics and topography of the western United States—An application of digital mapping: GSA Today, v. 6, p. 118–121.

Smoot, J. P., 1991, Sedimentary facies and depositional environments of early Mesozoic Newark Supergroup basins, eastern North America: Palaeogeography, Palaeoclimatology, Palaeoecology, v. 84, p. 369–423.

Stein, R. S., and Barrientos, S. E., 1985, Planar high-angle faulting in the Basin and Range: geodetic analysis of the 1983 Borah Peak, Idaho, earthquake: Journal of Geophysical Research, v. 90, p. 11,355–11,366.

Stein, R. S., King, G. C. P., and Rundle, J. B., 1988, The growth of geological structures by repeated earthquakes: 2. Field examples of continental dip-slip faults: Journal of Geophysical Research, v. 93, p. 13,319–13,331.

Stewart, I. S., and Hancock, P. L., 1991, Scales of structural heterogeneity within neotectonic normal fault zones in the Aegean region: Journal of Structural Geology, v. 13, p. 191–204.

Swanson, M. T., 1986, Preexisting fault control for Mesozoic basin formation in eastern North America: Geology, v. 14, p. 419–422.

Travis, C. J., and Nunn, J. A., 1994, Stratigraphic architecture of extensional basins: Insights from a numerical model of sedimentation in evolving half grabens: Journal of Geophysical Research, v. 99, p. 15,653–15,666.

Trudgill, B., and Cartwright, J., 1994, Relay-ramp forms and normal fault linkages, Canyonlands National Park, Utah: Geological Society of America Bulletin, v. 106, p. 1143–1157.

Versfelt, J., and Rosendahl, B. R., 1989, Relationships between pre-rift structure and rift architecture in Lakes Tanganyika and Malawi, East Africa: Nature, v. 337, p. 354–357.

Villemin, T., Angelier, J., and Sunwoo, C., 1995, Fractal distribution of fault length and offsets: Implications of brittle deformation evaluation—The Lorraine Coal basin, *in* Barton, C. C., and LaPointe, P., eds., Fractals in the earth sciences: New York, Plenum Press, p. 205–226.

Vreeland, J. H., and Berrong, B. H., 1979, Seismic exploration in Railroad Valley, Nevada, *in* Newman, G. W., and Goode, H. D., eds., Basin and Range symposium and Great Basin field conference: Denver, Colorado, Rocky Mountain Association of Geologists, p. 557–569.

Walsh, J. J., and Watterson, J., 1987, Distributions of cumulative displacement and seismic slip on a single normal fault surface: Journal of Structural Geology, v. 9, p. 1039–1046.

Walsh, J. J., and Watterson, J., 1988, Analysis of the relationship between displacements and dimensions of faults: Journal of Structural Geology, v. 10, p. 239–247.

Walsh, J. J., and Watterson, J., 1991, Geometric and kinematic coherence and scale effects in normal fault systems, *in* Roberts, A. M., Yielding, G., and Freeman, B., eds., The geometry of normal faults: London, Geological Society of London Special Publication 56, p. 193–203.

Walsh, J. J., and Watterson, J., 1992, Populations of faults and fault displacements and their effects on estimates of fault-related regional extension: Journal of Structural Geology, v. 14, p. 701–712.

Watterson, J., 1986, Fault dimensions, displacements and growth: Pure and Applied Geophysics, v. 124, p. 365–373.

Wernicke, B., and Burchfiel, B. C., 1982, Modes of extensional tectonics: Journal of Structural Geology, v. 4, p. 105–115.

Wheeler, R. L., 1987, Boundaries between segments of normal faults—Criteria for recognition and interpretation, *in* Crone, A. J., and Olmdahl, E. M., eds., Proceedings, Directions in Paleoseismology, 39th Conference: U.S. Geological Survey Open-File Report 87-673, p. 385–398.

Wu, D., and Bruhn, R. L., 1994, Geometry and kinematics of active normal faults, South Oquirrh Mountains, Utah: Implication for fault growth: Journal of Structural Geology, v. 16, p. 1061–1075.

Young, S. S., Schische, R. W., and Ackermann, R. V., 1995, Micro-normal fault populations in Mesozoic rift basins: Length-displacement scaling relations: Geological Society of America Abstracts with Programs, v. 27, no. 1, p. 94.

Zandt, G., and Owens, T. J., 1980, Crustal flexure associated with normal faulting and implications for seismicity along the Wasatch front, Utah: Seismological Society of America Bulletin, v. 70, p. 1501–1520.

Zhang, P., Slemmons, D. B., and Mao, F., 1991, Geometric pattern, rupture termination and fault segmentation of the Dixie Valley–Pleasant Valley active normal fault system, Nevada, U.S.A.: Journal of Structural Geology, v. 13, p. 165–176.

MANUSCRIPT ACCEPTED BY THE SOCIETY APRIL 21, 1995

Printed in U.S.A.

Index

[Italic page numbers indicate major references]

N

O

P

Q

R

S

W

Y

Z

Typeset and printed in U.S.A. by Johnson Printing, Boulder, Colorado